CAMBRIDGE LIBRARY COLLECTION

Books of enduring scholarly value

Physical Sciences

From ancient times, humans have tried to understand the workings of the world around them. The roots of modern physical science go back to the very earliest mechanical devices such as levers and rollers, the mixing of paints and dyes, and the importance of the heavenly bodies in early religious observance and navigation. The physical sciences as we know them today began to emerge as independent academic subjects during the early modern period, in the work of Newton and other 'natural philosophers', and numerous sub-disciplines developed during the centuries that followed. This part of the Cambridge Library Collection is devoted to landmark publications in this area which will be of interest to historians of science concerned with individual scientists, particular discoveries, and advances in scientific method, or with the establishment and development of scientific institutions around the world.

Conferences Held in Connection with the Special Loan Collection of Scientific Apparatus, 1876

In 1876 the South Kensington Museum held a major international exhibition of scientific instruments and equipment, both historical and contemporary. Many of the items eventually formed the basis of collections now held at London's Science Museum. In May 1876, organisers arranged a series of conferences at which leading British and European scientists explained and demonstrated some of the items on display. The purpose was to emphasise the exhibition's goal not merely to preserve archaic treasures (such as Galileo's telescopes or Janssen's microscope) but to juxtapose them with current technology and so inspire future scientific developments. Volume 2 of the proceedings covers chemistry, biology, and earth sciences including geology, mining, meteorology and hydrography. The contributors include Joseph Dalton Hooker, William Thiselton-Dyer, Andrew Crombie Ramsay and John Rae, all of whom have other works reissued in the Cambridge Library Collection, which also includes the full catalogue of the exhibition itself.

Cambridge University Press has long been a pioneer in the reissuing of out-of-print titles from its own backlist, producing digital reprints of books that are still sought after by scholars and students but could not be reprinted economically using traditional technology. The Cambridge Library Collection extends this activity to a wider range of books which are still of importance to researchers and professionals, either for the source material they contain, or as landmarks in the history of their academic discipline.

Drawing from the world-renowned collections in the Cambridge University Library and other partner libraries, and guided by the advice of experts in each subject area, Cambridge University Press is using state-of-the-art scanning machines in its own Printing House to capture the content of each book selected for inclusion. The files are processed to give a consistently clear, crisp image, and the books finished to the high quality standard for which the Press is recognised around the world. The latest print-on-demand technology ensures that the books will remain available indefinitely, and that orders for single or multiple copies can quickly be supplied.

The Cambridge Library Collection brings back to life books of enduring scholarly value (including out-of-copyright works originally issued by other publishers) across a wide range of disciplines in the humanities and social sciences and in science and technology.

Conferences Held in Connection with the Special Loan Collection of Scientific Apparatus, 1876

Chemistry, Biology, Physical Geography, Geology, Mineralogy, and Meteorology

VOLUME 2

VARIOUS

CAMBRIDGE
UNIVERSITY PRESS

CAMBRIDGE
UNIVERSITY PRESS

University Printing House, Cambridge, CB2 8BS, United Kingdom

Cambridge University Press is part of the University of Cambridge.

It furthers the University's mission by disseminating knowledge in the pursuit of
education, learning and research at the highest international levels of excellence.

www.cambridge.org
Information on this title: www.cambridge.org/9781108078146

© in this compilation Cambridge University Press 2015

This edition first published 1876
This digitally printed version 2015

ISBN 978-1-108-07814-6 Paperback

SOUTH KENSINGTON MUSEUM.

CONFERENCES.

SPECIAL LOAN COLLECTION OF
SCIENTIFIC APPARATUS,
1876.

SOUTH   KENSINGTON   MUSEUM.

CONFERENCES

HELD IN CONNECTION WITH

THE

SPECIAL   LOAN   COLLECTION   OF   SCIENTIFIC   APPARATUS.

1876.

CHEMISTRY, BIOLOGY, PHYSICAL GEOGRAPHY,
GEOLOGY, MINERALOGY, AND
METEOROLOGY.

Published for the Lords of the Committee of Council on Education

BY

CHAPMAN AND HALL, 193, PICCADILLY.

LONDON :

PRINTED BY VINCENT BROOKS, DAY AND SON,

GATE STREET, LINCOLN'S INN FIELDS.

CONTENTS

———◆———

SECTION—CHEMISTRY.

SECTION—BIOLOGY.

SECTION—PHYSICAL GEOGRAPHY, GEOLOGY, MINING AND METEOROLOGY.

CONTENTS.

CONTENTS.

SECTION—CHEMISTRY.

President : Professor E. FRANKLAND, Ph.D., D.C.L., F.R.S.

Vice-Presidents :

Professor ABEL, F.R.S.
Herr Professor Dr. VON BABO.
M. le Professeur BEILSTEIN.
Il Commendatore Professsore BLASERNA
M. le Professeur FREMY.
Dr. GILBERT, F.R.S.
Dr. GLADSTONE, F.R.S.
Herr Professor HEINTZ.
Herr Professoi HIMLY.

Herr Professor Dr. HOFMANN, F.R.S.
Herr Professor Dr. de LOOS.
The Right Hon. LYON PLAYFAIR, C.B., M.P., F.R.S.
Professor ROSCOE, Ph.D., F.R.S.
Herr Professor WAAGE.
Professor WILLIAMSON, Ph. D., F.R.S.
Herr Professor Dr. WOHLER, F.R.S.

May 18th, 1876.

DR. FRANKLAND'S ADDRESS.

The Conference which I have been requested to open to-day has for its object the discussion of the merits and defects of the various forms of chemical apparatus exhibited in these buildings; and the criticism of the original investigations which are here illustrated, partly by the instruments used in them, and partly by the chemical compounds, to the discovery of which they have led.

Various objects interesting to chemists have been displayed in former international exhibitions, but it may be safely asserted that such a collection as this, which has been brought together in these buildings, has never before been seen; neither has there before been the opportunity for discussion and criticism, by men eminent in science from all parts of Europe, which is now afforded.

B

Such a collection of apparatus and products, gathered from all part of Europe, is useful in disclosing, to chemical investigators and others, the best sources whence to procure apparatus; it is interesting, and as historically showing the improvements in chemical apparatus during the present century; and it is instructive in the comparisons it affords of the various forms of instruments used for the same purpose in different countries, and by different experimenters.

The entire novelty of such a collection as that belonging to this section has rendered the attainment of the object sought for, on the present occasion, exceedingly difficult. The workers in science have hitherto had no inducement to preserve the *instruments* with which they experimented. When an investigation was finished, the apparatus employed was dismantled and converted to other uses. Still less inducement has there been to preserve the *chemical compounds* resulting from research, although their creation required, in many cases, a great expenditure of time and labour. The chief object of preparing such compounds has hitherto been, in most cases, merely to ascertain their existence, to show their molecular relations to previously known bodies, and to ascertain a few of their leading properties such as colour, specific gravity, vapour density, melting point, boiling point, and chemical composition. They have been weighed and measured, and then dismissed out of existence. And thus the present collection of chemical preparations is but the merest skeleton of a complete exposition of all known chemical compounds.

It is, indeed, remarkable, that whilst *natural* chemical compounds are exhibited in almost endlessly multiplied specimens in the mineralogical collections of our national museums, the *artificial* compounds which have resulted from research, or have been the foundation of important theories and generalisations, have nowhere been honoured by admission into national collections. The neglect, not to say contempt, with which these productions of the laboratory have been treated, cannot be justified on the ground of their want of national utility. It is true that from an exclusively commercial point of view, no one of them can lay claim to the importance of coal, iron, silver, and gold. Still, many of them, such as the paraffins, the coal-tar colours, and many of the compounds of sulphur, potassium, sodium, and ammonium, have contributed, in an important degree, to the wealth and prosperity of this and other states. Had these artificial compounds remained undiscovered, how different would now have been the condition of the industries of bleaching, dyeing, calico-printing, glass-making, and the manufactures connected with the production of artificial light. Many of these artificial compounds have become of the most essential importance to the physician, the artist, the telegraphist, the engineer, and the manufacturer, and it cannot be doubted that many more would soon come into active service for similar purposes if they were better known.

But not alone on the ground of utility and incentive to further useful discovery of technical applications would I plead for the establishment of national museums of chemical preparations ; such collections would be of the highest interest both to the student and the investigator. They would call vividly before the mind the results of labours which can only otherwise become known by a tedious search through the transactions of learned societies. An intelligent study of a properly arranged collection of artificial chemical compounds would show the progressive triumph of mind over matter—not over masses moved by mechanical agencies—for monuments of this, the engineer and the architect need only bid the inquirer, in the language of Wren's tablet, to "look around him"—but over the ultimate atoms which, in these compounds, are compelled to submit themselves to the will of man, and to form new structures, seen only, in most cases, by the discoverer himself, and the qualities and uses of which are but very imperfectly ascertained. Nine-tenths of these compounds are no better known than islands which have been seen only from the deck of a ship and whose position has been accurately marked upon a chart. But a collection of them, if properly kept up, would represent the actual condition of our knowledge of chemical facts, and, if properly arranged, would suggest to the observant student the direction of future investigation.

I know of no other incentive to research which would be more likely to call original inquirers into existence. The student wishing to commence a chemical investigation is always confronted at the outset by the difficulty of finding the boundary line between the known and the unknown, and this difficulty must obviously increase from year to year, owing to the continued expansion of the circle of knowledge. It has led to a suggestion emanating from the British Association, that chemists who are intimately acquainted with particular departments of their science should suggest subjects of research for the benefit of students. Much may be said, no doubt, in favour of such a scheme; but it appears to me that the development of original talent in the young investigator would be more surely promoted by giving him the means of selecting for himself a subject for experimental inquiry, rather than by inducing him to follow the less invigorating plan of working out the suggestions of others. I venture, therefore, thus prominently to call attention to the non-existence, in any country, of a museum of artificial compounds, and to the great value, both economical, scientific, and educational, which such a museum would possess. I feel convinced that if such museums were established in the capitals of Europe, chemical investigators throughout the world would gladly contribute their new products to them, and thus keep them abreast of the discoveries of chemical science.

Amongst the groups of objects in the Chemical Section, not the least interesting is that which consists of *Apparatus and Contrivances employed in the Generation and Application of Heat.* The great advances which have been made in the modes of producing and applying heat for chemical purposes are strikingly conspicuous. The cumbrous furnaces of the earlier operators, constructed in fire-proof vaults, have gradually been replaced by simple and elegant contrivances, which would scarcely look out of place upon a drawing-room table. The time is still fresh in the recollection of many of us, when the fusion of a silicate for quantitative analysis, or the heating to redness of oxide of copper for the combustion of an organic compound, required in each case the expenditure of much time and trouble in the lighting of a coke or charcoal furnace. Now these operations are performed in small gas furnaces, with or without air blast. Conspicuous amongst these inventions are the gas-burners of Bunsen and Hofmann, the oxy-coal gas-furnaces of Deville, the blast gas-furnaces of Griffin, and the hot blast gas-furnaces of Fletcher. Of these fundamental inventions many ingenious modifications for special purposes have been devised, amongst which I may mention the valuable contrivances of Finkener, Mitscherlich, Wallace, and Müncke. The blast gas-burners of Hofmann and Bunsen, the blast gas-furnaces of Deville, Griffin, and Bunsen, and the furnaces for organic analysis by Hofmann, Bunsen, Finkener, Mitscherlich, and Müncke, are amongst the exhibits illustrating the application of heat in chemical operations.

These burners and furnaces command a range of temperature from the gentlest ignition up to the most intense heat procurable by chemical means; but the temperature produced by such combinations as those of oxygen and hydrogen, or oxygen and carbon, enormously high though it be, now no longer suffices, and recourse must be had to the still more intense heat of the electric discharge. The electric current and the stream of sparks are now not unfrequently called into requisition by the chemist, and from this point of view the electric lamp and the apparatus of Hofmann and others, for the decomposition of gases by the spark-stream must be classed with chemical furnaces.

To apparatus for the application of heat belong the various forms of water, steam, and air baths, or drying closets. Convenient contrivances of this class, invented by Bunsen, Mitscherlich, Habermann, and Müncke, are exhibited by Messrs. Warmbrunn, Quilitz and Co., Mr. Johann Lentz, and Mr. Julius Schober, all of Berlin, and by Mr. C. Desaga, of Heidelberg.

In the application of gas to chemical purposes, regulators of pressure and temperature are often of the utmost importance, in order that operations requiring the prolonged and regular action of heat may not require the constant attention of the operator. The ingenious and effective contrivances of Bunsen and Kramer, for this purpose are exhibited.

Closely connected with appliances for raising temperature are those intended for its reduction—the refrigerators or condensers. The Liebig's condenser is still the refrigerator almost exclusively used, but few pieces of apparatus have been so much modified and refined, as will be seen on comparing the original design with the present construction—the final light and convenient form having been given to it by my late friend Mr. B. F. Duppa. Most manufacturers of chemical apparatus exhibit various forms of this condenser.

Sprengel Pumps.—Of the comparatively recent appliances for facilitating chemical work, few can lay claim to higher merit than the invention of Dr. Hermann Sprengel, in the year 1865, for the production of vacua by the fall of liquids in tubes; and yet this invention remained for many years dormant, until the late Master of the Mint applied the mercurial pump to the extraction and collection of occluded gases, and Bunsen the water-pump to hastening the filtration of liquids. Without the mercurial pump the elements of the organic matter in potable waters could not be determined, and the highly interesting results which this pump has quite recently achieved in the hands of Mr. Crookes, come home to every one who has seen the various forms of the radiometer.

Bunsen's application of the water-pump to filtration has done much to shorten one of the most tedious and troublesome operations of gravimetrical analysis.

Dr. Sprengel's invention has, moreover, nearly abolished the use of the air-pump in chemical laboratories, and I need not therefore, perhaps, bring under the special notice of this Section the various improvements in air-pumps which are illustrated by the exhibits in the Physical Section.

Models, diagrams, apparatus, and chemicals used in the teaching of chemistry, include numerous exhibits of great interest. It is to be regretted, however, that models and plans of chemical laboratories are not more numerously represented. The important improvements which have been introduced of late years, and the numerous laboratories of truly palatial proportions which have been built, in almost every case at the cost of the State, would have rendered a complete exposition of their plans and fittings most instructive and interesting. Dr. de Loos has, however, sent us a model of the chemical laboratory in the secondary Town School of Leyden. And we have from Mr. Waterhouse plans of the Owens College laboratories in Manchester. The latter were devised after the Professor of Chemistry and the Architect had visited all the great laboratories of Europe; and for compactness, economy of space, appropriateness of fittings, and ventilation, they are unsurpassed.

In illustration of the permanent fittings of laboratories, we have from the Chemical Institute of the University of Strasburg a diagram showing elevation, section, and plan of a " digestorium," or iron closet, for use in dangerous operations in which explosions are liable to occur. This is a contrivance which ought never to be absent from a laboratory in which research is carried on.

Professor Roscoe exhibits a beautiful and effective series of diagrams ı d models illustrating the processes carried on in alkali works, and Mr. Henry Deacon a sectional model of his ingenious apparatus for exposing porous materials and currents of gases to mutual action. Dr. De Loos, of Leyden, has sent drawings of gas works used for teaching technical chemistry in secondary schools. We are indebted to Mr. Spence, of Manchester, for a series of specimens illustrating his process for the manufacture of ammonia-alum. To Messrs. Roberts, Dale and Co. for specimens illustrating the manufacture of oxalic acid. To Messrs. Calvert and Co. for similar illustrations of the manufacture of carbolic, cressylic and picric acids.

Messrs. Hargreaves and Robinson exhibit plans and specimens in connection with their new process of manufacturing sulphate of soda directly from sulphurous acid, steam, air, and salt ; whereby the intermediate production of sulphuric acid is avoided. A chemical factory is generally conspicuous in the landscape by a series of huge and ugly leaden vitriol chambers. Should the new process prove as successful as the inventors anticipate, these leaden chambers will almost entirely disappear, and the aspect of chemical factories will undergo a more profound modification than any which has occurred during the last half century.

The splendid platinum apparatus of Messrs. Johnson and Matthey for the concentration of sulphuric acid, will also contribute much to compactness in chemical works, by the abolition of cumbrous leaden pans and long ranges of glass retorts.

Not only is the sense of sight thus likely to be relieved, but that of smell, which, in the case of chemical works, is perhaps of even more importance, is also gradually being subjected to less offence by the adoption of Mond's process for the recovery of sulphur from soda waste. The vast mounds of this waste which surround alkali works not only pollute the air with sulphuretted hydrogen ; but also the neighbouring streams, with an offensive drainage which is very destructive to fish life. Herr Mond has succeeded in profitably extracting the sulphur—the offending constituent of the waste—and Messrs. John Hutchinson and Co., of Widness, exhibit specimens illustrating this important process.

Dr. Van Rijn, of Venlo, Netherlands, exhibits fine crystals of potash and chrome alums.   One of the Octohedrons of potash alum weighs no less than 11 lbs.

Messrs. W. J. Norris and Brother, of Calder Chemical Works, have sent specimens useful in teaching the technology of lichen colours, sulphate of alumina, and bichromate of potash.

Messrs. Brooke, Simpson, and Spiller contribute a fine series of specimens illustrating the technology of coal-tar colours.

Lastly, several magnificent series of specimens have been sent over by members of the German Chemical Society.   They comprise, firstly, some items of much historical interest.   Thus, we have from Professor Wöhler the first specimens of boron and aluminium ever prepared. And, from the same chemist, another historical specimen which, it is no exaggeration to say, is the most interesting now in existence ; for, after the discovery of oxygen, it marks the greatest epoch in chemical science.   I allude to this specimen of the first organic compound prepared synthetically from its elements by Wöhler, without the aid of vitality.   If the work of the army of chemists who have successfully attacked the problems of organic chemistry during the last quarter of a century were to be described in one word, that word would be SYNTHESIS.   In this specimen of urea we have then the germ of that vast amount of synthetical work which has done so much to dispel the superstition of vital force and to win for chemistry the position of an exact science.   In the absence of a specimen of the first oxygen from Priestley's laboratory in 1774, it seems to me that this specimen of the first sythesised urea made in 1828 is, historically, the most interesting chemical preparation the world has to show.

Secondly, we have a beautiful collection of all the compounds discovered by Liebig, but I need not dwell upon them, as they have been so recently described by their exhibitor, Professor Hofmann, in his Faraday lecture, delivered to the Fellows of the English Chemical Society.

And thirdly, there are several interesting series of specimens, illustrating the researches of Biedermann, Weltzien, Michaelis, Hübner, Hofmann, Lieberman, Oppenheim, Pinner, Wichelhaus, Tiemann, and others.

We come now to a review of that sub-division of the Chemical Section which illustrates original research, viz., chemical compounds discovered in certain specific investigations, and apparatus used in the prosecution of research. Whilst the sub-division which I have been describing illustrates for the most part the training of the young chemist in habits of observation and in the use of apparatus and processes, the one we are now considering aims at representing, so far as it can be objectively represented, the highest outcome of this training—the additions to our knowledge acquired through the accurate methods of observation and experiment, which it is the function of the chemical instructor to teach. I have already remarked on the interest and importance of exhibits of this class, and it is to be regretted that out of so many chemical investigators so few have exhibited. It is characteristic of the direction long taken by chemical research, that of about twenty-five exhibitors only two have contributed mineral as distinguished from organic products.

Professor Roscoe exhibits sixty-five compounds of vanadium discovered and investigated by himself. This classical research stands out as a model of thoroughness, and not only clearly discloses the habits of a comparatively rare metal, but brings to light some new and interesting facts in connection with the theory of atomicity. As Professor Roscoe has consented to deliver an address on these compounds, we shall have an opportunity of discussing the peculiarities and anomalies which have presented themselves in the course of this investigation.

The water of crystallisation of salts has been the subject of some controversy amongst chemists of late. It is generally considered to be present in atomic proportions, however complex these may sometimes be, and most chemists are inclined to regard the bond of union between this water and the salt proper in the light of a *molecular*, as distinguished from an *atomic*, attraction. Mr. Walcott Gibbs, however, has recently endeavoured to show that the union is strictly atomic, and subject to the ordinary laws of atomicity. The subject has attracted the attention of Professor Guthrie, who has attacked it from a new side, and obtained results which throw much light on this question. He has promised to give us an address on the subject at the next Chemical Conference. Professor Guthrie also exhibits—

NITROXIDE OF AMYLEN. Discovered by the exhibitor. Of historical interest as being the first instance in which nitroxyl NO_2 was shown to behave as a halogen in uniting directly with an olefine to form a body homologous with "Dutch liquid." The composition of the body is C_5H_{10} $(NO_2)_2$

SULPHIDE OF ŒNANTHYL.—Discovered by the exhibitor, and of historical interest as being the first instance in which a term of a higher alcohol series was made from terms of lower alcohols. It is formed by the action of zinc—ethyl on sulpho-chloride of amylene.

And NITRATE OF AMYL.—Discovered by M. Balard. Its therapeutic action discovered, and its introduction into the pharmacopœia recommended, by the exhibitor ; and it is now coming into use in tetanic and other nervous affections.

A series of twenty-three specimens of hydrocarbons derived from Pennsylvanian petroleum is exhibited by Prof. Schorlemmer. They form a striking record of the skill with which a most laborious and difficult investigation has been conducted.

Very interesting and important are the ethyl compounds derived from isolated radical methyl exhibited by Mr. W. H. Darling. The results of some experiments made by myself seemed to indicate that the products of the action of chlorine upon methyl were not ethyl compounds ; but the experiments of Schorlemmer and Darling, conducted with much larger quantities of material, show that my conclusion was erroneous. Mr. Darling exhibits ethylic chloride, ethylic alcohol, ethylidenic chloride and sodic acetate, all made from electrolytic methyl.

Mr. Perkin has sent a large collection of specimens illustrating his researches on mauneine, artificial alizarin, artificial coumarin, glyoxylic acid and other subjects. His investigation of glyoxylic acid seems to have at last put an end to the controversy as to the possibility of two semimolecules of hydroxyl being united with one and the same atom of carbon.

Amongst the other exhibits in this department are numerous and important contributions from the laboratories of St. Petersburg, Louvain and Edinburgh. For several years past, chemical research has been actively carried on in Russia.

The apparatus used in Research exhibited in the Chemical Section has suffered much from the depredations of the physicists, for although chemistry is essentially founded upon measurements of weight and volume

the instruments used for such determinations have been swept almost
en masse into the section of measurement ; nevertheless, the Chemical
Section contains several objects of unusual interest. The apparatus
with which chemists, both ancient and modern, prosecuted their
researches was generally of a simple description and often dismantled
as soon as the necessary operations were completed, consequently it
was far less likely to be preserved than the more expensive and elaborate
contrivances of the physicist. Here, however, is Black's balance presented to the Science and Art Museum of Edinburgh, by the Right
Hon. Lyon Playfair. Upon this balance Dr. Black ascertained in 1757,
the loss of weight suffered by carbonate of magnesia and limestone
when exposed to heat. Hales previously used a balance for this purpose, but the instrument before us was certainly one of the first
employed for quantitative chemistry. The balances used by Cavendish, Davy, Young, and Dalton are here, and each one of them has its
own historical interest for the chemist. The balance of Cavendish is
probably the instrument with which in 1783 or 1784 he first ascertained
that a globe filled with a mixture of oxygen and hydrogen gases
underwent no alteration in weight when the mixture was exploded.

From gravimetric instruments we are naturally led to volumetric
apparatus used in quantitative chemistry, and I will now, in conclusion,
briefly direct the attention of the Conference to apparatus used in the
analysis of gases, in the hope that a discussion of the merits and
defects of the numerous instruments now before me may have the
effect of directing a larger share of attention to eudiometric chemistry
than has hitherto been accorded to it. This branch of chemical
analysis originated in the attempts of Fontana, Landriani, Scheele,
Priestley, Cavendish, Gay Lussac, Dalton, and others, to determine the
volume of oxygen in samples of atmospheric air taken from various
localities. In these primitive instruments air was exposed to the
action of some substance either solid, liquid, or gaseous, which combined with the oxygen and left the nitrogen unacted upon. The chief
substances used were phosphorus, potassic sulphide, nitric oxide, a
solution of nitric oxide in ferrous sulphate, and a mixture of sulphur
and iron filings. Many of the instruments were of simple, or even
rude construction, and little calculated to inspire confidence in the
lurests. Nevertheless, the accuracy of a determination often depends

much more upon the skill of the operator than upon the construction of the instrument used; and thus Cavendish, with nitric oxide as his reagent and water as the confining liquid, made many hundred analyses of air, collected in various localities, in 1781, and found the percentage of oxygen to be invariably 20·83, a number nearly identical with those obtained by Bunsen and Regnault with much more perfect means. But the average chemist of that day obtained the most discordant results with the same apparatus and materials, and would doubtless also do so at the present day. By improved apparatus and methods, the work of the average chemist is made to equal, or nearly so, that of the most skilful.

Volta introduced a new reagent—hydrogen—for the determination of oxygen, and he was the first to employ the electric spark in eudiometry. The use of mercury, instead of water, for confining the gases eliminated, the source of fallacy caused by transfusion through the latter liquid, and, lastly, Bunsen, in the year 1839, brought Volta's eudiometer to its highest degree of perfection.

The PRESIDENT then proceeded to describe and criticise the various forms of apparatus for the analysis of gaseous mixtures invented or improved by Bunsen, Regnault, Frankland and Ward, Williamson and Russell, and McLeod. He concluded as follows :—

Every eudiometric chemist will agree that the following qualifications are very desirable in a good apparatus for the analysis of gases :—

1. The determination of the gaseous volumes should be made in a manner entirely independent of the pressure and temperature of the external atmosphere.

2. Such determinations of volume should also be self-correcting as regards the tension of aqueous vapour and the variations in the density of mercury.

3. Each change of volume should be expressed by a numerical difference as large as possible.

4. In order to avoid the inconvenience and loss of time occasioned by tedious calculations, references to tables, &c., the numerical expression of each volume actually read off should either be the true and corrected volume, or a number from which such volume can be at once obtained by the most simple arithmetical process.

The advantages of the 1st and 2nd of these qualifications are :

The apparatus does not need a room of nearly constant temperature, and the accuracy of the results is not affected by errors in the determination of expansion of gases by heat, the tension of aqueous vapour, or the expansion co-efficient of mercury. The 3rd secures delicacy of readings, and the 4th great economy of time.

Such are the modern developments of the eudiometer now at the disposal of chemists. For rapidity of working and delicacy of measurement they leave nothing to be desired; indeed, as regards delicacy, it may be doubted whether amongst all the apparatus for measurement in this exhibition, there is one which can, like some of these instruments, give a distinct value, in weight or volume, to the one fourteen-millionth part of a gramme of matter. Their drawback is their fragility, and any improvements tending to diminish this would doubtless be welcomed by chemists. I trust we may have some suggestions from those present, calculated to bring endiometry into more general use in chemical laboratories.

Professor ABEL, F.R.S. : I feel sure it will be the unanimous wish of all who have listened to the interesting address with which Dr. Frankland had favoured us, that we should record formally our gratitude to him for having placed before us in so lucid a manner, an account of the articles which are exhibited in the chemical section of this important collection, and that we should also thank him for the more than suggestive manner in which he has treated several portions of the very extensive field over which he has travelled. Subjects, such as the last one he has dwelt upon more particularly, to which, among many others, he has devoted with very great success, a very large share of his attention and study would also suffice for a long series of interesting addresses and discources. We can only admire the ease and succinctness with which Dr. Frankland has travelled on this occasion over so vast a ground as that which is included in the subjects he has placed before us. I ask you, therefore, to be so good as to record to Dr. Frankland a hearty vote of thanks for the valuable address with which he has favoured us.

The PRESIDENT : I will now ask Dr. Russell to give us a more explicit account of the apparatus contrived by himself and Dr. Williamson, than that which I have been able to put before you.

Dr. RUSSELL : I am sorry to hear there is no specimen of the apparatus referred to in the Exhibition. I thought there had been one. In the first place, as I happen to have had a certain amount of experience in working both the apparatus suggested by Dr. Williamson and myself, and the one which you, sir, have described, I should like to say a word or two with regard to what practically I have found to be the merits and demerits of these two forms of apparatus. It was an exceedingly interesting and graphic account which you have given us of this very important branch of chemical science, and I think it is worth while in passing, just to call attention again to one point. I think it is seldom that so great an advance in manipulatory processes is made in so short a time. We had in the apparatus of Bunsen, an admirable method of gas analysis ; the only point in which it failed was the length of time which it occupied. Now from Bunsen we pass to Regnault's method, which did away with that length of time ; but undoubtedly there was this defect, that although we gained so much in that way, we lost considerably in the accuracy of the process ; and it certainly is to your process that we owe this increase of accuracy, together with great rapidity of operation. I think very seldom has an operation been so improved in that way ; and I hardly think it is any exaggeration to say, that with your apparatus ordinary analyses can be made almost in as many minutes as it would have taken hours with a Bunsen apparatus. I believe that as now made there is no loss of accuracy, and yours gives results equal, if not superior, to the Bunsen apparatus. With regard to your apparatus, allow me to say one thing further, which is this,—I think you will remember that when it was first introduced I had the honor of working with you, and using the apparatus a good deal, and at that time there was a difficulty, which I will say one word upon. It did occasionally leak. There are stop-cocks necessary to the apparatus, and iron plates, which have to be united together, and in many cases, the stop-cocks did not hold perfectly tight, and even polished plates do occasionally let in air. But I am bound to say that, at the present time, those inaccuracies have been got over to a wonderful extent, certainly much more than I should have imagined was possible ; and, practically, now in my own laboratory I have that apparatus working, and it is the rarest thing to have any inconvenience arising from leakage. If used with ordinary tact and care, I believe it is practically, in almost all cases, quite tight. Formerly,

certainly, that was not the case, and it was with this inconvenience
strongly impressed upon us that Dr. Williamson and myself undertook
to try if we could make a gas apparatus not subject to this inconvenience.
We set before us as a standard that the apparatus should be without
any stop-cocks, and all sorts of leakage of that kind. The form we
ultimately recommended was the one you have alluded to. We took
the ordinary eudiometer of Bunsen. There you have an apparatus of
great simplicity and accuracy. The alteration we introduced was
simply this :—We eliminated entirely the observation of the thermometer and barometer, and the long and tedious calculations which
they necessitated. We eliminated again the necessity for having a
constant temperature. It signified not at what temperature we worked,
and it signified not whether any change of temperature took place
during the operation, or whether the barometer altered. To enable
us to do this we simply took a volume of air in a tube with a mark
upon it ; we kept that quantity of air always at the same tension; if
the thermometer altered, or the barometer altered, this tube was raised
or lowered, so that the volume was increased or diminished accordingly,
and the gas under manipulation was always brought to this particular
degree of elasticity. It did not signify what the temperature or
pressure was, the degree of elasticity being the same in all operations.
If you wanted to ascertain the exact weight of the gas operated upon,
all you had to do was to determine, once for all, what that temperature
and pressure was. This method was certainly exceedingly simple, and,
as far as it goes in that way, I believe it is one of the most accurate
and easily performed operations in gas analysis. The objection to our
apparatus is simply this, that the other manipulations connected with it
are far more cumbrous and troublesome than those with your apparatus.
The attaching of that laboratory tube, the introduction of the liquid re-
agents, the taking away the other tube, are all operations which occupy
only a few minutes. They are of the greatest simplicity, and can be
done very easily, far more so than the operations with the Bunsen
apparatus, and far more readily and quickly than with the apparatus of
Dr. Williamson and myself. At present, the special use of our
apparatus is for the measurement of large quantities, and then the
measurement with the pressure tube is very simple, and goes very well.
I think either in its present form, or some other one, it will be in future

still more used, but for simple measurements and analyses of small quantities of gas, I much prefer your apparatus to the one we have devised.

The PRESIDENT : I see Mr. Thorp in the room. He has worked with these apparatus, and it would be interesting to know what his opinion is about the desirability of improvements in these instruments and as to their working capacities in their present condition.

Mr. THORP : I am under the great disadvantage that I have only just this moment entered the room, and I really do not know what has been said. But after a great deal of practice, especially with the smaller form of the apparatus, I must say that I think for all gas operations, especially those which do not require explosion, that leaves very little to be desired, particularly for measuring very small quantities. I have not worked so much with Mr. McLeod's larger form, but from what I did do with it I was led to think that it was perhaps a little deficient in the measurement of very small quantities. In that case, the difference of pressure between two small quantities of gas, differing only slightly in volume, is very small and somewhat difficult to read, and errors of reading are rather important in that case. I have thought whether some modification involving the adoption of the form of tube used in the smaller apparatus to the large apparatus might not be introduced. It would involve some loss of strength, but I fancy that glass-blowers now could manage to give us that strength and yet retain the tube with a narrower diameter at the top. In that case, the difference of pressure, as read off the pressure tube, would be greater, and easier to read. The smaller apparatus can also be used very satisfactorily even for explosions by the adoption of a special laboratory vessel constructed for that purpose, the explosions taking place in the laboratory tube itself. The disadvantage is, that for safety's sake, the laboratory vessel has to be disconnected at every explosion, and unless you have perfect faith in the joints and clamp, that involves some risk of loss. The great advantage claimed for the large apparatus is, that it obviates some amount of calculation, the correction from the barometer being entirely avoided. This, however, is a matter of small moment, because after the barometer is carefully read with the smaller apparatus I do not think it leaves anything to be desired in point of accuracy and the saving of time in calculation is exceedingly small. I found after a great deal of practice that I could make the calculations in water analysis, correcting for barometric pressure and also for the thermometer, and calculating out the quantity of carbon and nitrogen from the readings of three distinct volumes of gas in about six minutes, so that the amount of saving in time possible in that must be exceedingly small, and for the rapidity of manipulation I take it the two forms are about on a par. I found I could make these three readings, in addition to two absorptions, one of carbonic anhydride and the other of oxygen, in about 25 minutes, so that the actual analysis and the calculation can be done in half an hour; but really in something less, because the time occupied by the absorption may be used to calculate out previous analysis, so that I think the time may be put down as about 25 minutes. I do not know whether the improvement suggested of the adoption in the larger apparatus of the form of tube used in the smaller one is possible. That would be worthy of consideration, but really, if you do not want too many explosions, the small one leaves very little to be desired.

Professor McLeod : I must rather object to the suggested improvement that Mr. Thorp has made in the explosion apparatus. The idea of using a narrow tube, was first suggested ; but it was supposed that there would be an imperfect combustion of the gases in the upper portion of the tube, and that was the reason it was not introduced into this apparatus. I fancy, however, that the difficulty of measuring small volumes might be overcome to some extent by using smaller divisions. The apparatus Mr. Thorp has been in the habit of using had a wider tube than this one, and if I mistake not, the distance between the top of the tube and the first division was rather longer, so that the volume contained there gave a smaller pressure than in this one. If the first division could be placed a little higher up, a sufficient pressure to measure even very small volumes would be obtained. There is one point with regard to this apparatus which gets over some of the difficulties Dr. Russell mentioned with regard to the first form ; all the connections are glass and india-rubber; there are no metal plates to bring together except at the junction above ; all the stop-cocks are carefully ground and polished with rouge and a solution of soda, so that there is not the slightest trace of leakage, even when there is a complete vacuum in the measuring tube. There is a point that I wish to press very

c

strongly on manufacturers of apparatus—with regard to whom, I must quite agree with the remarks of the President as to the improvements they often introduce. The stop-cocks must be most carefully made. I am told by a gentleman who lately bought an apparatus of this kind, that when he got it home, from every stop-cock there poured fountains of mercury. In that case it was useless for gas analysis, for the moment the slightest exhaustion was placed on the apparatus, the air leaked in, and the whole operation had to be commenced again. With this particular apparatus, I can say that when used some years ago there was not the slightest fear of leakage; the stop-cock at the top of the measuring tube was left open, reliance being placed on the tightness of the other stop-cock and on the joint, no leakage being noticed. The apparatus worked well, and some of the analyses which are described in the *Chemical Society's Journal* show that it gave very accurate results.

The PRESIDENT: Mr. Thomas has worked a good deal with this apparatus, and perhaps he may have some observations to offer.

Mr. THOMAS: The apparatus I have worked with myself is that modification of Dr. Frankland's, by Mr. McLeod. The objections which may be raised against the analysis by this method, are very few ; the readings are extremely accurate, and the analysis may be performed in a short time. I think, however, there are a few alterations which may be made with advantage. The connections in the bottom, which have not been referred to, are usually done by means of india-rubber, and there are three joints and two taps. I think if the glass connections were entirely done away with, and a steel bush introduced, which would be regulated by one tap of peculiar construction, a three-way tap, and a steel tube connected at the bottom with the flexible india-rubber tube, it would do away with the complications of the lower portion of the apparatus, and there would be less liability to leakage. Another objection I find is this: by the manner in which the clamp joints are separated, and the introduction of india-rubber for the purpose of keeping the tubes in their positions, and preventing the liability of fracture, the barometer tube, and also the eudiometer, are very apt to gradually get lower down, and so, in performing very accurate analyses, it is necessary to read off the level at the commencement, and also at the termination of the operation. By the introduction of a steel portion into the bottom, these objections are entirely done away with, because the glass tubes may be introduced into sockets in the steel portion with an india-rubber lining round them, and a small steel ring, screwed upon the india-rubber, holds it firm in place, in such a manner that it is impossible for the tubes to get lower down. The other connections, as Professor McLeod has remarked, are extremely good. If you grind the glass stop-cocks and the steel faces accurately, they will remain constant almost any length of time. I have worked with Mr. McLeod's apparatus six months at a time, and have never had the slightest difficulty in keeping the apparatus tight. I adopt, also, the same method he has mentioned, the using one only of the glass taps, during the time the apparatus was attached, for the purpose of absorbing by liquids. With regard to the analysis of compound gases, in Watt's Dictionary and other treatises on gas analysis, we find certain recommendations to add large volumes of air or oxygen to lessen or modify the force of explosion. The construction of this form of apparatus entirely removes this necessity. If you take five cubic centimetres of electrolytic gas, and place it under expansion until the mercury in the barometer tube shall only remain about 190 millimetres in height, or under an expansion of four times the volume, the gas can be readily and accurately exploded. The whole of the oxygen and hydrogen will have disappeared, and the explosion will be extremely light. Again, if you expand it about six times, under a pressure of only 127 millimetres of mercury, no explosion occurs at all; in fact, the spark will not ignite the gases. In working with explosive gases, for instance, marsh gas, with two and-a-half volumes of oxygen, may be safely and accurately exploded in the manner I have described, if you place it under a pressure of only 152 millimetres, or expand it to five times the volume. The contraction will be exactly twice the volume, and the carbonic acid formed will exactly represent the volume of gas used in the first instance. Again, with other explosive gases, hydride of ethyl, for instance, the necessary quantity of oxygen is three and-a-half times the volume, and with this, when expanded, to about 150 millimetres, it may be exploded safely and accurately. Ethyl under about 120 millimetres of mercury, may be exploded in the same manner. I made several quantitative experiments on various gases, and find it is quite as accurate as the analysis which can be made by the introduction of a large volume of air or oxygen. After the absorption of carbonic acid it is necessary to add hydrogen to explode the excess of oxygen, and you cannot work small quantities. You introduce such a volume of these gases as to make the volume unnecessarily large to work with. It is quite possible with this apparatus to use a much larger volume of explosive gases than could otherwise be employed. With ethyl, for instance, you require to add at least six and-a-half times the volume of oxygen for complete explosion, and it is necessary in order to work under ordinary atmospheric pressure, to add at least fourteen volumes of air, and three or four volumes of oxygen to modify the force of the explosion.

The PRESIDENT: I feel sure you will all agree with me that we have been fortunate in having at the Conference this morning gentlemen who have for so many years worked at eudiometry, and who have put before us their extensive experience. It has struck me during some of the remarks which have been made, that we have been rather too hard on the instrument makers. We must admit that one of the great improvements which Dr. Russell pointed out, was due to greater care on the part of the manufacturer in making stop-cocks, which, as Dr. Russell remarked, were a constant source of annoyance in the early use of instruments of the Regnault type. These stop-cocks are now manufactured of glass, and are usually so carefully ground that no leakage is experienced in the use of the apparatus. The leakage again in the steel plates, by which the laboratory tube and the eudiometer are connected together, which was also a constant source of annoyance in the early days of these instruments, has been obviated to a very great extent by the efforts of the manufacturers, although still more so by a very ingenious contrivance, for which we are indebted to Professor Bischoff. That is a contrivance which, as will be seen in one of the instruments exhibited by Messrs. Cetti and Co., consists in hollowing out in one of those steel faces a shallow ring, and of making a corresponding projection on the other steel face, this hollow, being filled with cerate, which is squeezed very tightly between the faces, when they are clamped together with the screw. This apparatus renders leakage from these plates almost impossible. I ought also, in conclusion, to direct the attention of the Conference to the beautiful examples of these eudiometric instruments which are exhibited by

Messrs. Cetti and Co. in the adjoining room. I will now ask Dr. Gilbert to read his paper "On some points in connection with Vegetation."

Shortly after the commencement of Dr. Gilbert's paper the Conference adjourned for luncheon, and on its re-assembling the chair was taken by Dr. Gladstone.

On Some Points in Connection with Vegetation.
By Dr. J. H. Gilbert, F.R.S., F.L.S., F.C.S.

The subject of vegetation is such a very wide one, and might be treated of in so many different ways, that it seems desirable to state at the outset what is the scope, and what are the limits, of the discussion which I propose to bring before you. I propose, then, to confine attention almost exclusively to the question of the sources of the nitrogen of vegetation in general, and of agricultural production in particular. I propose further to treat of this subject mainly in the aspects in which it has forced itself upon the attention of Mr. Lawes and myself during the now thirty-three years of our agricultural investigations; and, also, in so far as it illustrates, and is illustrated by, the objects contributed by Mr. Lawes to the Exhibition around us.

Before entering upon the special subject matter of my discourse, I must claim the indulgence of those present who are already well acquainted with the main facts of the chemistry of vegetation, whilst I call attention, very briefly, to some rather elementary matters, with a view of rendering what has to follow the more intelligible to any who may be less fully informed on the subject.

When a vegetable substance is burnt—as a familiar instance, let us say tobacco, for example—the greater part of it is dissipated, but there remains a white ash. The ashes of crude or unripe vegetable substances are found on analysis to contain most, or all, of the following constituents, namely:—

Oxide of iron, oxide of manganese, lime, magnesia, potass, soda, phosphoric acid, sulphuric acid, chlorine, and silica.

Rarer substances than these are also sometimes found. Now, much has of late years been established in regard to the occurrence, and the offices, of some of these substances in plants; but I do not propose to touch upon the questions herein involved. It will suffice further to say in regard to these incombustible, or "mineral" constituents, that the ash of one and the same description of plant, growing on different soils, may, so long as it is in the growing or immature state, differ very much in composition. Again, the ashes of different species, growing on the same soil, will differ very widely in the proportion of their several constituents. But it is found that the nearer we approach to the elaboration of the final products of the plant—the seed, for example—the more fixed is the composition of ash of such products of one and the same species. In other words, there is very little variation in the composition of the ash of one and the same description of seed, or other final product, provided it be evenly and perfectly matured. This fact alone, independently of all that has been established of late years in regard to the office or function, so to speak, of individual mineral constituents of plants, would be sufficient to indicate the essentialness of such constituents for healthy growth; and it is obvious that they must be provided *within the soil.*

But now as to the combustible constituents—the carbon, the hydrogen, the oxygen, and the nitrogen. Leaving out of consideration such exceptional cases as those brought to light in Mr. Darwin's beautiful investigation on insectivorous plants, and also the sources of the organic substance of fungi, and perhaps of some forced horticultural productions, it may be stated that the source of the carbon of vegetation generally is the carbonic acid existing in very small proportion, but in large actual amount, in the atmosphere; that the source of the hydrogen is water; and that the source of the oxygen may be either that in carbonic acid, or that in water. With regard to the nitrogen the case is, however, by no means so simple. Not that there are no questions still open for investigation in regard to the assimilation by plants of their incombustible or mineral constituents, or of their carbon, their hydrogen, and their oxygen; but those relating to the sources, and to the assimilation, of their nitrogen, are not only in many respects of more importance, but seem to involve greater difficulties in their solution.

What, then, are the sources of the nitrogen of vegetation? Are, they the same for all descriptions of plants? Are they to be sought entirely in the soil? or entirely in the atmosphere? or partly in the one, and partly in the other?

As the combined nitrogen coming down from the atmosphere in rain, hail, snow, mists, fog, and dew, does undoubtedly contribute to the annual yield of nitrogen in our crops, let us first briefly consider what is known as to the amount of it annually so coming down over a given area of the earth's surface; and as we are here discussing the subject in England, I will adopt the English pound as the unit of weight, and the English acre as the unit of area. The following table shows the amount of nitrogen coming down as ammonia and nitric acid in the total rain, hail, snow, and some of the minor deposits during the years 1853, 1855, and 1856, at Rothamsted (Herts), the nitric acid being in all cases determined by Mr. Way, and the ammonia in some cases by him, and in others by ourselves :—

TABLE I.

Combined Nitrogen in Rain and Minor Aqueous Deposits at Rothamsted.

	Nitrogen per acre, per annum, lbs.			
	1853.	1855.	1856.	Mean.
As ammonia	5·67	5·86	7·85	6·46
As nitric acid	(not determined)	0·77	0·73	0·75
Total		6·63	8·58	7·21

Numerous determinations of the ammonia and nitric acid in rain, and the other aqueous deposits, have been made in various parts of France and Germany, some in the vicinity of towns, and some in the open country. Of the latter, which are the most to our purpose, it may be stated that those of Boussingault at Liebfrauenberg, in Alsace, generally indicate a larger proportion of the nitrogen existing as nitric acid, and less as ammonia than our own; but, upon the whole, the observations in the two widely separated localities mutually confirm one another. Of the results of others, in other localities, some show about the same amount of combined nitrogen so deposited as our own, some, however, much more, and some much less, than ours; but the determinations on the Continent generally show a higher proportion of the total combined nitrogen to exist as nitrates than those in this country. It may be added, that numerous determinations of the combined nitrogen in rain, dew, &c., collected at Rothamsted, have much more recently been made by Professor Frankland, and his results, which are published in the " Sixth Report of the Rivers Pollution Commission," are substantially confirmatory of the earlier determinations, summarized in the foregoing Table, but upon the whole they indicate lower amounts. Lastly, M. Marié-Davy determined the ammonia in the rain, &c., collected at the Meteorological Observatory at Montsouris, Paris, during the last six months of 1875; and the amount of ammonia so coming down, even within the walls of Paris, only represented 5·25 lbs. of combined nitrogen per acre, or only 10·5 lbs. per acre, per annum. M. Marié-Davy did not make a complete series of determinations of the nitric acid in the meteoric waters, but his initiative results agree with the experiments of others in showing the amount of combined nitrogen so existing to be comparatively small.

Thus, the determinations hitherto made of the amount of combined nitrogen coming down in the measured aqueous deposits from the atmosphere, do not justify us in assuming that the quantity available from that source will exceed 8 or 10 lbs. per acre, per annum, in the open country, in Western Europe. It should be observed, however, that the amount of ammonia especially is very much greater in a given volume of the minor aqueous deposits than it is in rain; and there can be little doubt that there would be more ammonia deposited from them within the pores of a given area of soil, than on an equal area of the non-porous even surface of a rain gauge. How much, however, would thus be available to the vegetation of a given area beyond that determined in the collected and measured aqueous deposits, we have not the means of estimating with any certainty. On the other hand, numerous independent determinations, by both Dr. Voelcker and Dr. Frankland, of the nitric acid in the drainage-water collected from land at Rothamsted which had been many years unmanured, lead to the conclusion that there may be a considerable annual loss of nitrogen by the soil in that way.

The next point to consider is, what is the amount of nitrogen annually obtained over a given area, in different crops, when they are grown without any supply of it in manure. This point may be illustrated by the results obtained in the field experiments on Mr. Lawes' farm at Rothamsted, which have now been in progress for about a third of a century. Table II., which follows, shows the yield of nitrogen per acre, per annum, in wheat, in barley, and in root crops, each grown for many years in succession on the same land, either without any manure, or with only a complex mineral manure, that is supplying no nitrogen.

TABLE II.

Yield of Nitrogen per acre, per annum, in Wheat, Barley, and Root Crops, at Rothamsted.

Crop, &c.	Condition of Manuring, &c.	Duration of Experiment.	Average Nitrogen per acre, per annum.
			lbs.
Wheat.	Unmanured	8 yrs. 1844-'51	25.2
		12 yrs. 1852-'63	22·6
		12 yrs. 1864-'75	15·9
		24 yrs. 1852-'75	19·3
		32 yrs. 1844-'75	20·7
	Complex Mineral Manure ...	12 yrs. 1852-'63	27·0
		12 yrs. 1864-'75	17·2
		24 yrs. 1852-'75	22·1
Barley.	Unmanured	12 yrs. 1852-'63	22·0
		12 yrs. 1864-'75	14·6
		24 yrs. 1852-'75	18·3
	Complex Mineral Manure ...	12 yrs. 1852-'63	26·0
		12 yrs. 1864-'75	18·8
		24 yrs. 1852-'75	22·4
Root Crops.	Complex Mineral Manure. { Turnips ...	8 yrs. 1845-'52	42·0
	Barley... ...	3 yrs. 1853-'55	24·3
	Turnips ...	15 yrs. 1856-'70[1]	18·5
	Sugar-beet...	5 yrs. 1871-'75	13·1
	Total	31 yrs. 1845-'75	26·8

Bearing in mind what has been said as to the amount of combined nitrogen known to be annually deposited from the atmosphere, the figures in Table II. have great interest and significance. Thus, over a period of 32 years, the wheat has yielded an average of 20·7 lbs. of

---

[1] Thirteen years' crop—two years failed.

nitrogen, per acre, per annum, without manure. But if we look at the quantities yielded during the first 8, the next 12, and the last 12 years of that period, it is seen that there has been a gradual, but at the same time a considerable decline in the annual yield. From this it would appear probable that the nitrogen of the soil, derived from previous accumulations, is being gradually reduced. Whether or not the whole of the excess of yield over that available from the rain, and other measured aqueous deposits from the atmosphere, is due to previous accumulations within the soil, and is therefore inducing a gradual exhaustion of its stock of nitrogen to that extent, we have not conclusive evidence to show. Determinations of nitrogen in samples of the soil taken at different times during the course of the experiments do, indeed, show an appreciable reduction. It is probable, however, that a part of the excess of yield is due to condensation of ammonia within the pores of the soil, beyond that which would be deposited in rain, and in the dew and other minor deposits condensed on the non-porous even surface of a rain-gauge, as already referred to.

Excluding the first 8 years of the growth of wheat, it is seen that whilst over the next 24 years, 1852-1875, the wheat yielded 19·3 lbs. of nitrogen, per acre, per annum; the barley yielded an average of 18·3 lbs. over the same period. Again, during the first 12 of the 24 years, the wheat yielded 22·6 lbs., and the barley 22 lbs.; whilst, during the second 12 years, the yield in wheat was reduced to 15·9, and that in the barley to 14·6 lbs. The similarity in the yield of nitrogen over the same periods in these two closely allied crops, growing in different fields, is very striking, though, upon the whole, the indication is that the autumn-sown wheat has accumulated more than the spring-sown barley.

It is next to be observed that the annual use of a complex mineral manure has but very slightly increased the yield of nitrogen in either of these gramineous crops; and it is probable that the increased yield, such as it is, is derived from the previous accumulations within the soil, and not from atmospheric sources.

To sum up the evidence in regard to the sources of the nitrogen of these two typical gramineous plants, when none of it is supplied to them by manure, though it is not conclusively shown whence the whole of it is derived, it would at any rate appear probable, that it may be accounted for by the combined nitrogen coming down in rain, and in the other measured aqueous deposits from the atmosphere, by the condensation of the ammonia of the air within the pores of the soil, and by the previous accumulations within the soil.

Let us now consider what is the yield of nitrogen by plants of other natural families, and first of all by certain so-called "root-crops"—turnips of the natural order cruciferæ, and sugar beet of the order chenopodiaceæ. On this point we have the experience of 31 years, excepting that during three of those years barley was grown without any manure in order to equalise the condition of the land as far as possible before re-arranging the manuring, and during 2 other years the turnips failed and there was no crop.

It should be premised that when root-crops are grown without manure of any kind, there is after a few years scarcely any produce at all; and hence the results recorded in the table are those obtained by the use of mineral manures, but without any supply of nitrogen. It is seen that during the first 8 years of turnips, there was an average yield of 42 lbs. of nitrogen per acre per annum. During the next 3 years barley yielded 24·3 lbs. annually. During the next 15 years, 13 with Swedish turnips, and two without any crop, there was a yield of 18·5 lbs. per acre annually. During the last 5 years sugar-beet yielded 13·1 lbs. per acre per annum. Lastly, over the whole 31 years, during which there were 3 crops of barley, 2 years without any crop, 21 years of turnips, and 5 of sugar-beet, the average annual yield was 26·8 lbs. of nitrogen.

Here, then, we have a reduction to less than one-third during the later compared with the earlier years, and to a lower point than even with either wheat or barley; though, during the whole period, the annual yield is higher than with either of the two gramineous crops. It may be mentioned that we have other experimental evidence showing that the so-called "root-crops" exhaust at any rate the superficial layers of the soil of their available supplies of nitrogen, more completely than perhaps any other crop. It may further be added that the surface soil has shown during recent years a lower per centage of nitrogen than that of any of the other experimental fields. We have fair grounds for concluding, therefore, that if in the cases of the wheat and the barley the nitrogen yielded beyond that retained by the soil from the direct measurable aqueous deposits, together with that condensed within the pores of the soil, from the atmosphere, be derived from previous accumulations within the soil, so also may the excess of yield by the so-called "root-crops" be accounted for.

We now come to the consideration of the yield of nitrogen when plants of the *leguminous* family are separately grown, or when they, and plants of some other families, are grown in alternation, or in association, with the gramineæ. Table III. shows the results obtained with beans, and with clover ; with clover and barley grown in alternation ; and with turnips, barley, clover or beans, and wheat, grown in an actual course of rotation.

TABLE III.

Yield of Nitrogen per Acre per Annum in Beans, in Red Clover, and in Rotation.

Crops, &c.	Conditions of Manuring, &c.	Duration of Experiment.	Average Nitrogen per acre, per annum.
			lbs.
Beans	Unmanured	12 yrs. 1847-'58	48·1
		12 yrs. 1859-'70(1)	14·6
		24 yrs. 1847-'70	31·3
	Complex Mineral Manure ...	12 yrs. 1847-'58	61·5
		12 yrs. 1859-'70(1)	29·5
		24 yrs. 1847-'70	45·5
Clover	Unmanured	22 yrs. 1849-'70(2)	30·5
	Complex Mineral Manure ...	22 yrs. 1849-'70(2)	39·8
Barley Clover	} Unmanured ... ... {	1 yr. 1873	37·3
		1 yr. 1873	151·3
Barley	Unmanured ... { After Barley	1 yr. 1874	39·1
	{ After Clover	1 yr. 1874	69·4
	Barley after Clover more than } after Barley ... ...}		30·3
Rotation 7 Courses	(1 Turnips ⎫ ⎧Unmanured ⎨2 Barley ⎬ ⎨ ⎪3 Clover or Beans⎪ ⎨Superphos- ⎩4 Wheat ⎭ ⎩ phate ...	28 yrs. 1848-'75 28 yrs. 1847-'75	36·8 45·2

(1) 9 years Beans, 1 year Wheat, 2 years Fallow.
(2) 6 years Clover, 1 year Wheat, 3 years Barley, 12 years Fallow.

Referring first to the results obtained with beans, the table shows that without manure there was an annual yield over the first 12 years, 1847—1858, of 48·1 lbs. of nitrogen. Over the next 12 years, 1859—1870, it was reduced to 14·6 lbs. per acre per annum. Still, over the whole period of 24 years, we have an annual yield of 31·3 lbs., or more than one and a half time as much as in either wheat or barley.

In the case of wheat and barley it was seen that a mixed mineral manure increased the yield of nitrogen to a very small degree only. Not so in the case of the leguminous crop, beans. During the first 12 years a complex mineral manure, containing a large amount of potass —I call attention to this fact because we have abundant evidence that it is the potass chiefly that is effective—gave 61·5 lbs. of nitrogen per acre per annum against 48·1 lbs. obtained over the same period without manure. During the next 12 years, the potass manure gave 29·5 lbs. against scarcely half as much, or 14·6 lbs. without the potass manure. And finally, during the whole period of 24 years, the potass manure has given 45·5 lbs. of nitrogen per acre per annum, against 31·3 lbs., or only about two-thirds as much, without manure ; and we have more than twice as much yielded by a potass manure over a period of 24 years with beans than with either wheat or barley.

Before calling attention to the figures relating to another leguminous crop—red clover—it should be mentioned that leguminous crops generally are, and clover in particular is, extremely sensitive to adverse climatal circumstances ; but clover is pre-eminently sensitive to soil conditions also. Indeed, it is a fact well recognised in agriculture, that few soils can be relied upon to grow a good crop of clover oftener than once in about 8 years ; and many soils will not yield it so frequently. It will not excite surprise, therefore, that in attempting to grow clover year after year on the same land, we have only succeeded in getting any crops, and some of those poor ones, in 6 years over a period of 22. Indeed, the plant failed seven times out of eight during the winter and spring succeeding the sowing of the seed ; when, in some cases a crop of wheat or barley was taken, and in others the land was left fallow. Hence, over a period of 22 years we have had only 6 years of clover, one of wheat, three of barley, and twelve of fallow. Still, the annual yield of nitrogen over the 22 years was 30·5 lbs. without any manure, and 39·8 lbs., or nearly one-third more, by mineral manure containing

potass. Unfavourable as was this experiment in an agricultural point
of view, still it is seen that the influence of the interpolation of this
leguminous crop has greatly increased the yield of nitrogen compared
with that obtained in either wheat or barley grown continuously ;
and that, unlike the result with those crops, a potass manure has here
again, as with beans, greatly increased the yield.

Without attempting for the moment to discuss the probable source
or sources of this greatly increased yield of nitrogen by leguminous as
compared with gramineous crops, I will simply here remark in
passing that we have no evidence leading to the conclusion that
this increased assimilation is at the expense of the nitrogen existing
at any rate in the upper layers of the soil. In fact, such initiative
results as we have relating to the nitrogen in the soil of the experimental bean field, would rather lead to the conclusion that the better
the crop has grown, and the more nitrogen it has assimilated, the
richer rather than the poorer in nitrogen (as indicated by the soda-
lime method) has the surface soil become. To this point, however,
we shall have to recur presently; but in the meantime let us first refer
to the yield of nitrogen in other cases in which leguminous crops have
been interpolated with others.

It is, indeed, well known that the growth and removal of a highly
nitrogenous leguminous crop is one of the best possible preparations
for the growth of a gramineous corn crop, which characteristically
requires nitrogenous manuring. A striking illustration of this apparent
anomaly is afforded in the results next in order recorded in the Table III.

After the growth of six corn crops in succession by artificial manures
alone, barley was grown without manure in 1873 on one portion of the
same land; and on another portion clover was grown. It is calculated
that there were taken off in the barley 37·3 lbs. of nitrogen, and in the
three cuttings of clover 151.3 lbs. Yet, in the next year, 1874, barley
succeeding the barley gave 39·1 lbs., and barley succeeding the clover
gave 69·4 lbs. of nitrogen; or 30·3 lbs. more after the removal of
151·3 lbs. in clover than after the removal of 37·3 lbs. in barley.
Nor was this remarkable result to be explained by either accident
or error. For, determinations of nitrogen in four separately taken
samples of the soil, in the mixture of the four, and in the mixture of
six others, taken from each plot, and at different depths, all concurred

in showing an appreciably higher percentage of nitrogen, especially in the surface soil, 9 inches deep, of the land from which the clover had been removed than in that from which the barley had been taken; and this was so, although, in every case, all visible vegetable debris had been carefully picked out. Here, then, the surface soil at any rate was positively enriched in nitrogen (determinable by soda-lime) by the growth and removal of a very highly nitrogenous crop. It may be mentioned that Dr. Voelcker has obtained results of a similar character.

The results next to be considered are those obtained in an actual four-course rotation of crops—namely, turnips, barley, clover or beans, and wheat. The experiments have been conducted through seven such courses; that is to say, over a period of twenty-eight years. One portion of the land, the results relating to which are given in the table, has been entirely unmanured during the whole of that period, and the other has received super-phosphate of lime alone, once every four years—that is to say, for the turnips commencing each course; but it has received no other manure throughout the 28 years, either mineral or nitrogenous.

Under these conditions—that is with a turnip crop and a leguminous crop interpolated with two gramineous crops—we have, without manure of any kind, an average of 36·8 lbs. of nitrogen yielded per acre, per annum; or not far from twice as much as was obtained with either of those cereal crops, wheat or barley, grown consecutively. With super-phosphate of lime alone, which, in a striking degree increased the yield of nitrogen in the turnips, reduced it in the succeeding barley, increased it greatly in the leguminous crops, and slightly in the wheat immediately following them, we have the average annual yield of nitrogen raised to 45·2 lbs. per acre, per annum, over the 28 years; or to more than double that obtained by wheat or barley grown continuously by mineral manures alone. And it may be observed that where, in adjoining experiments, no leguminous crop was grown between the barley and the wheat, but the land was fallowed instead, the total yield of nitrogen in the rotation was very much less: the wheat succeeding the fallow yielding very little more nitrogen than that succeeding the leguminous crops which had removed so much of it. In other words, the removal of the most highly nitrogenous crops of the rotation—beans or clover—has been succeeded by a growth of wheat, and assimilation of nitrogen by it, almost as great as when it has succeeded a year of fallow—that is to say, a period of accumulation from external sources, and no removal by crops.

One other illustration must be given of the power of plants of the leguminous and some other families to assimilate more nitrogen over a given area than those of the gramineous family. But before entering upon the bearing of the results in question on this particular point, it will be necessary to digress a little to call special attention to the conditions of the experiments under which the results were obtained; and it is the more desirable to do this, since the most important of Mr. Lawes' contributions to this Exhibition is an illustration of the results I am about to refer to.

I must here forestall a little what I shall have to refer to more fully further on, as to the effects of characteristically different manuring substances on crops belonging to different botanical families. I will say briefly, then, that it is found that nitrogenous manures have generally a very striking effect in increasing the growth of gramineous crops grown separately on arable land, such as wheat, barley, or oats, all of which contain a comparatively small percentage of nitrogen, and, as has been illustrated, assimilate a comparatively small amount of it over a given area when none is supplied to them in manure. The highly nitrogenous leguminous crops, on the other hand, such as beans, peas, clover, and others, are by no means characteristically benefited by the use of direct nitrogenous manures, such as ammonia-salts or nitrates, though nitrates act much more favourably than ammonia-salts. Again, whilst, under equal conditions of soil and seasons, mineral without nitrogenous manures increase comparatively little the poor-in-nitrogen gramineous crops that are grown separately, such manures, and especially potass-manures, as has been seen, increase in a striking degree, the growth of crops of the leguminous family grown separately, and coincidently the amount of nitrogen they assimilate over a given area.

Such, then, is the result obtained in the separate growth, on arable land, of individual plants of the different families. Now, in the mixed herbage of permanent grass land, we may have fifty, or even many more species growing together, representing nearly as many genera, and perhaps eighteen or twenty natural orders or families. Of these,

the gramineæ generally contribute the largest proportion of the herbage; and, on good grass land, if the leguminosæ do not come second,
they are at any rate prominent. The degree in which other orders
are represented may be very various indeed, according to soil, locality,
season, and other circumstances. In Mr. Lawes' park, at Rothamsted,
nearly eighty species have been observed; but of many only isolated
specimens, and it may be stated generally that about fifty species are
so prominent as to be found in a carefully averaged sample of the hay
grown without manure.

Experiments on the influence of different manures on this mixed
herbage were commenced in 1856; at which time the herbage was
apparently pretty uniform over the whole area selected. About twenty
plots, from one-quarter to one-half an acre each, were marked out, of
which two have been left continuously without manure, and each of
the others has received its own special manure, and as a rule the same
description year after year—and the experiments have now been conducted over a period of twenty years.

Under this varied treatment, changes in the flora, so to speak,
became apparent even in the first years of the experiments; and three
times since their commencement, at intervals of five years—namely,
in 1862, 1867, and 1872—a carefully averaged sample of the produce of
each plot has been taken and submitted to careful botanical separation,
and the percentage. *by weight*, of each species in the mixed herbage
determined. Partial separations have also been made in other years.

Mr. Lawes has contributed a large case of specimens to the exhibition, which shows the botanical composition of the herbage on twelve
selected plots in the seventeenth season of the experiments (1872).
The quantities of the different plants there exhibited represent the
relative proportion, by weight, in which each species was found in
the mixed produce of the different plots; and the whole illustrates in a
striking manner the domination of one plant over another, under the
influence of different manures, applied year after year on the same plot.

The general results of the experiments may be briefly summarised as
follows:—

The mean produce of hay per acre per annum has ranged, on the
different plots, from about 23 cwt. without manure to about 64 cwt. on
the plot the most heavily manured.

D

The number of species found has generally been about 50 on the unmanured plots, and has been reduced to an average of only 20, and has sometimes been less, on the most heavily manured plots.

Species belonging to the order *Gramineæ* have, on the average, contributed about 68 per cent. of the weight of the mixed herbage grown without manure; about 65 per cent. of that grown by purely mineral manures (that is, without nitrogen); and about 94 per cent. of that grown by the same mineral manures, with a large quantity of ammonia-salts in addition.

Species of the order *Leguminosæ* have, on the average, contributed about 9 per cent. of the produce without manure, about 20 per cent. of that by purely mineral manures (containing potass), and less than 0·01 per cent. of that by the mixture of the same mineral manures and a large quantity of ammoniacal salts.

Species belonging to various other orders have, on the average, contributed about 23 per cent. of the produce without manure; about 15 per cent. of that by purely mineral manures, and only about 6 per cent. of that by the mixture of the mineral manures and a large amount of ammonia-salts.

Not only the amounts of produce, but the number and description of species developed, have varied very greatly between the extremes here quoted, according to the particular character or combination of manure employed, and to the character of the seasons, as is strikingly illustrated by the arrangement of the specimens in the case, which, however, it should be borne in mind, show the composition of the herbage on the selected plots in one particular season only—namely, in 1872.

Obviously, these few remarks can only very inadequately indicate the interest of these curious illustrations of the domination of one plant over another in the mixed herbage of permanent grass land. Nor do we pretend to be able to give a satisfactory explanation of the variations induced, founded on the obvious or recorded difference in above-ground or under-ground character or habit of growth of the individual species. The whole of the results—agricultural, chemical, and botanical— obtained during the twenty years of the experiments are, however, now in course of arrangement for publication; and that we may not overlook such explanations as might be suggested from the point of view of the botanist and vegetable physiologist, as well as that of the chemist, we have associated with ourselves Dr. Masters in working up the botanical part of the inquiry; and I think Dr. Masters will agree with me in saying that much more has yet to be known of the difference in the physiological capability, so to speak, of the leaves of plants of different species, genera, and orders, and of the difference in the distribution, and in the feeding power, of the roots, before satisfactory explanations of the facts observed can be given. Surely, a wide field of investigation for the botanist and vegetable physiologist is here opened up to view!

Let us now recur to the question of the various amounts of nitrogen assimilated over a given area by plants of different natural orders, and call attention to the facts bearing upon the point which these experiments on the mixed herbage of grass land have supplied.

In Table IV. is shown the average produce (in the condition of hay) in lbs., per acre, per annum, over 20 years, of herbage of the gramineous family, of herbage of the leguminous family, and of herbage of other orders, calculated according to the mean percentage of each of these, determined in separations at six periods, namely in 1862, 1867, 1871, 1872, 1874, and 1875, in samples of the produce of four of the plots which have received no nitrogenous manure from the comm mencement; and there is also given, by the side of these results, the average annual yield of nitrogen per acre over the first 10, the second 10, and the total period of 20 years, in each case.

TABLE IV.

Yield of Nitrogen in the Mixed Herbage of Permanent Grass-land at Rothamsted.

Plots.	Conditions of Manuring.	Average Produce per acre per annum, 20 years, 1856-1875, according to Mean per cent. at 6 periods 1862, '67, '71, '72, '74, '75.			Average Nitrogen per Acre per annum.		
		Grami- neæ.	Legumi- nosæ.	Other Orders.	10 years 1856- 1865	10 years 1866- 1875	20 years 1856 1875
		lbs.	lbs.	lbs.	lbs.	lbs.	lbs.
3	Unmanured	1635	219	529	35·1	30·9	33·0
4-1	Superphosphate[1] ...	1671	149	673	35·7	31·5	33·6
8	Complex Min. Man.[2]	2442	296	639	54·4	38·1	46·3
7	Complex Min. Man.[3]	2579	8.6	573	55·2	56·0	55·6

[1] Mean of 4 Separations only, namely 1862, 1867, 1872, and 1875.
[2] Including potash 6 years, 1856-1861 ; without potash, 14 years 1862-1875.
[3] Including potash 20 years, 1856-1875.

The quantities of nitrogen yielded are calculated from the results of actual determinations of the nitrogen in the mixed produce of the respective plots; but the estimates of the quantity of the produce referable to the different Natural Orders must be taken as only giving a general indication or an approximation to the truth; for, whilst the amount of the total mixed produce is the average of that of the twenty years, the amount of it referred to the different orders is calculated upon their percentage d termined in six years only, four of which are among the last five, and the fluctuation according to season is in some cases very considerable, whilst in others there is a progression in the changes, which render an accurate estimate of the average botanical composition of the herbage over the whole period impossible. The figures do, however, undoubtedly represent the truth sufficiently nearly for our present purpose. But before referring to the yield of nitrogen, it may be remarked, in passing, how much greater is the increase of gramineous produce by the use of purely mineral manures in this mixed herbage than in the case of gramineous crops grown separately. The interesting question arises, how far the result is due to the direct action of the mineral manures in enabling the grasses to form much more stem and seed—that is, the better to mature—which, as a matter of fact, they are found to do? or how far the increased growth is to be explained by an increased accumulation of combined nitrogen available for the grasses in the upper layers of the soil, as the result of the increased growth of the leguminosœ induced by the potass manure, as already illustrated by the results obtained in alternating clover and barley, and in an actual course of rotation?

Referring to the yield of nitrogen. it is seen that, without manure, it has diminished during the last as compared with the first 10 years; but that the average is 33 lbs. per acre, per annum, or considerably more than with a gramineous crop grown separately.

With super-phosphate of lime alone, the yield of nitrogen over the first 10, the second 10, and the 20 years, is very nearly the same as without manure. It is slightly higher, as also is the total amount of produce; but whilst the quantity contributed by gramineous species is rather more, that yielded by leguminous species is less, and that by species belonging to other orders more than without manure.

With super-phosphate of lime, and sulphates of potass soda, and magnesia, during the first 6 years, but no potass during the last 14 years (plot 8), the amount of both gramineous and leguminous herbage is very much increased; and that of the leguminous produce was especially so during the earlier years. The result is a yield of 55·4 lbs. of nitrogen per acre, per annum, over the first 10 years, of only 38·1 lbs. over the second 10 years, and of 46·3 lbs. over the 20 years.

With the complex mineral manure, including potass each year throughout the period of twenty years (plot 7), leguminous species contribute about one-fifth of the whole produce, or very much more than in either of the other cases. The result is an annual yield of 55·2 lbs. of nitrogen over the first 10 years; of even slightly more, or 56 lbs., over the second 10 years; and of 55·6 lbs. over the whole period of 20 years—that is, considerably more than twice as much as would be yielded by a gramineous crop grown separately on arable land. It may here be observed that, whilst in the case of each of the first three plots referred to, the produce of the mixed herbage diminished over the second as compared with the first 10 years, that of plot 7, with the potass manure, and so much leguminous herbage, increased slightly over the second compared with the first 10 years. Finally, it may be remarked on this point, how comparatively uniform is the average yield of produce by all other species other than the gramineous and the leguminous on the four very differently manured plots.

Here again, then, the results relating to the growth of species of many different natural orders growing together, like those relating to the growth of individual species grown separately, show that those of the leguminous family, and probably those of various other orders also, have the capacity of assimilating much more nitrogen over a given area than species of the order gramineæ.

Assuming for the sake of argument that the yield of nitrogen by the gramineæ grown separately may be explained, as already suggested, by reference to the amount of combined nitrogen acquired from the measured aqueous deposits from the atmosphere, together with that condensed within the pores of the soil, and that derived from previous accumulations within it, the question arises, can the greatly increased yield by other plants be so accounted for? or, if not, how otherwise may it be explained? We will endeavour to weigh the evidence bearing upon this point.

It so happens that the plants which do gather, or which have been supposed to gather nitrogen more readily than the gramineæ, have obviously a different character of foliage; as, for instance, the "root-crops"- turnips and the like; and the leguminous crops—beans, peas, clover, &c. An obvious explanation, therefore, which will be found in books of authority, is that these so distinguished "broad-leaved plants" have the power of taking up nitrogen in some form from the atmosphere, in a degree, or in a manner, not possessed by the narrow-leaved gramineous plants. It is true that Adolph Mayer in Germany, and Schlösing in France, have experimentally shown that plants can take up nitrogen by their leaves from ammonia supplied to them in the ambient atmosphere. But I think I am right in saying that the conclusion of both of these experimenters is that this action takes place in a very immaterial degree in natural vegetation.

In reference to this subject, I may observe that the results of the determinations of the ammonia in the atmosphere by different experimenters, and in different localities, vary very greatly; and it may be concluded that a shower of rain will wash out much of it. According to M. Schlösing's statement of the results of his recent determinations of the ammonia in the air of Paris (Compt. Rend. lxxxi. p. 1252 et seq.), it ranges from one part in about 12, 500,000, to one part in about 260.000,000 of air by weight. If, for the purpose of illustration, we assume that, on the average, the ambient atmosphere in the open country—in Europe, at any rate—will contain one part of ammonia in about 60,000,000 of air, or one part of nitrogen as ammonia in about 50,000,000 of air, the atmosphere would thus contain more than 8000 times less nitrogen as ammonia than carbon as carbonic acid. But cereal crops contain 1 part of nitrogen to about 30 of carbon, and leguminous crops, 1 of nitrogen to 15, or fewer, of carbon. On these assumptions, the ambient atmosphere would contain a proportion of nitrogen as ammonia, to carbon as carbonic acid, about 267 times less than that of nitrogen to carbon in cereal produce, and about 534 times (or more) less than that in leguminous produce. It is true that water would absorb very much more nitrogen as ammonia, or dissolve very much more as carbonate or bi-carbonate of ammonia, than it would of carbon as carbonic acid under equal circumstances. Hence, there would appear to be a compensating quality for the small actual and relative amount of nitrogen as ammonia in the atmosphere, in the greater solubility or absorbability of the compounds in which nitrogen exists, than of the carbonic acid in which the carbon is presented. Further, it can hardly be to a greater mere extent of leaf or above-ground surface that the result could be attributed. Thus, though a bean and a wheat crop may yield about equal amounts of dry matter per acre, the bean produce would contain from two to three times more nitrogen, and approximate measurements show that a wheat plant offers a greater external superfices in relation to a given weight of dry substance than a bean plant, and greater still therefore in relation to a given amount of nitrogen fixed. If, then, the bean can in some way take up more nitrogen from the atmosphere than the wheat, the result must be due to character and function, rather than to mere extent of surface above ground. It may, however, be observed that, as a rule, even those of the leguminous crops which are grown for their ripened seed, maintain their green and succulent surface, over a more extended period of the season of active growth, than do the gramineous corn crops.

It may safely be asserted, then, that neither direct experimental evidence, nor a consideration of the chemistry and the physics of the subject, would lead to the conclusion that the plants which assimilate more nitrogen over a given area than others, do so by virtue of a greater power of absorbing by their leaves combined nitrogen from the atmosphere in the form of ammonia. And here it may be said in passing that the argument would be still stronger against the supposition that nitric acid in the atmosphere supplies directly to the leaves of plants any important amount of the nitrogen they assimilate.

But apart from the more purely scientific considerations bearing upon the question, we believe that our statistics of nitrogen-production are themselves sufficient to justify the conclusion that, at any rate, the "broad leaved" *root-crops*, turnips and the like, to which the function has with the most confidence been attributed, do not take up any important proportion of their nitrogen by their leaves from combined nitrogen in the atmosphere. Thus, it has already been shown, that the yield of nitrogen in these crops, even with the aid of complex mineral manures, was in the later years reduced to a lower point than that in any other crop; the percentage of nitrogen in the upper layers of the soil was also reduced to a lower point than with any other crop. The evidence of this kind is, however, admittedly not so conclusive in regard especially to plants of the leguminous family.

But as about four-fifths of the atmosphere which surrounds the leaves of plants consist of free nitrogen, why should not this be a source to them of the nitrogen they require? To assume that it is so, is such an obvious and easy way out of so many difficulties, that this assumption has from time to time been freely made, and much experimental investigation has been undertaken on the point with the most conflicting results. It is now nearly 40 years ago since Boussingault showed that there was a greater assimilation of nitrogen over a given area in a rotation of crops than he could well account for; and almost from that time to this he has been occupied with investigations of very various kinds, sometimes on the atmosphere, sometimes on meteoric waters, sometimes on plants, and sometimes on soils, the main object of which has obviously been to throw light on the question of the sources of the nitrogen of vegetation. And almost for as long a period as Boussingault, Mr. Lawes and myself have devoted much thought and investigation to the same end.

On this point, of whether or not plants assimilate the free nitrogen of the atmosphere, leaving out of view, for lack of time and space, the experiments and conclusions of several others who have worked on the subject on a less comprehensive scale, I will first briefly direct attention to the most comprehensive series of experiments, the results of which led the author to conclude that the free nitrogen of the atmosphere is taken up and assimilated by the leaves of plants.

During the years 1849, 1850, 1851, 1852, 1854, 1855, and 1856, M. G. Ville, of Paris, made numerous experiments on this subject. His plants were generally enclosed in a glass case, and his soils consisted of washed and ignited sand, sand and brick, or sand and charcoal. They were sometimes supplied with a current of unwashed air, sometimes with a current of washed air, and they were sometimes in free air; sometimes a known quantity of ammonia was supplied to the air of the apparatus, and sometimes known quantities of nitrate were supplied to the soil. Lastly a great variety of plants was experimented upon. M. G. Ville's results are summarised in Table V. below.

TABLE V.

Summary of the Results of M. G. VILLE'S Experiments, to determine whether Plants assimilate free Nitrogen.

Plants.	In Seed, and Air; and Manure, if any.	Nitrogen – Grammes. In Products.	Gain or Loss.	Nitrogen in Products to 1 Supplied.
1849 : *Current of unwashed air supplying 0.001 grammes Nitrogen as Ammonia.*[1]				
Cress	0·0260	0·1470	0·1210	5.6
Large Lupins ...	0·0640	0·0640	0·0000	1·0
Small Lupins ...	0·0640	0·0470	—0·0170	0·7
	0·1550	0·2580	0·1030	1·7
1850 : *Current of unwashed air supplying 0·0017 grammes Nitrogen, as Ammonia.*[1]				
Colza (plants) ...	0·0260	1·0700	1·0440	41·1
Wheat	0·0160	0·0310	0·0150	1.9
Rye	0·0130	0·0370	0·0240	2·8
Maize	0·0290	0·1280	0·0990	4·4
	0·0357	1·2660	1·1803	14·8
1851 : *Current of washed air.*[1]				
Sunflower	0·0050	0·1570	0·1520	31·4
Tobacco	0·0040	0·1750	0·1710	43·7
Tobacco	0·0040	0·1620	0·1580	40·5
1852 : *Current of washed air.*[1]				
Autumn Colza ...	0·0480	0·2260	0·1780	4·7
Spring Wheat ...	0·0290	0·0650	0·0360	2·2
Sunflower	0·0160	0·4080	0·3920	25·5
Summer Colza ...	0·1730	0·5950	0·4220	3·4
Summer Colza ...	0·1050	0·7010	0 5960	6·7

[1] Recherches Expérimentales sur la Végétation, par M. GEORGES VILLE. Paris, 1853

TABLE V.—*continucd.*

Plants.	Nitrogen—Grammes.			Nitrogen in Products to 1 Supplied.
	In Seed, and Air and Manure, if any.	In Products.	Gain or Loss.	
1854: *Current of washed air (under superintendence of a Commission).*[1]				
Cress...	0·0099	0·0097	—0·0002	1·0
Cress...	0·0038	0·0530	0·0492	13·9
Cress...	0·0039	0·0110	0·0071	2·8
1854: *Current of washed air (closed, under superintendence of a Commission.*[1]				
Cress...	0·0063	0·0350	0·0287	5·6
1855 and 1856: *In free air, with* 0·5 *grammes Nitre =* 0·069 *Nitrogen.*[2]				
Colza	0·0700	0·0700[3]	0·0000	1·0
Colza	0·0700	0·0660[3]	—0·0040	0·9
Colza	0·0700	0·0680[3]	—0 0020	1·0
1855 and 1856: *In free air, with* 1 *gramme Nitre =* 0·138 *Nitrogen.*[2]				
Colza	0·1400	0·1970[3]	0·0570	1·41
Colza	0·1400	0·3740[3]	0·2340	2·67
Colza	0·1400	0·2160[3]	0·0760	1·54
Colza	0·1400	0·2500[3]	0·1100	1·79
1856: *In free air, with* 0·792 *grammes Nitre =* 0·110 *Nitrogen.*[2]				
Wheat	0·1260	0·2180[3]	0·0920	1·7
Wheat	0·1260	0·2240[3]	0·0980	1·8
1855: *In free air, with* 1·72 *grammes Nitre =* 0·238 *Nitrogen.*[2]				
Wheat	0·2590	0·3080[3]	0·0490	1·2
1856: *In free air, with* 1·765 *grammes Nitre =* 0·244 *Nitrogen.*[2]				
Wheat	0·2650	0·2170[3]	—0·0480	0·8
Wheat	0·2650	0·3500[3]	+0·0850	1·3

We have already discussed the results of M. G. Ville, as well as those of others, in a paper published in the *Philosophical Transac-*

---

[1] Compt. rend , 1855.   [2] Recherches Expérimentales sur la Végétation, 1857.
[3] In Plants only.

tions for 1859, and in a somewhat condensed form in the *Journal of the Chemical Society*, Vol. xvi. 1863 ; and we can only very briefly refer to them in this place. The column of actual gain or loss of nitrogen is seen to show in one case a gain of more than 1 gram of nitrogen ; the amount of it in the products being more than 41 fold that supplied as combined nitrogen in the seed, and air. This result was obtained with colza. Those obtained with wheat, rye, or maize, showed very much less of both actual and proportional gain. Experiments with sunflower and tobacco showed a less actual gain than that with colza ; but still it amounted in one case, with sunflower, to more than 30, and in two, with tobacco, to more than 40 fold of that supplied. In M. G. Ville's later experiments (as a glance down the last two columns in the Table will show), although he still had generally some gain, it was usually both actually and in proportion to the quantity supplied considerably less than in his earlier ones.

M. G. Ville attributed the gain, in some cases, to the large leaf-surface. In explanation of the assimilation of free nitrogen by plants, he calls attention to the fact that nascent hydrogen is said to give ammonia, and nascent oxygen nitric acid, with free nitrogen, and he asks—Why should not the nitrogen in the juices of the plant combine with the nascent carbon and oxygen in the leaves ? He refers to the supposition of M. De Luca, that the nitrogen of the air combines with the nascent oxygen given off by the leaves of plants, and to the fact that the juice of some plants (mushrooms) has been observed to ozonize the oxygen of the air, and he asks—Is it not probable, then, that the nitrogen dissolved in the juices will submit to the action of the ozonized oxygen with which it is mixed, when we bear in mind that the juices contain alkalies, and penetrate tissues, the porosity of which exceeds that of spongy platinum ?

The experiments of M. Boussingault, and of ourselves, on the other hand, have not given an affirmative answer to the question whether plants, by their leaves, take up and assimilate the free nitrogen of the air.

M. Boussingault commenced his experiments on this subject in 1837, and Table VI., which follows, summarises his results, obtained at intervals from that date up to 1858.

TABLE VI.

Summary of the Results of M. BOUSSINGAULTS' *Experiments, to determine whether Plants assimilate free Nitrogen.*

Plants.	Nitrogen Grammes.			Nitrogen in Products to 1 Supplied.
	In Seed, or Plants; and Manure, if any.	In Products.	Gain or Loss.	

1837 : *Burnt soil, distilled water, free air, in closed summer-house.*[1]

Trefoil...	0·1100	0·1200	+0·0100	1·09
Trefoil...	0·1140	0.1560	+0·0420	1·37
Wheat...	0·0430	0·0400	—0·0030	0·93
Wheat...	0·0570	0·0000	+0 0030	1·05

1838 : *Conditions as in* 1837.[2]

Peas	0·0460	0·1010	+0·0550	2·20
Trefoil (Plants)	0·0330	0·0560	+0·0230	1·70
Oats (Plants)	0·0590	0·0530	—0·0060	0·90

1851 *and* '52: *Washed and ignited pumice with ashes, distilled water, limited air, under glass shade, with Carbonic Acid.*[3]

Haricot, 1851	0·0349	0.0340	—0·0009	0·97
Oats, 1851	0·0078	0·0067	—0·0011	0·86
Haricot, 1852	0·0210	0·0189	—0·0021	0·90
Haricot, 1852	0·0245	0·0226	—0·0019	0·92
Oats, 1852	0·0031	0·0030	—0·0001	0·97

1853: *Prepared pumice, or burnt brick, with ashes; distilled water, limited air, in glass globe, with Carbonic Acid.*[3]

White Lupin	0·0480	0·0483	+0·0003	1·01
White Lupin	0·1282	0·1246	—0·0036	0·97
White Lupin	0·0349	0·0339	—0·0010	0·97
White Lupin	0·0200	0·0204	+0·0004	1·02
White Lupin	0·0399	0·0397	—0·0002	1·00
Dwarf Haricot	0·0354	0·0360	+0·0006	1·02
Dwarf Haricot	0·0293	0·0277	—0·0021	0·93
Garden Cress	0·0013	0·0013	0·0000	1·00
White Lupin	0·1827	0·1697	—0·0130	0·93

[1] Ann. Ch. Phys.,[2] lxvii, (1838).   [2] Ibid, lxix.   [3] Ann. Ch. Phys. [3] xli. (1854).

TABLE VI.— *continued.*

Plants.	Nitrogen—Grammes.			Nitrogen in Products to 1 Supplied.
	In Seed, or Plants; and Manure, if any.	In Products.	Gain or Loss.	

1854: *Prepared pumice with ashes, distilled water, current of washed air, and Carbonic Acid, in glazed case.*[1]

Plants	In Seed	In Products	Gain or Loss	Nitrogen
Lupin	0·0196	0·0187	—0·0009	0·95
Dwarf Haricot	0·0322	0·0325	+0·0003	1·01
Dwarf Haricot	0·0335	0·0341	+0·0006	1·02
Dwarf Haricot	0·0339	0·0329	—0·0010	0·97
Dwarf Haricot	0·0676	0·0666	—0·0010	0·99
Lupin	0·0180 } 0·0334		—0·0021	0·94
Lupin	0·0175 }			
Cress	0·0046	0·0052	+0·0006	1·13

1851, '52, '53, *and* 54: *Prepared soil, or pumice with ashes; distilled water, free air, under glazed case.*[1]

Plants	In Seed	In Products	Gain or Loss	Nitrogen
Haricot (dwarf), 1851	0·0349	0·0380	+0·0031	1·09
Haricot, 1852	0·0213	0·0238	+0·0025	1·12
Haricot, 1853	0·0293	0·0270	-0·0023	0·92
Haricot (dwarf), 1854	0·0318	0·0350	+0·0032	1·10
Lupin (white), 1853 ...	0·0214	0·0256	+0·0042	1·20
Lupin 1854...	0·0199	0·0229	+0·0030	1·15
Lupin 1854...	0·0367	0·0387	+0·0020	1·05
Oats, 1852...	0·0031	0·0041	+0·0010	1·32
Wheat, 1853	0·0064	0·0075	+0·0011	1·17
Garden Cress, 1854 ...	0·0259	0·0272	+0·0013	1·05

1858 : *Nitrate of Potassium as Manure.*[2]

Plants	In Seed	In Products	Gain or Loss	Nitrogen
Helianthus ... ... {	0·0144[3]	0·0130	—0·0014	0·90
{	0·0255[3]	0·0245	-0·0010	0·96

M. Boussingault's soils consisted of burnt soil. washed and ignited pumice, or burnt brick ; his experiments were sometimes in free air, sometimes in a closed vessel with limited air, sometimes with a current of washed air, and sometimes in free air, but under a glass case. When the plants were enclosed, a supply of carbonic acid was provided, and in a few cases known quantities of nitre were supplied as manure.

1 Ann. Ch. Phys., Sér. [3] xliii. (1855).   2 Compt. rend., xlvii, (1858).
3 Nitrogen in Seed and Nitrate.

The last two columns of the Table (VI.) show the actual and proportional gain of nitrogen in M. Boussingault's experiments. It will be observed that in his earliest experiments, those in free air, in a summer house, the leguminous plants, trefoil and peas, did indicate a notable gain of nitrogen; but in all his subsequent experiments there was generally either a slight loss, or, if a gain, it was represented in only fractions, or low units, of milligrams. After twenty years of varied and laborious investigation of the subject, M Boussingault concluded that plants have not the power of taking up and assimilating the free nitrogen of the atmosphere.

Our own experiments on this subject were commenced in 1857, and the late Dr. Pugh, of the Pennsylvania State Agricultural College, devoted between two and three years to the investigation at Rothamsted. Mr. Lawes has contributed one complete set of the apparatus employed to this exhibition. The arrangement, and the results obtained up to that date, are fully described in the papers already referred to, published in the *Philosophical Transactions* for 1859, and in the *Journal of the Chemical Society* in 1863. They may be briefly described as follows:—

The soils used were ignited, washed, and re-ignited, pumice, or soil. The specially made pots were ignited before use, and cooled over sulphuric acid under cover. The pots, with their plants, were enclosed under a glass shade resting in the groove of a specially made hard-baked glazed stone-ware lute-vessel, mercury being the luting material. Under the shade, through the mercury, passed one tube for the admission of air, another for its exit, and another for the supply of water or solutions to the soil; and there was an outlet at the bottom of the lute-vessel for the escape of the condensed water into a bottle affixed for that purpose, from which it could be removed and returned to the soil at pleasure. A stream of water being allowed to flow into a large stone-ware Wolff's bottle (otherwise empty), air passed from it through two small glass Wolff's bottles containing sulphuric acid, then through a long tube filled with fragments of pumice saturated with sulphuric acid, and lastly through a Wolff's bottle containing a saturated solution of ignited carbonate of soda; and, after being so washed, the air enters the glass shade, from which it passes, by the exit tube, through an eight bulbed apparatus containing sulphuric acid, by which communication with the unwashed external air is prevented. Carbonic acid is supplied as occasion may require, by adding a measured quantity of hydrochloric acid to a bottle containing fragments of marble, the evolved gas being passed through one of the bottles of sulphuric acid, through the long tube, and through the carbonate of soda solution, before entering the shade.

It will be observed that, by the arrangement described, the washed air is forced, not aspirated, through the shade, and the pressure being thus the greater within the vessel, the danger of leakage of unwashed air from without inwards is lessened. In 1857, twelve sets of such apparatus were employed ; in 1858 a larger number, some with larger lute-vessels, and shades ; in 1859 six, and in 1860 also six. The whole were arranged, side by side, in the open air, on stands of brickwork, as described in the papers referred to, and shown in the apparatus exhibited. Drawings of some of the plants grown were also exhibited, and the published results are summarised in Table VII.

TABLE VII.

Summary of the Results of Experiments made at Rothamsted, to determine whether Plants assimilate Free Nitrogen.

			Nitrogen—Grammes.			Nitrogen in Products to 1 Supplied.
			In Seed, and Manure if any.	In Plants, Pot and Soil.	Gain or Loss.	
With NO *combined Nitrogen supplied beyond that in the seed sown.*						
Gramineæ . .	1857.	Wheat .	0·0080	0·0072	—0·0008	0·90
		Barley .	0·0056	0·0072	+0·0016	1·11
		Barley .	0·0056	0·0082	+ 0·0026	1·46
	1858.	Wheat .	0·0078	0·0081	+0·0003	1·04
		Barley .	0·0057	0·0058	+0·0001	1·02
		Oats . .	0·0063	0·0056	—0·0007	0·89
	1858. A [1]	Wheat .	0·0078	0·0078	0·0000	1·00
		Oats . .	0·0064	0·0063	—0·0001	0·98
Leguminosæ.	1857.	Beans. .	0·0796	0·0791	—0·0005	0·99
	1858.	Beans. .	0·0750	0·0757	+0·0007	1·01
		Peas . .	0·0188	0·0167	—0·0021	0·89
Other Plants.	1858.	Buck Wheat.	0·0200	0·0182	—0·0018	0·91

1 These experiments were conducted in the apparatus of M. G. Ville.

TABLE VII.—*continued.*

			Nitrogen—Grammes.			Nitrogen in Products to 1 Supplied.
			In Seed, and Manure if any.	In Plants, Pot and Soil.	Gain or Loss.	

WITH *combined Nitrogen supplied beyond that in the seed sown.*

			In Seed, and Manure if any.	In Plants, Pot and Soil.	Gain or Loss.	Nitrogen in Products to 1 Supplied.
Gramineæ . .	1857.	Wheat .	0·0329	0·0383	+0·0054	1·16
		Wheat .	0·0329	0·0331	+0·0002	1·01
		Barley .	0·0326	0·0328	+0·0002	1·01
		Barley .	0·0268	0·0337	+0·0069	1·25
	1858.	Wheat .	0·0548	0·0536	—0·0012	0·98
		Barley .	0·0496	0·0464	—0·0032	0·94
		Oats . .	0·0312	0·0216	—0·0096	0·69
	1858. A[1]	Wheat .	0·0268	0·0274	+0·0006	1·02
		Barley .	0·0257	0·0242	—0·0015	0·94
		Oats . .	0·0260	0·0198	—0·0062	0·76
Leguminosæ.	1858.	Peas . .	0·0227	0·0211	—0·0016	0·93
		Clover .	0·0712	0·0665	—0·0047	0·93
	1858. A[1]	Beans. .	0·0711	0·0655	—0·0056	0·92
Other Plants.	1858.	Buck Wheat.	0·0308	0·0292	—0·0016	0·95

The upper part of the Table shows the results obtained in 1857 and 1858 in the experiments in which no combined nitrogen was supplied beyond that contained in the seed sown. The drawings show how extremely restricted was the growth under these conditions, and the figures in the Table show that neither with the gramineæ, the leguminosæ, nor with buckwheat, was there in any case a gain of three milligrams of nitrogen indicated. In most cases there was much less gain than this, or a slight loss. There was in fact nothing in these results to lead to the conclusion that either the gramineæ, the leguminosæ, or the buckwheat had assimilated free nitrogen.

The lower part of the Table shows the results obtained in 1857 and 1858, in the experiments in which the plants were supplied with known

---

[1] These experiments were conducted in the apparatus of M. G. Ville.

quantities of combined nitrogen in the form of a solution of ammonium sulphate applied to the soil. The gains or losses range a little higher in these experiments, in which larger quantities of nitrogen were involved, but they are always represented by units of milligrams only, and the losses are higher than the gains. Further, the gains, such as they are, are all in the experiments with the gramineæ, whilst there is in each case a loss with the leguminosæ, and with the buckwheat. On this point it should be stated that the growth was far more healthy with the gramineæ than with the leguminosæ, which are even in the open fields very susceptible to the vicissitudes of heat and moisture, and were found to be extremely so when enclosed under glass shades. It might be objected, therefore, that the negative results with the leguminosæ are not so conclusive as those with the gramineæ. However this may be, taking the results as they stand, there is nothing whatever in them to lead o the conclusion that either the gramineæ or the leguminosæ can take up and assimilate the free nitrogen of the atmosphere. We, indeed, do not hesitate to conclude from our own experiments, as Boussingault did from his, that the evidence is strongly against the supposition that plants can so avail themselves of the free nitrogen of the atmosphere.

Independently of the action suggested as possible by M. G. Ville, that is between free nitrogen and nascent or ozonized oxygen within the plant itself, it has been supposed that the free nitrogen of the atmosphere may unite with the nascent oxygen, or ozone, as the case may be, evolved by the plant, and so yield nitric acid. In our papers above referred to we have given reasons for supposing that such actions are not likely to take place ; but whether they do or do not, it is at any rate certain that in our own experiments we have not been able to persuade plants to avail themselves of this happy faculty of producing their own nitrogenous food. With regard to the action supposed possibly to take place externally to the plant itself, if it were in any material degree operative, we should expect some, at least, of the resulting combined nitrogen to be collected in the aqueous deposits from the atmosphere ; but we have seen how inadequate is the amount of combined nitrogen in those deposits to account for the yield of nitrogen, even of the gramineæ, and still less can it satisfactorily explain the yield in the leguminosæ and other plants.

E

But if the plant itself cannot either assimilate free nitrogen, or effect its combination so as to bring it into a state for its use, may not such combination take place under the influence of the soil?

More than 30 years ago, Mulder argued that in the last stages of decomposition of organic matter in the soil, hydrogen was evolved, and that this nascent hydrogen combined with the free nitrogen of the air, and so formed ammonia.

A few years ago Dehérain substantially revived this view. He maintained that at a certain depth the air of the soil is poor in, or destitute of, oxygen ; that hydrogen is evolved from the decomposing organic matter ; that it unites with free nitrogen to form ammonia ; and, that so, combined nitrogen increases in the soil in spite of the growth and removal of crops. This view he supports by some laboratory experiments.

It is obvious that if the reality of this action in soils were unquestionably established, it would greatly aid the solution of the question we are discussing. There are, indeed, results of others on record which would seem to lend it probability.

Thus, Bretschneider found, on exposure of a mixture of humic acid and quartz sand to the air for a whole year, under conditions in which it was protected from rain and insects, that there was a gain of combined nitrogen which would represent an increase of more than 40 lbs. per acre.

Again, Boussingault exposed a moist garden soil for three months, and found a small gain of nitrogen. His explanation, was, however, different. He supposed it possible that ozone might be evolved in the oxydation of organic matter in the soil, and unite with free nitrogen, and so nitric acid be produced, and the soil gain in combined nitrogen. In other experiments Boussingault put mixtures of vegetable mould and pure sand in small quantities in large glass vessels which he perfectly closed and preserved in a dark cellar for a whole year. At the end of that period oxydation of organic matter had taken place, nitric acid was formed, but there was upon the whole a small loss of combined nitrogen. Lastly in regard to Boussingault's results bearing upon this point, it has already been shown that in all of his experiments with plants in which his soils consisted of ignited pumice, ignited brick, or the like, without organic matter, he found no gain of combined nitrogen in soil and plant. In 1858 and 1859, however, he made a number of experiments on growth, in which part of the soil consisted of rich garden mould ; and in two cases with lupins growing in confined air, and in one with haricot growing in free air, his results showed a notable gain of combined nitrogen ; and although the quantity of garden mould employed was not the same in the three cases, the gain of nitrogen was approximately in proportion to the amount of soil used. The gain was, indeed, in the soil rather than in the plant. In the other experiments, however, either much less, or no gain was indicated.

Much more recently, Boussingault has published the results of experiments which showed that when a garden soil was confined for about 11 years in closed glass vessels in an atmosphere containing oxygen, the free nitrogen did not serve for the formation of nitric acid within it ; but, on the contrary, the soil lost a portion of its combined nitrogen.

Since the delivery of this lecture, M. Berthelot (Compt. Rend, T. lxxxii. p. 1,357) has stated that in experiments in which he exposed moistened cellulose to an electric current in an atmosphere of nitrogen, he found nitrogen taken up, and a fixed nitrogenous body formed. Referring to the last mentioned experiments of M. Boussingault, and his conclusions from them, M. Berthelot objects that the soils being in closed glass vessels, the intervention of atmospheric electricity was excluded, and the conditions of the experiments were, so far, unlike those of a natural soil.

Being very desirous to know the present opinion of M. Boussingault on the various points involved in this important question of the sources of the nitrogen of vegetation, I wrote to him shortly after undertaking to give this address, and asked whether he would be kind enough to favour me with a statement of his views on certain points. Unfortunately his reply did not reach me until after the delivery of the lecture ; but, with his permission, I am now enabled to contribute a very valuable addition to the discussion in the form of a translation of the more essential parts of M. Boussingault s letter. He says :—

"(1.) In confined stagnant air, or in air moving through a closed apparatus, after previous purification, but still containing carbonic acid, plants growing in a soil destitute of nitrogenous manure, but containing the mineral substances indispensable for the vegetable organism, do not assimilate the nitrogen which is in a gaseous state in the atmosphere."

" (2.) In the open air, in a soil destitute of nitrogenous manure, but containing the mineral substances necessary for the vegetable organism, plants acquire very minute quantities of nitrogen, arising, no doubt, from minute proportions of fertilising nitrogenous ingredients carried by the air, ammoniacal vapours, and dust, always containing alkaline or earthy nitrates."

" (3.) In confined stagnant air, or in air renewed in a closed apparatus, a plant growing in a soil containing a nitrogenous manure, and mineral substances necessary for the vegetable organism, or in fertile vegetable earth, does not assimilate free nitrogen."

" (4.) In field culture, where dung is applied in ordinary quantities, analysis shows that there is more nitrogen in the crops than was contained in the manure applied."

" This excess of nitrogen comes from the atmosphere, and from the soil."

" (A.) From the atmosphere, because it furnishes ammonia in the form of carbonate, nitrates or nitrites, and various kinds of dust. Theodore de Saussure was the first to demonstrate the presence of ammonia in the air, and consequently in meteoric waters. Liebig exaggerated the influence of this ammonia on vegetation, since he went so far as to deny the utility of the nitrogen which forms a part of farm-yard manure. This influence is, nevertheless, real, and comprised within limits, which have quite recently been indicated in the remarkable investigations of M. Schlösing."

" (B.) From the soil, which, besides furnishing the crops with mineral a kaline substances, provides them with nitrogen, by ammonia, and by nitrates, which are formed in the soil at the expense of the nitrogenous matters contained in diluvium, which is the basis of vegetable earth ; compounds in which nitrogen exists in stable combination, only becoming fertilising by the effect of time. If we take into account their immensity, the deposits of the last geological periods must be considered as an inexhaustible reserve of fertilising agents. Forests, prairies, and some vineyards, have really no other manures than what are furnished by the atmosphere, and by the soil. Since the basis of all cultivated land contains materials capable of giving rise to nitrogenous combinations, and to mineral substances, assimilable by plants, it is not necessary to suppose that in a system of cultivation the excess of nitrogen found in the crops is derived from the free nitrogen of the atmosphere.

As for the absorption of the gaseous nitrogen of the air by vegetable earth, I am not acquainted with a single irreproachable observation that establishes it; not only does the earth not absorb gaseous nitrogen, but it gives it off, as you have observed in conjunction with Mr. Lawes, as Reiset has shown in the case of dung, as M. Schlösing and I have proved in our researches on nitrification."

" If there is one fact perfectly demonstrated in physiology, it is this of the non-assimilation of free nitrogen by plants ; and I may add by plants of an inferior order, such as mycoderms, and mushrooms."

Numerous experiments of Schlösing indicate a similar result to that last quoted of Boussingault. He selected a soil rich in humus, containing about 16 per cent. of moisture, and 0·263 per cent. of combined nitrogen. Known quantities of it were placed in large wide glass tubes, and during a period of about four months, he aspirated over them air containing respectively from 1·5 to 21 per cent. of oxygen. He determined the carbonic acid in the air passing off, and the nitric acid in the soil before and after the experiment. He found that both the combustion of the organic matter, and the formation of nitric acid, were very considerable, even with the lowest proportion of oxygen in the air ; but that the formation of the nitric acid in particular was very much the greater, the larger the proportion of oxygen in the air.

In a second set of experiments, he used the soil in a moister condition ; and instead of the experiment in which the air contained only 1·5 of oxygen, he employed pure nitrogen ; and the experiments extended over a period of about six months. In the case in which the aspirated air contained no oxygen, the whole of the nitric acid previously existing in the soil disappeared ; but in the other cases there was a considerable formation of nitric acid.

In a third set of experiments, Schlösing determined the nitric acid in the soils, and added known quantities of potassium nitrate in a dilute solution. The mixture was enclosed in a flask of several times the capacity of the volume of soil. At the conclusion of the experiment only traces, if any, of gas containing hydrogen and carbon were present in the air of the vessel. The amount of ammonia in the soil increased considerably, but in only small proportion to that which the nitric acid would yield. At the end of the first experiment more potassium nitrate was added, and an atmosphere of known volume and composition supplied. At the conclusion of this experiment the soil contained no nitric acid ; the amount of ammonia was increased, but again in only small proportion to the amount which the nitrate would yield. There was indeed a loss of total nitrogen in the soil.

Schlösing concludes that the combustion of organic matter in the soil is accompanied by a loss of nitrogen ; that the combustion may be at the cost of the air as in the experiment of Boussingault, or at the cost of nitrates, of ferric oxide, or of the oxygen of the organic matter, as in his own experiments.

It will be seen that on this important point of whether or not the soil may acquire combined nitrogen either in the form of ammonia by the combination of free nitrogen with nascent hydrogen evolved in the decomposition of organic matter in defect of oxygen, or in the form of nitric acid by the oxydation of free nitrogen, the evidence is, to say the least, conflicting. The more recent results of Boussingault, and those of Schlösing, would, however, indicate a greater probability of a loss of combined nitrogen, and·evolution of free nitrogen.

Judging of the probabilities by reference to some of the results of our own investigations, we think that they are rather against than in favour of the supposition that there is any material gain of the kind assumed by Mulder and Dehérain. It may be well, however, briefly to call attention to some few facts which seem to bear upon the point, whether in favour, or otherwise, of the view in question.

The action assumed by Mulder and Dehérain, if it have place at all in soils in their natural condition, would be supposed, and is assumed by Dehérain, to occur in layers sufficiently deep to be poor in oxygen. In the lower layers of the soil there is, however, a deficiency of carbonaceous organic matter also. Again, if such formation of ammonia do take place, it is probable that some at any rate of it must be oxidated into nitric acid ; a condition which, on the other hand, implies an atmosphere not poor in oxygen. Thus, numerous results of analysis of the drainage water from many of the experimental plots at Rothamsted, to which further reference will be made presently, show that nearly the whole of the combined nitrogen in the drainage collected at a depth of about 30 inches, exists as nitrates and nitrites ; which, obviously, would hardly be the case if the solution passed through a considerable layer of soil, the interstices of which contained an atmosphere poor in, or destitute of, oxygen.

Again, assuming such formation of ammonia to take place in the upper layers of the soil, where there is the most organic matter, and much oxidation of it, the supposition would be that the conditions would favour oxidation rather than the formation of ammonia from free nitrogen ; and the fact of the formation of a good deal of nitric acid by the oxidation of nitrogenous organic matter, or ammonia, in the surface soil, is sufficiently established.

Further, if it were to the action assumed by Mulder and Dehérain taking place in the upper layers of the soil that we owe the supplies of combined nitrogen available to leguminous and other plants which assimilate so much more of it over a given area than the gramineæ, the question may be asked—why cannot the gramineæ avail themselves of this superficial supply ? On this point it may be mentioned that, on some parts of the experimental wheat and barley fields at Rotham-sted, farm-yard manure has been applied year after year, for a quarter of a century or more, in quantity containing perhaps six or seven times as much nitrogen as is removed in the increase of crop, and that thus the percentage of nitrogen in the surface soil has been more than doubled. Yet, as large a produce of barley, and a larger produce of wheat, is annually obtained by the use of very much smaller quan-tities of nitrogen, as ammonia-salts or nitrate. It would thus appear that the nitrogen of the farm-yard manure was only available to the cereals after its transformation into ammonia or nitric acid. Unfortu-nately, we are not at present able to adduce direct experimental evidence as to the condition in which the large amount of inefficient nitrogen exists in the soil, or as to whether a leguminous crop would or would not grow luxuriantly in it, but there is little doubt that it would do so. On the other hand, a good crop of clover would appear to be attainable in soil comparatively poor in nitrogen in its upper layers and comparatively poor in organic matter also ; for, in the experiments already referred to in which barley was grown after barley and after clover, the large amount of clover obtained, and nitrogen assimilated in it, was after six corn crops grown by artificial manure alone ; conditions under which the amount, both of available nitrogen, and of organic matter, in the upper layers of the soil, would be supposed to be comparatively small.

The answer of Dehérain would probably be, that under the circumstances supposed, the nitrogen would be in a condition of combination not favourable for assimilation by the gramineæ; that, in fact, the ammonia formed would combine with organic acids in the soil, yielding compounds specially favourable as food for the leguminosæ. An objection to this view is, that if the accumulation in the soil by time, of nitrogen in a condition specially favourable for the leguminosæ were such as is here assumed, we should expect the amount of nitrogen in the soil, determinable by the soda-lime process, to be higher before than after the growth of a leguminous crop ; whereas, on the contrary, after the growth of a leguminious crop, the amount of nitrogen so determinable in the upper layers of the soil is very appreciably increased.

The evidence in favour of the supposition that the special source of nitrogen to the leguminosæ is ammonia, or other compounds than nitric acid. in the upper layers of the soil, is then, to say the least, inconclusive. It remains to consider whether it may not be nitric acid, either in the soil or in the subsoil ?

As already said, there is abundant evidence of the formation and existence of a considerable amount of nitric acid in surface soils ; even in such as contain a relatively high amount of carbonaceous and nitrogenous organic matter. For example, a soil at Rothamsted which has been under garden cultivation, and as such probably manured almost every year for centuries, has successfully grown clover every year for more than twenty years. This soil was shown by the late Dr. Pugh, and has been again recently by Mr. Warington, to contain a considerable amount of nitric acid. But such a soil would, there is no doubt, grow large crops of gramineæ also ; which direct experiments show to attain great luxuriance under the influence of artifically applied nitrates. But such a rich garden soil contains an abundance of every thing—mineral constituents, carbonaceous organic matter, and combined nitrogen in various forms, and thus the exact conditions which it supplies favourable to the leguminosæ cannot at once be discriminated. The fact of the comparatively little, or at least uncertain, action of directly applied nitrates on the growth of the leguminosæ, would seem to be inconsistent with the supposition that it is the nitric acid in such a surface soil that has given it its special adaptation for the growth of clover for so many years—unless, indeed, it be the case, that it is much more available to such crops when in combination with some bases than with others.

The next point to consider is, whether there are any facts in favour of the supposition that clover, and leguminous crops generally, acquire any material proportion of their nitrogen in lower layers, and in a more extended range of the soil, than the gramineæ. As an element in the discussion of this question it will be well in the first place to call attention to the effects of direct nitrogenous manures, such as ammonia-salt, or nitrates, on the growth of some of our crops.

In Table VIII. is shown the estimated amounts of carbon, yielded per acre per annum, in wheat over twenty years, in barley over twenty years, in sugar-beet over three years, and in beans over eight years ; each with a complex mineral manure alone, and each with the same mineral manure and given quantities of nitrogen in addition, supplied in some cases in the form of ammonia-salts, and in others as nitrate. The gain of carbon by the use of the nitrogenous manure is also given.

TABLE VIII.

Estimated yield and gain of Carbon per acre, per annum, in experimental Crops at Rothamsted.

Manuring, Quantities per acre, per annum.	Average Carbon per acre, per annum.	
	Actual	Gain
Wheat 20 *years,* 1852-1871.		
	lbs.	lbs.
Complex Mineral Manure	988	
Complex Min. Man. & 41 lbs. Nitrogen, as Ammonia...	1590	602
Complex Min. Man. & 82 lbs. Nitrogen, as Ammonia...	2222	1234
Complex Min. Man. & 82 lbs. Nitrogen, as Nitrate ...	2500	1512

TABLE VIII.—*continued.*

Manuring, Quantities per acre, per annum.	Average Carbon per acre, per annum.	
	Actual	Gain.
Barley 20 *years,* 1852-1871.		
Complex Mineral Manure	1138	
Complex Min. Man. & 41 lbs. Nitrogen. as Ammonia...	2088	1150
Sugar-Beet 3 *years,* 1871-1873.		
Complex Mineral Manure	1136	
Complex Min. Man. & 82 lbs. Nitrogen, as Ammonia...	2634	1498
Complex Min. Man. & 82 lbs. Nitrogen, as Nitrate ...	3081	1945
Beans 8 *years,* 1862 *and* 1864-1870.		
Complex Mineral Manure	726	
Complex Min. Man. & 82 lbs. Nitrogen, as Nitrate ...	992	266

It is quite evident that in the case of the gramineous crops, wheat and barley, which contain a comparatively low percentage of nitrogen, and assimilate a comparatively small amount of it over a given area, and also in that of the sugar-beet, there was a greatly increased amount of carbon assimilated by the addition of nitrogenous manure alone. In the case of the wheat, there is much more effect from a given amount of nitrogen supplied as nitrate, which is always applied in the spring, than from an equal quantity as ammonia-salts, which are applied in the autumn, and are subject to winter drainage. There is also more effect from ammonia-salts applied to barley than to wheat ; the application being made for the former in the spring and for the latter in the autumn. There is again more effect from the nitrate than from the ammonia-salts when applied to sugar-beet, the application being made in both cases at the same date, in the spring.

On the other hand, the effect of the nitrogenous manure upon the highly nitrogenous bean crop is seen to be, comparatively, very insignificant.

In reference to this point it should be observed that there has been this greatly increased assimilation of carbon in the wheat and in the barley for more than twenty years, without the addition of any carbon to the soil. It is indeed certain that, in the existing condition of our soils, the increased growth of our staple starch-yielding grains is greatly dependent on a supply of nitrogen to the soil. It is equally certain that the increased production of sugar in the gramineous sugar-cane, in the tropics, is likewise greatly dependent on the supply of nitrogen to the soil.

In reference to the great increase in the assimilation of carbon in the sugar-beet by the use of purely nitrogenous manures, it may be of interest to observe that over the three years of the experiments with sugar-beet, the increased production of sugar per acre per annum was about 20 cwts. by the use of 82 lbs. of nitrogen per acre per annum as ammonia-salts, and about 28 cwts. by the use of 82 lbs. of nitrogen as nitrate of soda.

It is then our characteristically starch and sugar producing crops that are the most characteristically benefited by the application of nitrogenous manures ; whilst our highly nitrogenous leguminous crops are comparatively little benefited by such manures.

But now let us consider what is the proportion of the nitrogen supplied in manure that we get back in the increase of the crops that are the most specially benefited by its use?

In Table IX. is shown the amount of nitrogen recovered, and the amount not recovered, in the increase of crop for 100 supplied in manure, to wheat, and to barley, respectively ; the result being in each case the average over a period of twenty years.

TABLE IX.

Nitrogen recovered, and not recovered, in the increase of produce, for 100 supplied in Manure.

Manuring, Quantities per acre, per annum.	For 100 Nitrogen in Manure.	
	Recovered in Increase.	Not Recovered in Increase.
Wheat 20 years, 1852-1871.		
Complex Min. Man. & 41 lbs. Nitrogen, as Ammonia...	32·4	67·6
Complex Min. Man. & 82 lbs. Nitrogen, as Ammonia...	32·9	67·1
Complex Min. Man. & 82 lbs. Nitrogen, as Nitrate ...	45·3	54·7
Barley 20 years, 1852-1871.		
Complex Min. Man. & 41 lbs. Nitrogen, as Ammonia...	48·1	51·9

Speaking generally, it may be said that, notwithstanding the great effects produced by the nitrogenous manures, two-thirds of the nitrogen supplied were unrecovered in the increase of crop when the ammonia-salts were applied to wheat ; the application being made in the autumn. When, however, nitrate of soda was used, which is always applied in the spring, the quantity left unrecovered was not much more than half that supplied. With barley also, the manuring for which takes place in the spring, there is again nearly half the nitrogen supplied in the manure recovered in the increase, and therefore little more than half left unrecovered.

It may be observed that, in the case of root-crops, when the supply of nitrogen is not excessive, the proportion of the nitrogen of the manure recovered in the increase may be much greater than in the case of the cereals ; whilst in the case of the leguminosæ the effects of such direct application of soluble nitrogenous manures to the surface soil is comparatively so small, and so uncertain, that it would be useless to give an estimate of the amounts recovered and not recovered respectively.

But what becomes of the one-half or two-thirds of the nitrogen supplied for the increased growth of the cereals, but not recovered in the increase of crop? Dr. Frankland and Dr. Voelcker have made numerous analyses of the drainage water from the experimental wheat plots which have yielded the results above referred to, and a summary of their results is given in Table X.

TABLE X.

Nitrogen as Nitrates and Nitrites, per 100,000 *parts of Drainage Water from Plots differently manured, in the Experimental Wheat Field at Rothamsted, Wheat every year, commencing* 1844.

Manuring, Quantities per acre, per annum.	Nitrogen as Nitrates and Nitrites, per 100,000 parts Drainage Water.					
	Dr. Frankland's Results.		Dr. Voelcker's Results.		Mean.	
	Experiments.		Experiments.		Experiments.	
Farm-yard Manure	4	0·922	2	1·606	6	1·264
Without Manure	6	0·316	5	0·390	11	0·353
Complex Mineral Manure. .	6	0·349	5	0·506	11	0·428
Complex Min. Man. & 41 lbs. Nitrogen, as Ammonia . .	6	0·793	5	0·853	11	0·823
Complex Min. Man. & 82 lbs. Nitrogen, as Ammonia . .	6	1·477	5	1·400	11	1·439
Complex Min. Man. & 123 lbs. Nitrogen, as Ammonia . .	6	1·951	5	1·679	11	1·815
Complex Min. Man. & 82 lbs. Nitrogen, as Nitrate	5	1·039	5	1·835	10	1·437

The figures in the Table conclusively show that the quantity of nitrogen as nitrates per 100,000 parts of the drainage water, increased in very direct proportion to the increase in the amount of ammonia or nitrate supplied, and it is obvious that there has been a considerable loss of the nitrogen of the manures by drainage. But as the subsoil rests upon chalk not many feet below the surface, and there is, therefore, natural drainage constantly going on, even when there is no flow from the pipes, it is impossible accurately to estimate the total amount of drainage, and therefrom the total amount of loss. Other experiments at Rothamsted, however, lead to the conclusion that, according to season, from one-quarter to nearly one-half of the annual rainfall may pass below 40 inches. Now, supposing drainage water to contain one part of nitrogen as nitrates per 100,000 parts of water, an inch of rain passing beyond the reach of the roots would carry with it $2\frac{1}{4}$ lbs. of nitrogen per acre; and it is obvious that if from seven to ten inches passed annually of that average strength, the loss would be very great. In reference to this point it is of much interest to observe, that in the Report of the River's Pollution Commission already referred to, Dr. Frankland gives a series of analyses of land drainage waters collected at Rothamsted, at depths of twenty, forty, and sixty inches, respectively; and those collected at twenty inches, almost invariably show much more nitrogen as nitric acid than those taken at either forty or sixty inches. It would thus appear to be indicated that a considerable amount of nitric acid has been arrested in the soil below the depth of twenty inches. Further, determinations of nitrogen in the soils do show some accumulation. Indeed, it would appear probable, that the whole of the nitrogen applied to the wheat as ammonia salts or nitrate of soda, was either recovered in the increase of crop, or may be accounted for by determinable accumulation within the soil, or by loss by drainage.

In ordinary agriculture, the amounts of soluble nitrogenous manures applied would generally be much less than in some of these special experiments; and the losses by drainage would from that cause alone be proportionately less than that shown above. Much, obviously, would also depend upon the character of the soil and of the subsoil. Again, in an ordinary rotation of crops, more of the supplied nitrogen would probably be gathered up before it reached the lower layers, than

in the case of a cereal corp grown year after year on the same land
It may be safely concluded, however, that whenever cereals were
grown, a material proportion of the nitrogen specially applied to, or
existing in the soil, which would be available to other crops, would not
be so to them ; but would in the first instance accumulate in the surface
soil, and gradually pass into the lower layers in the form of nitrates, to
be eventually lost by drainage if not arrested by some other crop.

The question obviously arises, whether we have not here a source of
some at least of the nitrogen available to leguminous or to other plants
having possession by their roots of a greater range of subsoil than the
gramineæ. We have evidence enough that although wheat and barley
send roots down very deep into the subsoil, and pump up moisture
from the deeper layers, they nevertheless derive much of their nitrogen
within the surface soil. If the leguminosæ do not so readily do so, or
at any rate naturally depend more upon the nitrogen in the lower
layers for a considerable proportion of that which they require, and
moreover are able to avail themselves of the residue from the manuring for other crops, what is the nature of the problem that we may
have to solve to elucidate this point ?

By way of illustration it may be mentioned that, supposing a leguminous crop to acquire 100 lbs. of nitrogen per acre from a layer of subsoil
three feet in thickness, weighing approximately 10,000,000 lbs. (exclu-
sive of stones and water), this would represent only ·001 per cent. of
nitrogen so acquired in such subsoil ; 200 lbs. of nitrogen per acre so
available would represent ·002 per cent., and so on. Now, even supposing that the nitrogen existed in the subsoil in such a condition as to
be converted into ammonia in the process of combustion with soda-
lime, the difference between one subsoil containing this, or even a
larger amount of nitrogen, more than another, could not with certainty
be determined by that process ; for, in taking say 15 or 20 grams of the
subsoil for combustion, the difference between two or more determinations could not be expected to be less than some units in the third decimal
place (per cent.) ; that is, in fact, equal to the total amount that may be
in question as between two subsoils to be compared. Further, if this
available nitrogen exist in the subsoil as nitrates, it may be a question
whether there would be a sufficient amount of organic matter present
to insure the evolution as ammonia of the nitrogen of the nitric acid.

It has been shown, then, that there are many questions still open for investigation in regard to the relations of the surface soil to combined nitrogen ; and there are obviously also equally important points to investigate in regard to the nitrogen of the sub-soil, before we can hope to arrive at a satisfactory solution of some of the problems which the consideration of the facts of vegetable production which have been adduced, suggest for enquiry. Nor are the problems still open connected with the amount, and the condition, of the mineral food of plants within the soil, either few, or without special, and independent interest. And although those relating to the nitrogen seem to call for the first attention, the marked effects, so far, of potass manures, in increasing the amount of nitrogen assimilated over a given area by the leguminosæ, seem to indicate the probability that even the difficulties connected with the sources of the nitrogen of our crops may not be solved without further knowledge as to the required conditions, or the actions, of the incombustible or mineral constituents in soils.

Our results in regard to the variations in the amount of nitrogen in the soils and subsoils of our different experimental plots, obtained by the soda-lime process, together with the results already referred to, relating to the composition of the drainage water from plots variously manured, as well as others of quite a different kind, have shown the absolute necessity for an extended investigation of the soil question by more exact methods ; and Mr. Warington is about to devote, probably some years, to this enquiry at Rothamsted. It is proposed that the questions relating to the nitrogen in subsoils should be the first considered, as, if the results do throw light upon some of the points at present in doubt, a definite step in advance will be so gained ; and should they not do so, the ground will thus be cleared of certain obvious suppositions, and the course of further research will be the more clearly indicated. But if the amount of nitrogen to be discriminated should prove to be represented by only units in the third decimal place per cent, say ˙002 for example, it is obvious that to get as little as four milligrams involved in the analysis, 200 grans of soil would have to be operated upon. The difficulties of the problem are thus sufficiently obvious. But, by the aid of the processes of water and gas analysis which have been explained by the President in his opening address, there is little doubt that they can be overcome, at any rate so far as the nitrogen existing as nitric acid is concerned ; and by the kindness of Dr. Frankland, Mr. Warington is at the present time gaining experience in the use of those methods, in the laboratory of the College of Chemistry, before entering upon this special investigation at Rothamsted.

But even supposing we arrive at a satisfactory solution of the, at present, unsettled points in regard to the sources of the nitrogen yielded in agricultural production, when, as in the experiments to which attention has been directed, we have a soil to work upon which already contains accumulations of combined nitrogen amounting to several thousands of pounds per acre within the range of the roots of our crops, further questions in regard to the nitrogen may still be left open, namely,—to what actions a large proportion of the existing combined nitrogen may be attributed ; and what in particular is the exact source of the accumulations of it in our soils and subsoils? And here it may be observed, in passing, that determinations made at Rothamsted have shown approximately the same percentage of nitrogen in the Oxford-clay obtained in the recent Sub-Wealden exploration boring at a depth of between 500 and 600 feet, and in the subsoil at Rothamsted, taken a depth of about 4 feet only.

It is not within the scope of the present discourse to discuss fully what is known of the actual or possible sources of the already existing combined nitrogen, the special object of the enquiry being, as intimated at the commencement, to bring to view the facts relating to the yield of nitrogen in agricultural production, which the extended period of the investigations of Mr. Lawes and myself have enabled us to establish, and to point out the relation of this to the various known or supposed sources of present periodic supplies, so as to indicate what points seem the most urgently to demand further investigation. In the papers already referred to, we have more fully considered what was known of the various actual or possible sources of the combined nitrogen which we know to exist, and to circulate, in land and water, in animal and vegetable life, and in the atmosphere, and we have pointed out how little was established of either the actual or the relative importance, in a *quantitative* sense, of the various actions by which it is admitted that free nitrogen may in nature be brought into combination. I may, however, observe that M. Boussingault and M. Schlösing nave quite recently made interesting contributions to the discussion of this subject. (Compt. Rend. T. lxxvi, lxxx, lxxxi, and lxxxii).

But whatever may be the origin of the existing combined nitrogen, or whether or not the agencies of its formation are more or less active now than during the earlier history of the earth and its atmosphere, the question arises whether, assuming the origin to be independent of the direct action of vegetation, the large accumulations within our soils and subsoils, admits of any reasonable explanation? On this point it may be remarked, that ages of forest growth, or of the growth of natural herbage only grazed by animals, would doubtless leave the soil richer year by year ; as the amount annually lost to it would probably be less than even the small amount known to be annually deposited from the atmosphere, in temperate regions, at the present time ; and the accumulation would probably be greater still, were the amounts of combined nitrogen in the atmosphere, and brought down from it, greater than with us at the present time. Then, again, the influence on aqueous deposits of ages of submarine vegetation, and of the subsistance of animal life upon it, has to be considered. But a soil once broken up, and under arable culture, it is difficult to conceive of any system of agriculture by which so little nitrogen as that hitherto quantitatively determined to be annually deposited from the atmosphere would be annually exported from the land.

And now, to summarise in a few words the results of the whole discussion, I think the balance of the evidence points to the conclusion, that the answer to the question—what are the sources of the nitrogen of vegetation in general, and of agricultural production in particular, is more likely to be found in the relations of the atmosphere, and of the plant to the soil, than in those of the atmosphere to the plant itself.

One word more in conclusion. I have, as explained at the outset, confined attention almost exclusively to one aspect of the great subject of vegetation ; but it will not be supposed that I have done so from any want of appreciation of the interest and importance of other lines of enquiry ; and allow me, before closing to allude to a point which can hardly fail to suggest itself on an inspection of the numerous organic compounds, made by transformation in the laboratory, which are collected in the Chemical Section of this Exhibition. Without in the least degree disparaging such work, I would ask whether some of

F

those who have become masters of such transformations, might not with advantage, armed with the experience thus gained, now devote themselves to the study of the transformations going on within the plant and the animal? In other words, whether it would not be desirable, that some of the thought and labour now expended on transformations in the chemical laboratory should be transferred to the laboratory of nature?

The CHAIRMAN: I am sure you will give your most hearty thanks to Dr. Gilbert for having brought before us the result of these many lengthened and careful experiments, which have been conducted by him on the estate of Mr. Lawes. The work, in fact, belongs to them jointly. We have had brought before us a great deal that would occupy our minds, were we to go fully into the subject and to do justice to it. But there are other communications to come, and therefore we must be brief. Still there may be some questions to be asked, and if so, I am sure Dr. Gilbert will be happy to reply to them.

Mr. WILKINS: I think a table of analyses of the ash of weeds would be a very useful guide to the public, and would enable one to some extent to tell the nature of the soil and what it is capable of growing. I have never seen one giving analyses of the different weeds.

Dr. GILBERT: We have already published very many analyses of the ashes of wheat, grain and straw grown under the special circumstances which have been here described, and we have some 200 analyses not yet published. As I mentioned to you, when you get a perfectly ripe product—and a grain should be so, though with different seasons it is not always so, and with different conditions of manure it is sometimes a little more backward and sometimes a little less backward, according to the characters of any particular season, so that you do not get the seeds always equally ripe ; otherwise you might say that, under ordinary circumstances, the same description of wheat would give the same composition of ash if equally and fairly ripened ; but we have found in these very experiments under the conditions which we were enabled to introduce—in some cases the great exhaustion of one constituent, in others, the great exhaustion of another, which we do for the purpose of investigation—that thus we have been able to get a greater variation in the composition of the ash, even of comparatively ripe grain, than would ordinarily be the case.

Mr. WILKINS : I should like an analysis made, particularly of the weed called " May-weed," which is very troublesome. We know that where lime is deficient in the soil, that weed grows; but if you add lime to the soil it disappears. That is the sort of information which I should think would be very useful to the public.

Mr. LIGGINS : As Dr. Gilbert has been good enough to make some remarks about the sugar-cane, I may say a word or two, not upon the chemistry of the subject, but on a question of fact. I am, unfortunately the possessor of large sugar estates in the West Indies, and at various times I have sent out, at the request of those who manage the estates, chemical manures, amongst others, nitrate of soda and sulphate of ammonia. They have been extensively used. Perhaps £1,000 worth at a time has been sent there, and, I admit under favourable circumstances of weather, the appearance of the crop has been very greatly improved. Of course, the sugar-cane is of the grass tribe, and the grass has been increased, perhaps, from 17 or 18 feet to 25 or even 30 feet, and the cane has been increased from the size of my umbrella to, perhaps, double that diameter. But there is one principal feature which we observe with great regret, which I think militates against the use of these chemical manures, viz., that the sugar made, although greater in quantity when it first goes to the scale in the manufacturing house, diminishes in the form of drainage on the voyage home, so that really in England we get less sugar from an acre of land than we did without those chemical manures. It appears to give to the sugar a deliquescent nature. I think if chemists would show us the way to preserve that which we have already created, by the introduction of some other chemical, or by a modification of that which they have recommended for our use, they will be doing the British West Indies a very great service.

Mr. WARINGTON : It may perhaps be of interest to mention one of the preliminary results obtained by Dr. Frankland's exceedingly delicate method in the examination of soils. Dr. Frankland has told us, we can if we measure gases, determine far smaller quantities of matter than by any other process, and we find as the result of preliminary experiments, that it is possible to determine nitrogen in the sub-soil to the extent of one part in a million of soil if we determine the nitrogen in the form of nitric acid by the method Dr. Frankland has recommended for water analysis. I think you will admit that that holds out to us some hope, that by these more refined modern methods we may be able to answer some of those questions which Dr. Gilbert has brought before us to-day.

Dr. GILBERT : In answer to the remarks made with regard to the effect of different manures on the sugar cane, I may say, that we know perfectly well, if we have an excess of nitrogenous manures in relation to other matters, and in relation to what the season will ripen, the characters spoken of, are those sure to arise. I may say, at the same time, that when judiciously used in fairly average seasons, you do get an enormous increase of good sugar by the application of these manures. By these samples of sugar-beet which I have before me, I may illustrate what difficulties there may be in this question. I do not know whether there are any German gentlemen here, but if there are, they would doubtless be very angry if I said that this small root was a pattern of the German beet growing in the best districts, but it would not be very far from the truth. In Germany, because the duty is levied on the weight of roots submitted to manufacture, and for other reasons, because land can be held in large quantities, and the makers of sugar become enormous landholders, it happens that it is to their interest to produce roots containing the highest percentage of sugar possible. So they have a great many restrictions preventing the application of such manures as give a very active growth, in order that they may get a root which will yield them in the manufacture as high a percentage of sugar as possible. In France, on the other hand, the circumstances are quite different. There is comparatively little land held by sugar makers, and they have to buy the roots of the small proprietors. Then again they pay their duty upon the juice, and not upon the root. Thus the object of the growers is to produce large crops of large roots, whilst the object of the manufacturers is to buy small ones. By the application of a certain amount of nitrogenous manure, ammonia salt, you get one of those large roots; and by the application of the same quantity as nitrate you get the other kind, but you get a smaller percentage of sugar and a larger percentage of water, and there is also a larger percentage of saline matter where you use the nitrate, the more soluble manure, than where you apply the ammonia salt. The ammonia salt would, I have no doubt in this point of view, be much more beneficial than the nitrate.

Just at this moment a great controversy is going on in France about this. The growers will grow large roots, and the consequence is there is not only a larger percentage of nitrogen in the roots but also a higher percentage of other matter which destroy the sugar in the making, so that they get out a lower proportion of what they do produce. They have been having a Conference at Lille, and are going to try to enforce conditions under which manure shall be limited, the seeds selected, the plants grown near enough to one another, so as not to grow too large, and not to give them these very nitrogenous and over-saline juices. This is, in fact, an exactly parallel case to that which Mr. Liggins referred to as occurring in the use of large quantities of certain nitrogenous manures in the West Indies.

The CHAIRMAN : I have no doubt that we might still obtain if we had time, still further information from the great stores of knowledge Dr. Gilbert possesses on these very important matters—the application of Chemistry to agriculture and to botany—but time presses and I must now call upon Mr. W. F. Donkin to give us an account of Sir Benjamin Brodie's Ozone Apparatus.

ON THE OZONE APPARATUS.

Mr. W. F. DONKIN : The apparatus exhibited by Sir Benjamin Brodie, and which, in his unavoidable absence, I have the honour to describe to-day, is that by means of which he carried out those researches on Ozone which are described in his paper "On the action of Electricity of Gases—Part I," in the *Royal Society's Transactions* for 1872. By means of this apparatus he was enabled to generate and collect the electrised oxygen in sufficient quantities for examination, to submit successive portions of the same gas to the action of different re-agents, and to estimate the variations in volume which the gas undergoes under the influence of these re-agents with facility and precision.

The apparatus as shown, is complete, with the exception of the arrangement for supplying pure and dry oxygen. This was usually prepared by electrolysis, and the arrangement being one in ordinary use it was not considered necessary to exhibit it. The remainder of the apparatus consists essentially of five distinct parts, besides the Ruhmkorff's induction coil and battery. They may be described briefly as follows :—

1. The instrument by which the oxygen is submitted to the electric action. This is called an induction tube, and is fundamentally of the kind originally devised by Dr. Siemens, and first described by him in Poggendorff's Annalen. It consists of two tubes of thin glass sealed at one end, like long test-tubes, one inside the other, and so fused together at the open ends as to leave a narrow annular space between them. The gas is made to traverse this narrow space, entering and leaving it by small tubes fused to the outer tube at its two ends. The instrument is suspended vertically in a tall glass jar, which is filled with water, and water is also poured into the inner tube, so that when the wires from the terminals of the Ruhmkorff's coil are immersed in the water in the jar and in the inner tube respectively, the gas passing through the induction tube is exposed to two oppositely electrified glass surfaces of comparatively large area. In order to dry the oxygen thoroughly, which is an essential condition for the production of a considerable percentage of ozone, it is first passed through sulphuric acid, and then over phosphoric oxide in three glass bulbs attached to the induction tube. The temperature may be kept at zero by filling the jar and inner tube of the induction tube with fragments of ice, or at lower temperature by means of a freezing mixture. The production of ozone is thus considerably increased.

2. A gas-holder in which the electrised gas is stored. It consists of a glass bell jar, of about three litres capacity, suspended in a large glass cylinder containing concentrated sulphuric acid, this liquid fulfilling the double purpose of confining the gas and keeping it dry. After being in use for these experiments for some time, the sulphuric acid becomes singularly free from colour and of a peculiar brightness, and the electrised gas may then be kept over it for many hours without either absorption or appreciable alteration. In the apparatus as shown, water is substituted for the sulphuric acid, for obvious reasons.

Since ozone rapidly corrodes india-rubber, the connection between the induction tube and the gas-holder is effected by slipping over the ends of the tubes to be connected, and which are placed close together, a short piece of glass tubing into which they exactly fit, and then running melted paraffin into the narrow space between the tubes after gently warming them. The paraffin, on solidifying, forms a strong and perfectly tight joint which will resist great pressure and is quite unaffected by ozone. All the joints between the remaining portions of the apparatus are made in the same way.

3. The third portion of the apparatus consists of a pipette, by means of which successive known portions of the electrised gas may be taken from the gas-holder. This is effected by means of sulphuric acid, which is admitted into and removed from the pipette at its lower end. Both in this case and in that of the gas-holder the necessary alterations of level in the acid are made by means of reservoirs of the acid placed at a higher level, and connected by means of syphons with the gas-holder and pipette respectively. The syphons are provided with stop-cocks to regulate the flow of the acid, which may be reversed when required and drawn up again into the reservoirs, by partially exhausting the air in them by means of an air-pump or syringe connected with them by means of india-rubber tubing. The capacity of the pipette was accurately determined by calibration with mercury, and is 290·8 cc. The pipette is suspended in a jar of water, and a thermometer in the water gives the temperature of the gas in the pipette. The acid is admitted into the pipette by a vertical tube connected with it at the lower end, and the leg of the syphon from the reservoir above passes to the bottom of this tube. By filling the pipette with gas to a point a little below the mark on the stem, while the acid stands in the vertical tube a little above the same mark, and then raising for a moment one of the stop-cocks attached to the upper end of the pipette, the excess of gas escapes, and the pipette is exactly filled with gas at atmospheric pressure.

4. From this pipette an accurately measured quantity of the electrised gas may thus be delivered, and may be sent through tubes or bulbs containing any required re-agents. In this case a bulb is shown, of the form used in the experiments with solution of iodide

of potassium; and this constitutes the fourth distinct portion of the apparatus.

5. In order to measure any change in the volume which the electrised gas may undergo by the action upon it of re-agents, it passes finally into a measuring apparatus, the principle of which is identical with that of Regnault's and of Frankland's apparatus for gas analysis, namely, the determination of the pressure which it is necessary to put upon the gas in order to make it occupy a known volume at a known temperature. This measuring apparatus is called the aspirator. It consists of a strong glass cylinder into which the gas may be admitted through a steel three-way cock at the top. Parallel to, and close beside it, is fixed a vertical pressure-tube, also of glass, graduated in millimetres and open to the air above. These tubes communicate permanently with each other below, being cemented into the iron foot of the apparatus, and, practically, constitute one vessel. Mercury is admitted into the tubes from an iron reservoir below, from which it may be forced up into them by means of compressed air pumped into the reservoir by a small force-pump, connected with it by strong caoutchouc tubing; a steel stop-cock between the reservoir and the tubes allows of exact adjustment of the quantity of mercury admitted. Inside the cylinder is fixed a thin piece of glass rod, to which seven points are attached, also of glass, turned downwards; these points give the means of adjusting the mercury to a definite level with great exactness, as in the cistern of Fortin's barometer. The capacity of the cylinder from the top down to each of these points is determined by mercurial calibration; and the accuracy of the measurements based on the numbers so obtained is shown by comparing the ratio of the volumes at each successive pair of points, thus determined, with the ratio of the pressures which one and the same quantity of air exerts, when measured at the same pair of points. The numbers agree very closely, the differences amounting on the average to about one in 1000. Now the capacity of the cylinder at one of the points, say the fifth, may also be determined by drawing over into it a pipette of air, and ascertaining the pressure which that volume of air exerts at that point. Assuming this to give the true capacity of the cylinder at that point, and applying the ratios previously obtained, we get a series of numbers which represent the capacity of the cylinder down to each of the points.

These are given in the following table ; and in the last column are given the capacities as determined by weighing the mercury run out between the successive points.

			By air.			By mercury.
At point 1	...	...	34·96 cc	...	...	34·93 cc
„ 2	...	...	72·00 „	...	...	71·87 „
„ 3	...	...	119·04 „	...	...	118·88 „
„ 4	...	...	172·19 „	...	...	171·92 „
„ 5	...	...	240·89 „	...	...	240·51 „
„ 6	...	...	353·75 „	...	...	353·45 „

The two methods of calibration are quite independent of each other ; and considering the number of observations, of different kinds, necessary for any one determination, the concurrence of the numbers is remarkable, and affords a sufficient guarantee of the accuracy of the method employed for the determination of the volumes of gases in the experiments made with the apparatus.

As it was unfortunately impracticable to remove the apparatus into the conference room, a diagram of it was shown ; and after the conference the course of manipulation involved in the experiment of passing a pipette-full of electrised gas through a bulb of iodide of potassium solution, and measuring it in the aspirator, was gone through with the original apparatus.

The CHAIRMAN: I am quite sure you will give your best thanks to Mr. Donkin for having, in this able manner, brought before us the apparatus of Sir Benjamin Brodie, and I will then call upon Professor Andrews.

The vote of thanks was passed unanimously.

EXPERIMENTS ON GASES.

Professor ANDREWS, F.R.S. : I wish to mention before proceeding further that it was not my intention on coming here to make any communication to the Chemical Section, and that it is only at the request of Dr. Frankland I appear before you on the present occasion. I do not propose to make more than an indirect reference to the subject on which some here may perhaps expect me to speak—I allude

to the properties of matter under high pressure—because this subject
will be brought before another section of the Conference.

I am glad, however, to have an opportunity on the present occasion
of describing some experiments, which many years ago, and also in
later times, have occupied my attention, and which bear a little on the
subject we heard this morning so elaborately and well described
by the President. I allude to certain novel methods of examining
gases—methods which do not interfere at all with the use of the large
apparatus you have seen this morning. In the first place, with regard
to these methods, in order that my remarks may not be entirely dis-
jointed, I wish to say that what led me to enter upon the subject was
an experimental difficulty. Reference has been made to-day to an
interesting and beautiful experiment, and it is one which can be made
with great effect, more effectually, indeed, than in any other way, by
means of a Holtz' machine which I have of late used in exhibiting it
to the students of my class. I allude to the experiment made originally
by the Dutch chemists, Paets Van Troostwyk and Deiman, and afterwards repeated by Wollaston, in which electric sparks were produced
in water from fine metallic points, and those sparks had the property
of decomposing the water. As Faraday, who gave great attention to
the subject, and examined it with his usual ability, ascertained in the
clearest manner,—this is not a case of electrolysis or electro-
chemical decomposition. You place in distilled water, for the experiment does not succeed in acid water, two glass tubes exposing very
fine platinum points, and when sparks are drawn from the points you
get hydrogen and oxygen at both, so that it is not a case of polar
decomposition. In fact what occurs here is that the water is split up
into its elements at the two points, and you obtain oxygen and hydrogen
together at each point. Passing from this experiment, there comes in
the next place the question,—how to decompose water by a current of
electricity from the common machine, so as to collect and measure the
gases evolved. This problem, although partially solved by Davy, had
long baffled the efforts of chemists to resolve it completely. If you
moisten a small piece of white blotting paper with a solution of iodide
of potassium, and bring a platinum wire from the prime conductor of
an electrical machine into contact with it, connecting at the same
time the moistened paper with the ground, you will obtain at once a

deposit of free iodine at the platinum wire, no doubt arising from true polar decomposition. But if you immerse two fine platinum wires, or platinum plates, in acidulated water, or if you have a plate on one side and a wire on the other, and pass a silent current from the electrical machine, or a current of atmospheric electricity through the liquid, not a trace of gas will appear. There can be no doubt, however, that decomposition actually occurs, because the plates become what is called polarized, or acquire the peculiar condition which is always produced when electrolytic decomposition takes place. Considering this question some years ago, it occurred to me that the gases might be obtained, and even measured, if we arranged the experiments so as to prevent their absorption by the liquid electrolyte. Accordingly, I succeeded in overcoming the difficulty by operating in a tube, in which the volume of gas to be obtained, and the amount of liquid to which it was exposed, were in some degree comparable; in other words, I took two fine tubes, such as I have here—for these are the very tubes I worked with—fused into one end of each tube a platinum wire, which projects into the tube half an inch, or, it may be, for the whole length. The capillary tubes were filled with acid water in a very simple way by introducing them into a large flask containing the liquid, which was repeatedly boiled and cooled so as to expel the air from the tubes. In order that the problem may be more clearly understood, I may mention that with a galvanometer constructed by Gourjon, with 2000 coils of carefully insulated copper wire, you can examine the amount of electricity derived from the machine or from the electric kite. If you take an ordinary electrical machine in good order, and connect one end of the wire of the galvanometer with the prime conductor and the other end with the ground, you will obtain a deflection of the needle amounting to about 20°. To compare this with the quantity of electricity produced by a voltaic arrangement, you have only to take a clean zinc plate and lay upon it a small speck of white blotting paper moistened with saliva, completing the mimic cell by touching the paper with a fine copper wire. The current obtained from this feeble source will produce upon the galvanometer at least triple the effect of an ordinary electrical machine. Its action upon the galvanometer corresponds to the power of the electrical machine to decompose a certain amount of water. These small tubes being filled with acid water are placed in a wine glass or other similar vessel ; one being connected with the prime conductor and the other with the ground. On working the machine, fine bubbles of gases will appear in two or three minutes, and at the end of an hour the hydrogen tube will be filled with gas. On measuring the gases they were found to be more exactly in the proper proportion than in the ordinary decomposing cell when a large voltaic pile is used, because the exposure to the liquid was less. In the next place, this sort of electricity, as every one knows, has immense tension, or power of overcoming resistances. Accordingly with even a powerful voltaic battery, if you place two or three decomposing cells in succession you diminish the effect enormously, but it is quite different here. I placed 100 such couples in series, and the amount of decomposition in each was the same as if one couple only had been taken. I did not go beyond 100 couples, but up to that number there was no difference whatever when acidulated water was employed. With pure water, on the contrary, a marked diminution of effect occurred with fifty or sixty couples. Another remarkable fact is this, that if you fill a long thermometer tube with pure water and pass a current from the electrical machine through it, the slight asperities on the inner surface of the tube interrupt the current, and lead to the formation of a number of decomposing cells in the course of the fine column of water.

Having succeeded so far, I then applied my little tubes to the decomposition of water by means of atmospheric electricity, and for this purpose I flew an electrical kite. Having a small observatory not in use at the time, I found the dome of it formed a very suitable place from which to fly the kite. The kite I employed was an ordinary one, with fine copper wire fastened to the string, and at the lower end the string and wire were wound round a glass bottle fixed on a stand, so that the whole was insulated. This was an experiment which any one may make with perfect safety in fine weather. I have never myself attempted it when there were thunder-clouds hanging about ; but in fine weather the experiment is a very interesting one and can be made without risk. You get very fine sparks about an inch or an inch and a half in length, which give a considerable shock, particularly if you touch a large mass of metal at the same time. Of course you can produce all the ordinary electrical phenomena. Applying this sort of electricity to the fine tubes, the water was decomposed, but not much more freely than by an ordinary machine, because the amount of electricity was not more—scarcely so much in fact—as from a machine in good order. Thus, then, the question was finally settled that you can decompose pure water by common or atmospheric electricity ; that is to say, you can obtain polar decomposition, and collect the oxygen and hydrogen gases separately.

In the next place, I may mention a few modes of experimenting with gases in fine tubes, which may be almost regarded as a distinct branch of eudiometry. Suppose you have a gas in one of these tubes, and you wish to ascertain whether it is oxygen. The mode of determining this is one which I do not think chemists have hitherto employed. You place the tube containing the gas is a vessel with a little distilled water, introduce it under the receiver of an air-pump, and exhaust until you bring the gas just to the end of the tube. Then repeat the operation two or three times until the vessel is pretty well washed. Now, remove your water and replace it with a solution of iodide of potassium. Repeat the experiment once or twice, and in this way, by exhausting to the point I have mentioned, so as not to remove any of the gas, you will have the liquid in the tube replaced by a solution of iodide of potassium. Now, pass the silent electrical discharge through the tube, and in the course of a few seconds, if the gas be oxygen, it will be converted into ozone and absorbed by the liquid.

As regards hydrogen, the red colour of the spark will indicate its presence, and the appearance of its characteristic lines, when the spark is examined by the spectroscope, will remove every doubt. Absorbents, such as caustic potash for carbonic acid can be easily applied by the manipulation already described. Indeed all the ordinary eudiometric experiments can be performed accurately on almost microscopic quantities of gases by the methods I have indicated. Ammonia can be readily decomposed in these tubes and the nature of the resulting gases determined by the spectroscope.

I will now exhibit a few of the remaining tubes with which I worked upon ozone some years ago in conjunction with Professor Tait. I had previously made many experiments myself on ozone, and discovered the mistake into which a distinguished chemist fell, who supposed that ozone was a compound of hydrogen and oxygen—a super-oxide of

hydrogen. To these I will not refer now ; but on continuing the inquiry with my friend Professor Tait, we found that oxygen was diminished in volume when it changed into ozone. M. Fremy had previously
established the fact that ozone is a form of oxygen. This was ascertained by passing an electric spark through dry oxygen gas, when ozone
was formed, and nothing else. Afterwards, in the experiments which
occupied my friend and myself a considerable time, we determined
that ozone is a condensed form of oxygen gas. We made several con-
jectures as to what the amount of condensation might be under certain
circumstances, and here are some tubes with which we operated, still
working on rather a small scale. In these cases we always adopted
the method referred to to-day, of having a second tube, which gave us
the correction for temperature and pressure. We took a pair of tubes
exactly alike, one filled with air and the other filled with oxygen gas,
which terminated at the other end with a U tube, containing sulphuric
acid. The acid at the beginning stood at a certain level. The two
tubes were placed in a large vessel, the exact position of the acid in
both tubes was ascertained, and then the one containing oxygen was
sealed. One tube was left merely as a test, so that we might ascertain
the change of volume arising from alteration of temperature or
pressure. The other tube was sealed, and a silent discharge, as we
call it—a faint electrical discharge was passed through the oxygen gas.
It was then placed again in the large vessel, and when such a tube as
the one I show you was opened, the sulphuric acid rose often about
three inches. A very interesting and remarkably instructive experiment was to seal the tube again and then introduce it into a chamber
heated to 300° or 400°. C. In this way the ozone was destroyed and
the oxygen restored to its original state, thus proving beyond all doubt
that ozone is a condensed form of oxygen. We did not succeed,
although we made many experiments with iodide of potassium,
except by conjecture, in ascertaining what the amount of condensation was. That point was afterwards determined by M. Soret
who ascertained definitively the density of ozone, and his results
have since been confirmed by the elaborate experiments of Sir Benjamin Brodie.

I wish now to show you a simple apparatus with which the latent
heat of vapour can be accurately determined on a small scale. This is

an object of importance in such experiments in order to enable the experimenter to operate upon liquids in the purest possible state.

Before concluding, allow me to say a few words regarding some details of the method of observing in experiments of precision on gases at high pressures and varied temperatures. The external vessel containing the water or vapour, by which a steady temperature is maintained, cannot in exact experiments be a glass cylinder. The inequalities or striæ in the most perfect glass cylinder by distorting the image of the object always interfere greatly with the accuracy of the readings ; and I confess it was with some surprise I observed that no attempt was made to rectify this source of error in any of the forms of apparatus for gas analysis which were brought under your notice this morning. Accurate measurements can only be made in a vessel having plate glass sides, and the plate glass ought to be of the best quality. The tubes in my pressure apparatus are enclosed in rectangular vessels of this kind, which admit of any temperature from 0° to 100° or even higher being steadily applied. For high temperatures I am in the habit of employing the vapours of methylic alcohol and of water. Even under a pressure of 300 atmospheres, it is easy to work at the temperature of boiling water. The vapour from a boiler in an adjoining apartment is introduced at the upper end of the containing vessel, and is carried off below by a tube passing through an outer wall. The extra pressure of the steam was ascertained by a small water guage. In experiments with steam there was at first great difficulty from the water having a tendency to condense in drops on the inner tube. This was at last overcome by the simple expedient of pouring boiling water into the vessel—an experiment which at high pressures appeared to be very hazardous, but was in reality attended with no real risk. With the vapour of methylic alcohol it was not possible to obtain a perfectly steady temperature, and the thermometer had to be used. The vapour of water gives an absolutely steady temperature, but other vapours, no doubt from the presence of impurities, do not act so well.

To measure the true volume of gas originally taken is a difficult experiment—where it is required that the results should be true to the three or four thousandth part. Of the calibration of the tube and the determination of its capacity by weighing the mercury it is capable of containing at a given temperature, I will not now speak ; but I may refer to a serious source of difficulty arising from the thin film of air which is always interposed between mercury and glass unless very special precautions are taken. Even if these precautions are originally taken, the accuracy of the observations may be interfered with from the imprisonment of the film of air in the subsequent experiments. To fill tubes with a known volume of gas at a given temperature and pressure, I am now in the habit of passing a current of the gas through the tube placed in a vertical position, and having the lower end enclosed in a test tube containing such a quantity of mercury that the end of the tube is about one-tenth of an inch above the mercurial surface. When the whole of the air has been expelled by the gas the upper end of the tube is hermetically sealed, and the current of gas is continued below for half an hour in an apartment where the temperature is as steady as possible. The test tube is then raised so as to bring the mercury into contact with the lower end of the tube, the temperature and height of the barometer being at the same time carefully noted.

I must apologise for these rambling observations, but they may perhaps be useful to persons occupied in delicate researches of this kind. In conclusion, I have here an apparatus which I showed in action two or three weeks ago, at a meeting of the Chemical Society, and which enables the condensation of oxygen, when converted into ozone, to be easily exhibited as a class experiment. It is a modified form of Siemens' tube, the delivery tube being a long vertical glass tube, and the entrance tube being sealed after the apparatus has been filled with oxygen. The end of the delivery tube dips under a surface of pure sulphuric acid. On passing the electrical discharge through the apparatus, the acid will be seen in a few minutes to rise through several inches. In this apparatus I have succeeded in reducing oxygen by one-tenth of its volume from the action of the discharge. The same apparatus has been employed under most favourable conditions to ascertain whether chlorine gas undergoes any change of volume under the action of electricity. The result was negative, so that neither chlorine, hydrogen, nor nitrogen, change into more active forms when subjected to the influence of the electrical discharge.

The CHAIRMAN : I am sure it was not necessary for Professor Andrews to apologise in any way for the account he has given us of these experiments which have so great interest. He has shown us the great care with which he worked at this subject and overcame very great difficulties in these experiments which have led in his hands to very important results. I have now only to express your thanks to him for his communication.

The Conference then adjourned.

SECTION—CHEMISTRY.

Tuesday, May 23rd.

The PRESIDENT : Ladies and gentlemen, the first communication
to the section of this morning ought to have been made by Professor
Fremy in person, but I regret to say that the very heavy occupations
which fall upon M. Fremy, and so many other professors, at this time
of the year, entirely prevent his coming over to make this communication in person. I regret this the more, because had he been here
we should have had an interesting discussion upon the proposition
which he makes in the short paper, which I shall read to you. The
diminution of scientific research in this country, as well as in France,
has been for many years deplored by those who pursue the various
branches of science. It is well known that the causes of the decline
of scientific research in this country have recently been investigated
in the most careful way by a Royal Commission, of which the Duke of
Devonshire was the Chairman. Professor Fremy, who has sent us
this communication, has been one of the most devoted apostles, of
the attempts to revivify original research in France ; and about two
years ago he instituted a Laboratory of Research, in which students
who had already completed their chemical education, were received
and allowed to work gratis. This plan has been in operation for
twelve years, and, therefore, although the causes of the decline in
research, and still more the remedies applicable to it, may be the
subject of a good deal of difference of opinion, I think you will all
agree with me that the utterances of a man in M. Fremy's position,
who has directed so much attention to the subject, and who has
endeavoured so earnestly to remedy these defects, deserve to be
listened to with attention.

M. Fremy says :—

Many thoughtful men who know the services which science can

render to humanity, have of late occupied themselves in considering the causes of the neglect which the higher scientific studies have suffered for a long time past, and they have attempted, both by encouragement given to scientific men, and by the further development of superior instruction, to revive those studies.

I do full justice to these generous efforts, and I have for the last twelve years associated myself with them in establishing, to the best of my ability, a laboratory for instruction and research in the Natural History Musuem, and a number of students attend this laboratory to pursue chemical investigations gratuitously.

But the long experience that I have had clearly shows that these endeavours will continue to miss their mark. They will not prove themselves capable of revivifying the cultivation of pure science ; and, moreover, the evil will not be eradicated until scientific careers are regularly organised and properly recruited.

It must not be forgotten that at the present day, more than ever every career without a future is rejected.

The most zealous aspirations are paralysed and arrested by want of the necessaries of existence.

Our laboratories are overflowing with students, but the latter leave as soon as they have acquired the minimum of knowledge by which pecuniary remuneration can be obtained ; they abandon pure science for its industrial applications; and who can blame them when science cannot ensure them a living?

I nevertheless believe it would be easy to bring the deserters back to science.

Before making known the means by which I propose to obtain this desirable end, I hope to be permitted to repeat here some of the remarks that I have made on former occasions.

I have said that for the last ten years science has been abondoned in France, because it offers only an uncertain income to those who pnrsue it.

Scientific education is both costly and laborious, and requires several years of uninterrupted study.

The greatest scientific discovery does not contribute any remuneration to its author, but often causes him ruinous expense.

Instances can be quoted of scientific men who, from want of means have been obliged to abandon important researches, and remain content up to the age of fifty with a modest assistantship.

The frequent pecuniary assistance given by the friends of science shows that often the most illustrious scientific men die leaving their families in extreme poverty.

Unlike all other professions, the scientific one gives no opportunity. for regular advancement.

The Professor of the Collège de France, of the Natural History Museum, of the Ecole Polytechnique, and of several other public educational institutes, receive after thirty years service, the same salary as on the day of their nomination.

A Professorship is the only remuneration given to scientific men, but even the most worthy do not always attain to that position, and a talented man may not always possess the special faculty of communicating his knowledge to others.

In short, the professors, although they are the fortunate ones among scientific men, live with difficulty ; whilst scientific men, NOT professors, die of hunger.   Thus, at a period like the present, when the expenses of life are increasing at a considerable rate, young men who would formerly have devoted themselves to science with much enthusiasm, finding no fixed career before them, seek lucrative posts, which science does not offer them, in manufactories or elsewhere.

Thus a scientific career is shunned, recruiting for it becomes daily more difficult, and the country loses every year mnch valuable scientific discovery. Such a loss is incalculable.

I know that these remarks harmonise with the sympathies of all those who are interested in the progress of science. Nevertheless, I regret to say that I have not been able to get a hearing from those who judge and dispense in these matters.   My publications have not succeeded in obtaining any of the measures that I asked for, and the career of a scientific man remains as before,—an uncertain and precarious one.   It is true that numerous laboratories have been built, but, as there are no scientific careers, they have been of more use to applied than to pure and higher science, and yet it is pure science that ought to be first thought of ; for is *it* not the source of all its useful applications ?

The men who extend to manufactures, the discoveries due to scientific investigators do not ask for encouragement, for they always find sufficient remuneration in the industrial branches which they carry on. But the scientific man, he who by his discoveries opens up new fields of knowledge, how is he rewarded ?

I know that there are men who never complain, their sacred zeal inspires and encourages them ; yet their means become exhausted, and I could quote several who have fallen before attaining to the end which they sought.

I do not want to suggest wealth for scientific men, but a modest progressive career, such as is offered to the soldier or state engineer.

I only demand that young students of science be treated in the same manner as young men entering the Ecole Polytechnique or the Ecole de St. Cyr.

Why not treat the scientific man and the soldier in the same manner ? Both serve their country with the same zeal and disinterestedness ; has not science, like the army, its wounded and its martyrs ?

The state chooses and adopts its officers and engineers after a severe course of study extending over several years ; it instructs them, never abandons them, and ensures their regular advancement, according to the services they have rendered.

This national measure secures excellent recruiting for the army and public service : I claim it for science also, which renders such invaluable services to all.

It is with this conviction that I suggest the following organization or scientific careers, of which I will here only enunciate the principles, leaving to others, more competent than myself, the task of working out the details.

The scientific career shall consist of five grades :

			Per ann.
The scientific man of the 5th grade shall receive			£120
,,	4th	,,	£200
,,	3rd	,,	£320
,,	2nd	,,	£600
,,	1st	,,	£800

The entrance upon a scientific career, *i.e.*, admission into the fifth grade will not be granted till decisive tests have proved with certainty the scientific capabilities of the candidate.

For this proof general education must certainly be taken into account, but above all inventive genius and originality in work.

The scientific vocation in a youth announces itself by unmistakable characteristics, and at a much earlier age than is generally supposed.

In the science of mathematics young men are often met with, who, even on quitting the Ecole Polytechnique, publish remarkable researches.

For the sciences of experiment and observation, the aptitude of young men can be easily tested in the laboratories, now so numerous and useful, in which their original work would be carried on under the supervision of professors.

I maintain, therefore, that the recruiting of science by the method here advocated, presents no difficulties whatever, and that it also offers many more guarantees than those which are made use of in other careers.

He who advances science by his discoveries, works in the interest of all ; the state ought, therefore, to reward him proportionaly to the scientific services he renders it.

It is this reward for scientific work in all its phases, which is the basis of the organization I propose.

The worth of men of science would be weighed by a jury, consisting of scientific men, who in official reports would make known the claims to advancement, the judge and candidate would thus be subject to public criticism ; everything would be done openly. Such a system would give no opportunity for intrigue or favour, for the scientific man would be judged by his peers, and it would be they who would determine the position he should fill.

The future of science depends upon the constitution of the jury ; its members, therefore, ought to be men of acknowledged scientific reputation, independence, and integrity ; and it is of course understood that their services would be entirely unremunerated.

Whilst the members of this jury ought not to neglect any of the claims of scientific men to advancement, they should be merciless in the case of mediocrity or idleness. Supposing, therefore, that a young man had been too easily admitted into the scientific course, and that his vocation had been misunderstood, such an error could not long remain undetected : the insufficiency of his scientific work would justify measures for his removal, which the jury ought never to hesitate to take.

The PRESIDENT : These are the remarks of a man who, as I have said, has devoted, at great self-sacrifice, twelve years of his life to attempts to revivify higher scientific studies—this original investigation in chemical and physical science—and he declares that his labour is nearly in vain, and cannot succeed without state aid. I am afraid with the press of matter before us to-day I cannot invite discussion upon this theme ; it is one, no doubt, that is of very great importance ; but if we once begin to discuss it I am afraid we should be led on to very considerable length. I do not think that in the absence of Professor Fremy it is advisable for us to criticise his proposals here, but I am sure you will all agree that our thanks are due to him for the communication which he has made to the section.

I will now ask Mr. Matthey to give us his communication

ON GAS BURNERS AND GAS FURNACES AS APPLIED TO PLATINUM MELTING.

Mr. G. MATTHEY : Ladies and gentlemen, having been requested by Dr. Frankland to describe the process of melting platinum, (a furnace which has been used for that purpose being exhibited in this collection,) I will merely premise that I must of necessity be brief, the subject being one which does not allow of any extended description.

Up to the time when Mr. Deville introduced the process of melting platinum in lime furnaces, by a combination of oxygen and common gas in the year 1858 or 1859, melted platinum was a thing of rare occurrence and only existed in very small specimens, and although the metal had been extensively used, it had always been prepared by welding the spongy platinum at a high heat.

It may be as well to remark that the quality of platinum carefully prepared in this manner is identical with that of melted metal, for although a trifling difference may exist in the density of the metal so prepared, we have abundant proof in the manufacture into apparatus or vessels for chemical use and more particularly in the process of autogenous burning instead of soldering two edges of the metal together, that no perceptible difference or change can be observed, and a skilful burner cannot detect which is melted or unmelted platinum. The process of melting therefore is chiefly useful for alloys of platinum, for as the pure metal is quite as soft as copper, it would be useless for many purposes unless hardened. This can be done, and it is extensively used slightly alloyed with copper. I have also alloyed it with gold, silver, palladium, rhodium, and iridium..

A very large piece of melted platinum was exhibited at the International Exhibition of 1862, weighing over 100 kilogrammes or two cwt. Also a great many castings were exhibited by my firm in the one held in Paris in 1867, but the largest operation of the kind took place in Paris during the spring of 1874 for the Commission Internationale du Metre, at which I had the honour of assisting.

The quantity of platinum then melted was 250 killogrammes or about five cwt., and it was alloyed with ten per cent. of iridium for the purpose of hardening it—and as this latter metal is extremely difficult to melt—requiring pure hydrogen and oxygen gasses for the purpose, this large mass of alloy required more heat than that of platinum alone.

The time taken to melt this quantity was seventy minutes, and 1,000 cubic feet of oxygen and 850 of common gas were consumed—the former at a pressure of seven inches of mercury whilst the latter was at the ordinary pressure of the main.

The size of the bar when melted was

3 feet 9 inches long
„ 7 „ wide
„ 4 „ thick

It resembled very much an ordinary cast pig of iron but naturally very much brighter.

The blow-pipes through which the gasses passed were similar in construction to that which I have here, only about half the size, and there were seven employed.

The blow-pipe, or burner, now exhibited, is capable of melting 100 killogrammes of platinum, and was used in the furnace exhibited here by my firm, in which a portion of the platinum is left as it cooled in the furnace.

The furnaces are made of ordinary lime stone, the large one used in Paris was lime stone containing a small portion of sand and answered the purpose very well.

They are made in two pieces—the cover which holds the burner and can be lifted to allow of fresh metal being added to the bath of melted metal, and the lower part or body of the furnace. When all the metal is melted and ready to be cast, the furnace, together with the cover, is grasped by a pair of tongues similar to those used by iron founders, and the melted metal poured into moulds of which several castings are here exhibited. It therefore becomes really an ordinary metallurgical operation, and there need be no limit to the size of the furnace or the quantity of metal to be cast, for it can be effected just in the same way, with the difference of applying the heat as in gold, silver, copper, or steel.

The oxygen gas, used for the large melting I have described as taking place in Paris, was made from chlorate of potash. That made from the ore of manganese requiring a long time to make besides containing impurities such as nitrogen and carbonic acid.

Up to the present time none of the various new processes for making oxygen gas that I have seen answer the purpose so well as chlorate of potash, for although it may be more expensive, it is rapidly and surely made. At the same time, I believe it will be superseded by another inexpensive process, which I have worked with Mr. Deville and made very pure gas indeed, and as soon as the apparatus is perfected I have no doubt that it will be extensively used whenever oxygen gas is required.

The PRESIDENT : I think you will all agree with me that our thanks are due to Mr. Matthey for bringing before us this powerful instrument for the metallurgy of metals of such very high melting points, as platinum and iridium. This improvement in the manufacture of platinum, has, I believe, contributed a good deal to the employment of this metal in the arts, and in processes to which formerly it was not so well applicable. We have present here this morning General Morin, who, I believe, has had considerable experience in the working of this process in Paris, and we are indebted for these novelties in the working of the platinum to M. Deville in the first instance, and for this method which has been carried out with singular success by

Messrs. Johnson and Matthey of London. I hope you will accord Mr. Matthey your thanks for his communication.

GENERAL MORIN, who apologised for speaking in French, said he was very happy to be present on this occasion, and to express publicly the acknowledgments of his countrymen to the talents, ability and goodwill of Mr. Matthey, and for the assistance which he had rendered to the Commission Internationale du Metre, of which he had the honour to be President. Mr. Matthey was able to produce for that commission by the process he had described an enormous block of platinum weighing 250 kilos at one operation, and from this block were obtained the materials for forming the exact measures which were of much interest to the scientific world. In thus assisting in the production of the new metric measures, of which the utility is so widely recognised, Mr. Matthey had rendered a great service for which he was happy to be the opportunity of publicly offering his thanks.

The PRESIDENT : If there are no further remarks upon Mr. Matthey's communication, I will call upon Professor Roscoe to favour us with his communication

ON VANADIUM AND ITS COMPOUNDS.

Mr. ROSCOE : Mr. President, Ladies and Gentlemen : I have the honour to exhibit to you a collection of specimens of the compounds of the rare metal vanadium, which at the present day possesses interest to the chemist, both from a scientific and a practical point of view. I may premise that the name vanadium is derived from Vanadis, a designation of the Scandinavian goddess Freia, and was given by the Swedish chemist Sefström to a new metal obtained by him for the first time in most minute quantities in the year 1830. The properties of this new metal were carefully investigated by the great Swedish chemist Berzelius, and although he had but a very small quantity of this rare substance in his possession he was able to give a full and particular account of the properties of this interesting substance. Up to the year 1868 all our knowledge concerning vanadium was derived from the experiments of Berzelius. In that year I was fortunate enough to obtain possession of a considerable quantity of this very rare substance, and carefully examined its properties. It was then found that in one important respect Berzelius had misunderstood the relations exhibited by vanadium, and it was only after experiments had been made working with larger quantities of material that this metal which had hitherto been wandering astray amongst her fellows was brought into the position which she now occupies as an acknowledged member of an important group of chemical elements.

The doctrine of isomorphism is one of the most important aids which the chemist possesses in the classification of the elements. Now the analogy between the elements phosphorus, arsenic and vanadium is pointed out in the isomorphism of three minerals which contain phosphoric, arsenic, and vanadic acids, and are known as pyromorphite, mimetesite and vanadinite. Each of these minerals crystallizes in hexagonal prisms, having the same angles and the same relative lengths of axes. In the year 1868 it was shown that the chemical constitution assigned to the vanadium compounds by their first investigator did not accord with that which the crystallographic relations required ; thus, whilst the highest oxides of phosphorus and arsenic, contained respectively in the isomorphous minerals pyromorphite and mimetesite, have the formula $P O_5$ and As_2O_5 given to them, the formula denoting the corresponding oxide of vanadium is, according to Berzelius, V_2O_3. Hence then we have here either to do with an exception to the law of isomorphism, or the experimental conclusions of Berzelius are incorrect.

The explanation of this is that Berzelius had overlooked the existence of two additional atoms of oxygen in the highest oxide of vanadium to which, therefore, the formula V_2O_5 must be given instead of V_2O_3. Thus the substance supposed by Berzelius to be the metal is in fact an oxide, and in order to obtain the true atomic weight of the metal we must subtract 16, the weight of one atom of oxygen, from 67·3, the weight of the metal according to Berzelius, and we thus obtain 51·3 as the true atomic weight of the metal vanadium. This then proves the analogy of vanadium (V=51·3) with phosphorus (P=31), and with arsenic (As=75), and thus the observed isomorphism in the case of the minerals above referred to is satisfactorily explained.

I will not detain you with a detailed description of the compounds of vanadium, but content myself with saying that the more we investigate the properties of these compounds the more we become convinced of the analogy of the vanadium compounds with those of phosphorus and arsenic. In proof of this I may refer you to the diagram showing the analogous composition of the phosphates and vanadates :—

	Phosphates.	Vanadates.
Ortho-Salts	$\begin{cases} Na_3\ PO_4 \\ Ag_3\ PO_4 \\ Pb_3\ (PO_4)_2 \end{cases}$	$Na_3\ VO_4$ $Ag_3\ VO_4$ $Pb_3\ (VO_4)_2$
Pyro-Salts	$\begin{cases} Na_4\ P_2\ O_7 \\ Ag_4\ P_2\ O_7 \\ Ca_2\ P_2\ O_7 \\ Pb_2\ P_2\ O_7 \end{cases}$	$Na_4\ V_2\ O_7$ $Ag_4\ V_2\ O_7$ $Ca_2\ V_2\ O_7$ $Pb_2\ V_2\ O_7$
Meta-Salts	$\begin{cases} Na\ PO_3 \\ NH_4\ PO_3 \end{cases}$	$Na\ VO_3$ $NH_4\ VO_3$

And as one other example I may state that this beautiful yellow solid body crystallising in bright golden plates, which have so bright a lustre and are so permanent in the air that they may possibly be used instead of gold leaf in gilding, is meta-vanadic acid, HVO_3 corresponding in composition with metaphosphoric acid HPO_3.

Another very interesting point of analogy between vanadium and phosphorus and arsenic is clearly shown by the very peculiar poisonous action which the vanadium compounds exert on the animal system, for vanadium turns out to be as active a poison as arsenic itself.

It is not, however, only in its relation to other elements that vanadium is of interest, for it possesses most remarkable properties of its own. Thus, for instance, the lowest stage of oxidation which the element can assume, viz., the oxide, V_2O_2, is a powerful reducing agent, bleaching indigo by reduction almost as quickly as chlorine does by oxidation. On the other hand, the highest oxide, V_2O_5, acts as a most powerful oxidising agent, in this resembling the highest oxide of chromium CrO_3.

This property of vanadic acid to oxidize organic matter enables this substance to be employed in photography.* If gelatine be mixed with sodium divanadate, and the film unequally exposed to light, the portions strongly insolated become slightly less soluble in warm water than the non-exposed portion, so that it is possible to print from such a film. Again if paper which does not contain any animal size is coated with a solution of sodium ortho-vanadate and then exposed to light, the portion insolated assumes a dark tint dependent upon the length of

---

* This proposal was first made by Mr. James Gibbons in 1874.

exposure and the strength of the solution employed. If the paper thus prepared be immersed, after exposure, in a solution of silver nitrate, the colour in the exposed part instantly changes to a dark brown or black colour, doubtless due to the reduction of the silver salt by the vanadous compound formed in the paper. Paper thus prepared may be used for photographic printing. The unexposed portions of the print are in this process coated with yellow silver ortho-vanadate; but this can be completely removed by ammonia or by sodium hyposulphite. Silver ortho-vanadate is capable of forming a latent image, like the chloride or bromide, and this may be developed by the ordinary ferrous developer. Two or three minutes' exposure to sunlight is needed. In the development little or no silver nitrate must be present.

By far the most important and interesting application of vanadium is, however, that recently suggested for the preparation of a permanent black which is now largely coming into use amongst dyers and calico printers, and is already extensively employed as a permanent marking ink.

Of the commercial value of a permanent rich black dye it is difficult for the uninitiated in such matters to form an idea. Suffice it to say that it is very great. We must, however, remember that this application of vanadium is only in its infancy, and whether the vanadium black will realize all the requirements of practice is a question which can only be settled by long and patient inquiry; still it has already so far proved a success that we may look forward with confidence to its future.

It is not at first sight easy to understand how a rare substance like vanadium, the price of which not long ago was 1s. 6d. per grain, and which even now cannot be obtained for less than ½d. a grain, can be employed for the production of a black colour, which, if not cheap and able to compete with the other common dyes, is of course useless in a practical and commercial point of view.

In order to understand the possibility of the technical application of one of Nature's rarest gifts, the history of the preparation of aniline black must be noticed. The splendid red, crimson, violet, green, and blue colours which are obtained from aniline are now universally known and appreciated, and their wide-spread manufacture serves as a striking illustration of the value of original scientific investigation. It is, however, not so generally known only that not the bright and gay colours, but also sombre browns and jetty blacks—by far the most valuable because by far the most generally used of colours—can be obtained from aniline.

In the year 1860, Mr. John Lightfoot, calico printer, of Accrington, applied to the processes of calico printing a black colouring matter which had been previously obtained in the manufacture of mauve from aniline by Messrs. Roberts, Dale and Co., of Manchester.

This black colouring matter is invariably formed when either aniline or toluidine, or mixtures of these two substances, are subjected to oxidising actions ; but in spite of several researches which have recently been published on aniline black, we are as yet unacquainted with its chemical formula, nor indeed can we say that it even possesses a constant chemical composition.

In order that a colouring matter shall be fixed or permanent, it must be fastened in some way to the fibre of the cloth. In the case of cotton this is generally effected (1) either by the precipitation of the soluble colouring matter in the fibre by means of a mordant which forms an insoluble compound termed a lake with the colour, as in madder dyeing and steam-colour printing ; or (2) by the fixation of the colour by means of albumin, as in pigment printing ; or (3) by the gradual oxidation and consequent precipitation of the colouring matter in the fibre, as in indigo printing. It is to this latter class of processes that aniline-black dyeing or printing belongs ; for the aniline salt under the action of certain oxidizing agents passing more or less quickly from the condition of a colourless solid readily soluble in water, into that of a black amorphous insoluble powder not to be distinguished at first sight from soot. Hence if the cloth can be impregnated with the aniline and with the oxidizing agent at the same time, and if the process of oxidation can be allowed to go on in the fibre, the black will be formed and will be permanently fixed in the fabric.

Many oxidizing agents, such as chlorine, ozone, or electrolytic oxygen, have the power of transforming aniline into this black pigment. In most cases a high temperature is needed for this purpose. Thus, for instance, if aniline is heated with chlorate of sodium, and if then hydrochloric acid be carefully added, a deep black almost solid mass s produced. In order, however, that the process may be employed in dyeing and calico printing, it is absolutely necessary to avoid high temperatures as well as the action of strong acids, because when exposed to these the cloth invariably is rotted or becomes "tender." If a mere mixture of aniline salt and chlorate of potash be heated strongly enough, the black is formed; but the heat necessary to produce the colour is sufficient, together with the hydrochloric acid which is at the same time liberated by the decomposition, to make the cloth rotten, and therefore to render this process useless.

It was found by Lightfoot that if an addition of four ounces of nitrate of copper solution was made to the pound of aniline and to the chlorate, the oxidation of the aniline went on at a lower temperature than when the copper salt was absent, and hence, when carefully worked, the black could be formed by this process without tendering the cloth. Certain technical objections to this process, however, soon arose; and in 1865 Lauth proposed to use the insoluble copper sulphide instead of the soluble nitrate, by which means he prevented the deposition of copper on the "rollers" and on the "doctors" which took place in Lightfoot's process. The method thus modified has been and is now extensively used for the production of black, and the chief, if not the only, objection which can be urged against it is that the black thus obtained is not perfectly permanent, but is liable to become green when exposed to reducing agents, such as the sulphurous acid contained in the impure air of our towns. This is, however, a serious drawback, and one which those practically engaged in solving such problems have not been able to remove. So much so indeed is this the case, that it is generally believed that the property of aniline black to become green when exposed to sulphurous acid, and to return to the black when treated with alkalies, is an essential property of the substance, which may be compared with the property of litmus to change colour in presence of acids and alkalies.

That the aniline black can not only be produced in presence of copper but also, as Mr. Lightfoot showed in the year 1871, in presence of vanadium salts, and that by vanadium alone can the black be obtained of the requisite permanent character, has now been proved beyond doubt. Moreover, the quantity of the vanadium necessary in order to produce the oxidation of the aniline is about one thousand times less than that of the copper. Thus if a piece of calico be dipped into a solution of 2·5 grains of vanadate of ammonia dissolved in a gallon of water and then dried, the cloth thus prepared is capable of producing an intense black if treated wtth the mixture of aniline salt and chlorate. In the same way if one gallon of colour be made containing twenty ounces of aniline hydrochlorate, ten ounces of chlorate of soda, and three grains of vanadate of ammonia, a mixture is obtained with which no less than from twenty to twenty-five pieces, or from 500 to 600 yards of cloth, can be thus printed of a permanent black.

In dyeing also, the vanadium will be extensively used; and in the same way only mere traces of this rare metal are requisite, whereas the copper black cannot be used for dyeing. Thus, for instance, one gallon of colour intense enough to dye forty pounds of cotton yarn black is obtained by mixing eight ounces of aniline hydrochlorate, four ounces of sodium chlorate, and eight grains of vanadate of ammonia. Cotton, wool, or silk dipped twice into this mixture and then aged, or allowed to oxidize, and "raised" in a solution of a carbonate of soda, is dyed a deep rich and permanent blue black. The goods may also be allowed to steep in a bath of the above strength for three days, then well washed in warm water, or boiled in a weak solution of acetic acid, to remove any bronze colour found on the surface of the silk or wool. The permanent black is then formed, and the fibre found to be quite strong.

The part played by vanadium in the formation of the black colour may be easily explained, when we remember the ease with which the metal passes from one degree of oxidation to another; thus from V_2O_5 the highest degree, to V_2O_4, and *vice versa*. In this way it doubtless acts, as M. Guyard has suggested, as a carrier of the oxygen of the chlorate to the aniline, being alternately reduced and re-oxidized, so that an infinitely small quantity of vanadium compound will convert an infinitely large quantity of aniline salt into aniline black, reminding one of the action of nitrous fumes in the leaden chamber.

Some time after the discovery of aniline black, Mr. Robert Pinkney, of the firm of Messrs. Blackwood and Co., of London, discovered, independently of Mr. Lightfoot, that vanadium can be most advantageously substituted for copper in the formation of aniline black; and he employed this re-action for the preparation of a permanent marking ink termed "Jetoline," of which many thousands of bottles have been sold. A few grains of vanadium—say from seven to twelve—being sufficient to produce, together with hydrochlorate of aniline and chlorate of soda, a gallon of marking ink.

The subject of the use of vanadium as a valuable dyeing agent was next taken up by the Magnesium Metal Company, of Patricroft, near Manchester; and, thanks to the unwearied exertions of Mr. Samuel Mellor, this firm have now succeeded not only in securing a very considerable supply of the rare element which occurs in the Keuper sandstone as the new mineral Mottramite, but are now in a position to produce a vanadium black for both calico printing and dyeing which is perfectly permanent. This is the more remarkable, as up to this time no aniline black made with copper has been produced in commerce which will withstand the reducing action of sulphurous acid.

As the result of a large number of experiments made with various qualities of commercial aniline, and by varying the strengths of solutions, proportions of aniline and sodium chlorate employed, and also by altering the temperature and the conditions of ageing, Mr. Mellor has found (1) that within certain limits the purer the aniline used, the deeper and more permanent is the black obtained. (2) That there is a maximum density of colour, beyond which if larger proportions of aniline salt and chlorate are used, corresponding advantages of colour are not obtained. This maximum colour is yielded by sixteen ounces to wenty ounces of hydrochlorate of aniline per gallon of colour. (3) That for the formation of a permanent black, the amount of aniline salt and sodium chlorate used for one gallon of colour must bear a definite relation to each other, the weight of sodium chlorate being about one-half that of the aniline hydrochlorate used. (4) That the permanency of the black depends very much upon the care and skill shown in " ageing " the cloth. If the cloth is aged in a moist atmosphere a blue-black is developed, which is very fleeting; but if aged in a dry air, and at a high temperature, a permanent black is obtained. It is also interesting to learn that for other colours also, the use of vanadium appears to be of value, as in the production of catechu browns as well as in some of the brighter aniline dyes.

It is indeed impossible to say what important technical functions

H

this rare and hitherto unapplied substance may not fulfil. Only the other day vanadium was accounted one of our greatest chemical curiosities, and the investigation of its properties would have been thought, by the practical Englishman, to be a mere waste of time.

Now, however, we have in vanadium a new example of the value of pure scientific research, which must carry conviction even to the most utilitarian of minds.

The PRESIDENT : Ladies and gentlemen, I am sure you will give your most hearty thanks to Professor Roscoe for bringing before us this interesting substance, vanadium. He has very justly remarked that it affords a very striking instance of the value of abstract research, which is by many thought to be comparatively useless. It is believed by some, that chemists and others may set about endeavouring to discover some particular utility that they may fix their minds upon. It cannot be too widely known, that such discoveries as far as I am aware, have never been made. Such useful applications as the one Professor Roscoe has glanced at, if not already a reality, is an immediate possibility always arising in this way. The application of chloroform to medicine and chloral hydrate, and numerous other compounds which I might mention have always been brought about in this way. They have not been discovered by the investigator setting to work to find out such a body, but have been the result of abstract research, lying dormant, it may be, for many years. Take for instance the paraffine industry which is now wide-spread in this country, and which contributes so much to the productiveness of England. This industry is also the result of purely abstract research, but discovery which was made by Reichenbach in 1831 or 1832, lay dormant for twenty years before it was resuscitated as a technical application by Mr. James Young, to whom we are indebted for the very extensive manufacture of paraffine which has gone far to revolutionise the production of artificial light.

I cannot invite a very long discussion on this paper, as my friend, Professor Guthrie, has consented to give us his ideas on cryohydrates and water of crystallization this morning in the place of a paper which we expected, but have been disappointed in. I shall be glad to devote a few minutes to any remarks you may have to make as to vanadium, its chemical properties and its application.

As there are no remarks I will ask Mr. F. Guthrie to be good enough to bring before us his communication

ON CRYOHYDRATES AND WATER OF CRYSTALLIZATION.

FREDERICK GUTHRIE: Mr. President, ladies and gentlemen, I owe you some sort of apology for endeavouring to compress into three quarters of an hour the substance of a research which has taken me the better part of three years. In connection with this subject to which I wish to direct your attention, it is my privilege to mention two names, the Right Honourable Lyon Playfair, and Dr. John Rae. The first amidst his onerous duties as a statesman keeps vigourously alive his interest in modern research, and the second while adding hundreds of miles to the survey line of what is now the Canadian Dominion, kept his keen eyes open to the manifold physical phenomenon presented to him. When I had the honour of being Professor Lyon Playfair's assistant, he suggested to me a possible explanation of that remarkable phenomenon the maximum density of water. Ice we know to be lighter than water. We know that water being gradually cooled from say this temperature to 0° centigrade, its freezing point, shrinks, till its temperature is four degrees and then swells until it freezes. Professor Playfair suggested that the phenomenon is continuous, that the water shrinks to four degrees centigrade, and that then ice is really formed, but that this dissolves in the water, and ice being lighter than water the solution is also lighter than the water at that temperature, and that accordingly when the water begins to freeze it is nothing more than the separation of ice from a saturated solution of ice. I brooded for many years over this point, and I believe it has had great influence in directing my attention to this research. Dr. John Rae in examining the ice floes found that the saltness of the ice depends upon the age of the floe, and concluded that in the first instance pure ice is separated from the salt water, and that when the floe is found to contain brine, it is mechanically enclosed between the crystals of pure ice, and gradually oozes away by gravitation.

In bringing the results of my experiments before you, of course I shall follow the logical rather than the historical sequence. I may state at once that the whole of this investigation depends upon the simple fact that all watery solutions whatever can be cooled below 0° centigrade without undergoing complete solidification, and it is the investigation of that portion which remains liquid that I wish to speak of now. Let me refer at once to this large diagram, which is a complex one, formed of fifteen diagrams. I have some small copies which will be distributed amongst you. Let us start with a weak solution of say such a salt as nitre. The abscissae are the weight per centages of the body in water. The ordinates are temperatures C. Every one of these dots represents one per cent. The central horizontal line is the freezing point of water. At the left end you have zero C, and nothing per cent., that is pure water. Supposing now you take a solution of nitre containing one per cent., cool that below zero, nothing will separate out until you get to —0·1 degree centigrade, and then pure ice separates out. If you take a little stronger solution, say five per cent., then you have to cool it to —1·5 before any ice separates out, and so on : a lower temperature is requisite as the solution is stronger and stronger until you come to eleven per cent. solution, and then the whole of the salt separates in combination with the remainder of the water at —3°. So if you take a solution of nitre of twelve per cent. strength saturated at zero, and cool it, then you do not get ice separated out but pure anhydrous nitre. If you take a little weaker solution the separation of the nitre will be at a little lower temperature until in the end you reach the same composition as before, when the same body will separate out, and also of course, at the same temperature. Let us take now this salt chloride of ammonium, and instead of examining successively solutions of different strength let us take a weak solution, say three per cent., and cool that, ice separates out, and by the separation of the ice of course the solution becomes enriched until it reaches a per centage of nineteen or twenty, and then the whole of the remaining water and the whole of the salt separate out together.

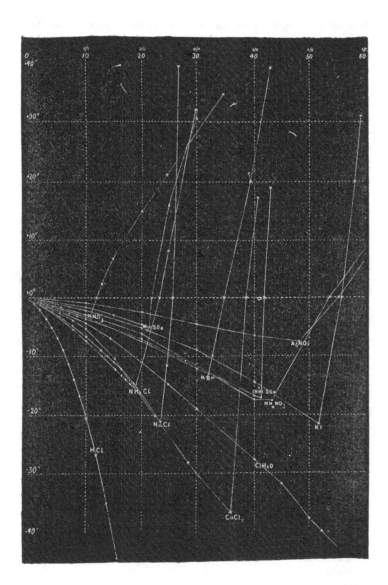

Now the body which separates out therefore at this temperature having this composition is a constant body, it freezes or solidifies and melts at a constant temperature, namely minus sixteen degrees. It can only exist in the liquid form below zero, and therefore, for an obvious reason I call it a cryohydrate, and I find that every body, whether previously known in combination with water such as sulphate of copper, or alum, or whether previously unknown to combine with water, such as nitre, or nitrate of ammonia, I find, I say, that all these bodies being soluble in water are capable of uniting in certain definite proportions with water to form these cryohydrates; and in this sense one may say that the number of known definite bodies has been doubled. Intimately connected with this point is the hitherto rather obscure theory of freezing mixtures, or cryogens. What determines the temperature of a freezing mixture? The quantity of heat absorbed when ice and a salt are mixed together may be considered as the sum of two things. The quantity of heat requisite to melt the ice, and the quantity of heat which the salt absorbs when it dissolves in water. This latter term, however, may be either positive or negative. Positive when we use a salt such as nitre, or nitrate of ammonia; negative when we use a salt such as anhydrous chloride of calcium. We must, therefore, consider it the algebraic sum of these two quantities. But the temperature or heat tension depends also upon the rapidity of solution, the rate of solution, and what governs this? Are those salts which are most soluble in water those most competent to depress the temperature? Are they the most powerful cryogens? No, neither those most soluable at the ordinary temperature of the air, nor even those most soluble at zero centigrade, the freezing point of water. But the heat tension, or temperature, of cryogens depends upon the temperature of these very bodies we have been studying, the temperature of solidification, and the temperature of melting of these cryohydrates. They as precisely govern the temperature of a freezing mixture as the melting point of ice governs the temperature of mixed ice and water when ice dissolves by the acceptation of heat. To show how fully this is carried out I must refer you to the table on the wall, and to the three first columns of that table. On the left hand are the chemical formulæ of the various salts, and the next column shows the temperatures of the cryogens.

TABLE.—Showing (1) the chemical formula of the salt, (2) the lowest temperature to be got by mixing the salt with ice, (3) temperature of solidification of the cryohydrate, (4) molecular ratio between anhydrous salt and water of its cryohydrate (water-worth or aquavalent), (5) percentage of anhydrous salt in portion of cryohydrate last to solidify.

(1)	(2)	(3)	(4)	(5)
Formula of salt.	Temperature of cryogen or freezing mixture.	Temperature of solidification of cryohydrate.	Molecular ratio or water-worth or acquavalent.	Percentage of anhydrous salt in last cryohydrate. M.L.
NaBr	$-28°$	$-24°$	8·1	41·33
$NH_4 I$	-27	$-27·5$	6·4	55·49
NaI	$-26·5$	-30	5·8	59·45
KI. . . .	-22	-22	8·5	52·07
NaCl	-22	-22	10·5	23·60
$SrCl_2 + 6H_2 O$ . .	-18	-17	22·9	27·57
$NH_4{}^2 SO_4$ . . .	$-17·5$	-17	10·2	41·70
$NH_4 Br$ . . .	-17	-17	11·1	32·12
$NH_4 NO_3$ . . .	-17	$-17·2$	5·72	43·71
$NaNO_3$ . . .	$-16·5$	$-17·5$	8·13	40·80
$NH_4 Cl$ . . .	-16	-15	12·4	19·27
KBr	-13	-13	13·94	32·15
KCl . . .	$-10·5$	$-11·4$	16·61	20·03
$K_2 CrO_4$ . . .	$-10·2$	-12	18·8	36·27
$BaCl_2 + 2H_2 O$ . .	$-7·2$	-8	37·8	23·2
$AgNO_3$ . . .	$-6·$	$-6·5$	10·09	48·38
$Sr2NO_3$ . . .	-6	-6	33·5	25·99
$MgSO_4 + 7H_2 O$ . .	$-5·3$	-5	23·8	21·86
$ZnSO_4 + 7H_2 O$ . .	-5	-7	20·0	30·84
KNO_3 . . .	-3	$-2·6$	44·6	11·20
$Na_2 CO_3$ . . .	$-2·2$	-2	92·75	5·97
$CuSO_4 + 5H_2 O$ . .	-2	-2	43·7	16·89
$FeSO_4 + 7H_2 O$ . .	$-1·7$	$-2·2$	41·41	16·92
$K_2 SO_4$. . . .	$-1·5$	$-1·2$	114·2	7·80
$K_2 Cr_2 O_7$ . . .	-1	-1	292·0	5·30
$Ba2NO_3$ . . .	$-0·9$	$-0·8$	259·0	5·30
$Na_2 SO_4 + 10H_2 O$.	$-0·7$	$-0·7$	165·6	4·55
$KClO_3$ . . .	$-0·7$	$-0·5$	222·0	2·93
$Al_2(NH_4)_24SO_4 + 24H_2O$	$-0·4$	$-0·2$	261·4	4·7
$HgCl_2$ . . .	$-0·2$	$-0·2$	450·0	3·24

The third column shows the temperature of the solidification of the cryohydrates, and there you see there are some discrepancies of one or two degrees. In the third on the list I at first imagined there to be a wide discord between the temperature of the freezing mixture and that of the corresponding cryohydrate, namely with iodide of sodium, but I found I had there deceived myself and that I had not then obtained a true cryohydrate, but that I was dealing with a sub-cryohydrate. I will not detain you by entering further into that point, but will merely point out that within the limits of experimental error (and dealing with a very low temperature one is more subject to incur these errors), those low temperatures are in strict accord with one another. And why? Supposing you have a freezing mixture, such as this mixture of ice and salt, the liquid portion of this freezing mixture is the liquid cryohydrate of common salt. If the temperature were to sink below the temperature of solidification of that cryohydrate, the cryohydrate would solidify, and would give up heat. Just as it is impossible to cool water below zero as long as there is any water left in the freezing mass, so here it is impossible for this temperature to sink below twenty-two centigrade, for if it did so a solid cryohydrate would be formed ; and so in all cases. By mixing a solution of nitre and ice or snow you can easily get a temperature of —3°, and there is a liberal margin in the proportion of the constituents of the nitre and the ice, with which you take to get this temperature. You can easily get it, but by no proportion can you get a fraction of a degree below, because if that were obtained then the solid cryodydrate of nitre would form and give up heat. Therefore it is, that the liquid portion of every freezing mixture is the molten cryohydrate of the corresponding salt. Direct experiments and analysis have proved this abundantly. In looking at the different quantities of water which the salt assumes, I have, for the sake of comparison, given on the extreme right of the table the simple percentages of the anhydrous salt, 100 minus each of these gives, of course, the percentage of water. Then dividing the percentage of the salt by the molecular weight of the salt, and the percentage of water by the molecular weight of the water, and dividing the greatest of these numbers by the least, that is, in all cases dividing the last by the first, you get what I call water-worths. I dared not call them water equivalents, because I am in very strong doubt whether these are true chemical ratios. And that doubt not only arises as you see obviously, from the disaccord of these numbers from whole numbers, there seems to be no tendency for them to drop their decimal places ; and really some of them are so easy to analyse that errors of this magnitude are beyond the errors of legitimate observation. Not only on this account did I hesitate to call them equivalents, and to write formulae for them, but, also, because as you will see towards the base of the table the water-worths or molecular ratios appear by their great magnitude to be what we may call ultra-chemical ratios. They are strictly gravimetric, but I cannot say whether they are what we call true chemical molecular ratios. If indeed we were wholly guided by what is commonly assumed to be the case, that chemical union is always accompanied by the development of heat, as a broad guide, then we might call into doubt the true chemical molecular ratio of other solid waters, such as the waters of crystallization. When anhydrous sulphate of copper is moistened it gives out heat, but when, however, some other salts are moistened they absorb heat. I remember at this moment one of the crystalline sulphates of sodium which, when it takes up more water takes up also heat, that is, it produces cold. Of course cases are known, as in the case of common salt to a small extent, and more in the case of nitrate of ammonium, where mere mechanical solution, without the formation of any definite hydrate, is accompanied by the absorption of heat. Having talked about these things, I should like, if you will allow me a few minutes more, to show you one or two of them ; and I will take, in the first instance, this chloride of ammonium, which solidifies at minus sixteen degrees. If I were to make a freezing mixture or cryogen, composed of chloride of ammonium and snow, I could get a temperature of minus sixteen degrees, but no lower, and, therefore, a temperature incompetent to solidify the cryohydrate of chloride of ammonium. If I take the ordinary freezing mixture, consisting of salt and ice, and cool the chloride of ammonium in that, then I must at last, if I employ a weak solution, by the separation of ice, or if I employ a strong solution, by the gradual separation of the anhydrous chloride of ammonium, reach the composition of the cryohydrate, and at that moment I reach the temperature of its solidification. To save time I have prepared a solution of chloride of ammonium which is, in fact, a cryohydrate melted in its own water ; the solution of chloride of ammonium contains 19·3 of the anhydrous chloride of ammonium. That when brought into contact with the ice-salt freezing mixture solidifies and gives rise to this very beautiful salt, which, of course, rapidly melts in the air as it absorbs heat. Others, such as the chlorate of potassium, do not require so low a temperature. In a few seconds I can reduce the temperature of a solution of chlorate of potassium to such a degree, that the cryohydrate solidifies, the molten hydrate being already of the right proportion. It freezes in a few seconds. Here, on the other hand, I have a solution of common salt in water in the right proportion ; that is, it is the molten cryohydrate. That you see has not solidified ; and it cannot solidify in its own cryogen. The temperature of this freezing mixture is just insufficient to cause solidification. Let me show you one branch of the investigation which I have not further pursued. You know that sulphuric ether dissolves in water ; water also dissolves in sulphuric ether to a slight exent. By shaking sulphuric ether up with water one gets a saturated solution of ether in the water. The quantity of the ether dissolved in the water in this bottle at this temperature is about one part in thirteen. If this be cooled, the water and the ether solidify together and you get a cryohydrate, the composition of which I have not yet determined ; but you will get a solid mass of ether and water which is combustible, that is to say, the ether is combustible, and you will see this remarkable phenomenon that when the ether burns it burns with a non-luminous flame, just as alcohol does. We know that the raising the temperature of combustible gases also increases their luminosity. Here we have something of the converse ; for this, although on fire, will never rise beyond the melting point of the cryohydrate of ether, and accordingly the ether is cooled, and that seems to diminish the luminosity. Then, finally, I should like to direct your attention to the peculiarity of these curves. In the first place you notice that it is by no means always the case that in these curves which consist of a valley with two slopes, the deeper the valley the wider it is open. What does that mean ? It is not always the case that the salt most soluble at zero gives the lowest freezing mixture, or as a cryohydrate, of the lowest solidifying temperature. Another thing which strikes you at once is that those salts which are metal-halogens, such as chloride of sodium, bromide of sodium, and so on, have all sharp points ;

their valleys are sharp at the bottom, whereas those salts which contain oxygen have a more rounded appearance. I should mention that I have not in any way rounded off this diagram. I have joined the points of observation by straight lines, and any apparent curvature there, is due to the continuity of the phenomena ; I have not drawn the intermediate curves approaching near to these polygons, but it appears from this that taking the same class of salts, the chloride of ammonium and the chloride of sodium, and chloride of potassium, then you have that continuity, that those which are most soluble at zero have also the lowest cryogens. These are the chief points to which I wished to direct your attention. I will not venture upon some other subjects connected with this, because if I did I should be led too far. I believe that perhaps one of the most important practical applications of these experiments will be the establishment of constant temperatures, because if I take one of these solid cryohydrates I can use it to keep a fixed temperature. If I take the cryohydrate of chloride of ammonium and plunge a body in it, I can maintain the body in it at a constant temperature for any time, and so these constant temperatures can be maintained as faithfully as we can maintain the constant temperature of zero by melting ice or the temperature of 100 degrees by boiling water. In conclusion I will point out the remarkable analogy between the phenomena of boiling and freezing salt solutions. When a solution of salt boils you know that its temperature is always above 100 degrees centigrade, as the vapour passes off. The salt soultion becomes richer and richer and its temperature of ebullition rises until it becomes saturated at its then boiling point, when there is a simultaneous separation of vapour above, and of salt below. These separate from one another, but one being gas and one being solid, they do not combine. In gelation you have a perfect analagous phenomenon. At this temperature at this strength the water separates, not as a vapour but as a solid. The salt separates as a solid, and these two being crystallizable bodies they unite together. There is some such sort of union in the case of a boiling liquid, in the case of boiling nitric and hydrochloric acids, although, as Professor Roscoe and others have pointed out, this boiling gives a variable compound depending upon the barometric pressure. One of the great reasons why I consider these to be true compounds and not to be mixtures, apart from their crystalline form, their constancy of composition and constancy of melting is this, that it is possible to get a solution of this kind supersaturated, as one may say, in three dimensions, and this has been one of the greatest difficulties I have had to encounter. That is to say, if I cool a melted cryohydrate down it will not solidify, nothing separates out, although it is four or five degrees below its proper solidifying point. If I throw into that a little crystal of ice nothing separates but ice, which comes to the surface. If I throw in a little anhydrous salt nothing but the anhydrous salt separates out, and that sinks to the bottom, but if I throw into it a crystal of a previous crop of cryohydrate then nothing but the cryohydrate separates out with its characteristic silvery opacity. You might imagine that these bodies were mere mixtures of the two substances, salt and ice ; but when we have this three-fold phenomenon of supersaturation, I think that is conclusive that there is some physical relation between these two constituents, even if you deny it the privilege of being a true chemical one. I have now, as you see, by dint of the freezing mixture of solid carbonic acid and ether solidified the cryohydrate of common salt.

Professor ANDREWS : We have had no great difficulty in following this extremely lucid and beautiful statement of these classical researches of Professor Guthrie, for what he has said has been so clear and so distinct that at least there is nothing more to say in the way of comment, and I only wish to make one or two slight observations. In the first place we have not here an example of an isolated chemical fact. It has long been known indeed that chloride of sodium will at low temperatures yield crystals containing water of crystallization, or, in the language of Professor Guthrie, will form a cryohydrate. But one or two cases of this kind only were known, till Professor Guthrie by his researches has opened a new field of chemical inquiry of great interest. Allow me just to mention, with regard to one of his last observations that these cryohydrates will give us a means of getting a number of constant temperatures, that this fact forms at least a justification, of the old thermometer of Fahrenheit, which however inconvenient it may be for certain purposes, still holds its ground. It turns out, after all, to have been strictly a scientific instrument, the zero of his scale being that of a cryohydrate. We have few facts in chemistry more interesting than the observation of Professor Guthrie, namely, that from the same supersaturated solution you can obtain three different bodies according to the foreign body you introduce. You can obtain either ice, or salt, or the cryohydrate, but it was not necessary, I am sure, for Professor Guthrie to make any apology for considering whether they were purely chemical compounds or not ; I should rather say that we have no definition in nature of a purely chemical compound ; that everything which is combined is a compound. It may not be a compound of the same firm order as oxygen and potassium form, but a compound it is, unless it be a mere mechanical mixture, such as sand and sugar. Wherever you have a solution, or anything of that kind, you have a certain form of combination, and, therefore, the fact that these cases are a little different from those we are accustomed to, is only increasing so much the extent of our knowledge. I should like to refer to one subject as pertaining to this question. An important observation was made not long ago by Professor James Thomson, who has examined with great care, not experimentally, but as he can do, with great ability, the question as to whether the tension of the vapour of the same body at the same temperature in the liquid state and in the solid state (as we know that bodies can be obtained at the same temperature in both states) is the same or not. Regnault came to the conclusion that they were the same, but Professor Thomson, from reasoning and from a careful examination of Regnault's own experiments, came to the conclusion that they are not the same. I only refer to this subject because these cryohydrates of Professor Guthrie will furnish a large number of bodies in which that very question can be examined, and I should think the question of the tension of the vapours of those substances in these different conditions would be one of some interest. I need say no more than to ask you to pass a vote of thanks to Professor Guthrie.

Professor GLADSTONE : I think this investigation which Professor Guthrie has been carrying out for the last two or three years with very great care, and with very fortunate results, promises to be a very large one, and it might be taken up in a great number of different directions. There is one point which struck me upon which I have not had the opportunity of speaking to him at all, and I should like to ask him one question ; it is this. He stated just now that he could hardly regard these cryohydrates as being compounds of the salt with a certain number of atoms of water, and he referred to the third column of the table in proof of that difficulty or doubt which he felt. No doubt if we considered that those numbers had been obtained with the greatest precision and accuracy he would be perfectly right ; and I want to ask him to what extent he thinks those analyses are to be thoroughly depended upon. I should like to know with regard, for instance, to bromide of sodium, if Dr. Guthrie thinks that it is possible by a more accurate analysis, the 8.1 would be simply eight, and, again, whether in the iodide of ammonium the 6.4 might not be 6.5 ? The iodide of potassium and chloride of sodium with which, I believe, he has worked very much, and which are substances easy to determine, we find 8.5 and 10.5, exactly intermediate between eight and nine and between ten and eleven, and altogether the figures in the third column seem rather to suggest to me that they might be whole numbers or intermediate between whole numbers, and that there may be some real atomic relation in the case.

Dr. GUTHRIE : I will answer Professor Gladstone's suggestion by saying, in the first place, that of course it has been my wish to reduce these proportions to simple atomic relations, but the more carefully I worked them out the less justified I felt in attributing the divergence to errors of observation. Of course when you have a small water-worth, such as those towards the head of the list, then you have in one sense the most favourable circumstances for analysis, but in another sense you have not, because you will see those require very low temperatures, and these cryohydrates in the solid form are, I will not say hygroscopic in the ordinary sense, but they assume water by reason of their extreme coldness. They condense the water of the atmosphere upon them, and that produces another element of difficulty. Of course by multiplying the figure for the water by one number, and that of the salt by another, it is possible to evolve formulæ strictly in accord with atomic weights, but I thought that was rather an unwarrantable proceeding and my reason for hesitation was the fact that the numbers towards the bottom of the list are so enormous, the water-worths are so great, that, of course, the errors of observation will then be vanishing. They seem to be ultra chemical ratios, even to what is usually called the water of crystallization, such as in that of alum which contains twenty-four molecules of water to one of salt ; these far exceed that, for some of them go up to 300 or 400. With regard to what Professor Andrews referred to about the hydrate and chloride of sodium, that reminded me that I had not time to mention the peculiarity of these sodium salts. Professor Andrews pointed out that there is a compound of common salt and water obtained at between four and ten degrees centigrade below zero. This is one of the sub-cryohydrates which have given me so much perplexity in the manipulation. It is only by approaching the cryohydrates from the side of dilute solutions and removing the ice that I have succeeded in obtaining the cryohydrates of sodium-salts. Iodide of sodium, chloride of sodium, and acetate of sodium—all assume sub-cryohydrates. With regard to the future which this research will take, I may be pardoned for saying that I have no doubt it will lead to very important results. Another point about which I have had some conversation and received some suggestions from my friend, Professor Dewar, is with regard to taking the latent heats of the solution of these cryohydrates ; thus perhaps we shall detect whether these are chemical compounds or merely physical ones, by finding discontinuity in the latent heat, if I may so call it, in water at this temperature, and the solution of the salt in water at the same temperature.

The Conference then adjourned for luncheon.

The PRESIDENT : I will now ask Professor Williamson to give us his communication

ON THE MANUFACTURE OF STEEL.

Professor WILLIAMSON, Ph. D., F.R.S. : I have been requested to give some account of the recent improvements in the manufacture of steel, and at the same time I have been told that the practice in these addresses is that the speaker should limit himself to half an hour. Under these conditions it would be clearly a mistake for me to give an amount of the fundamental facts of the case or an elementary description of the processes.

I shall, therefore, not go into any particulars of the processes by which pure iron used to be made into that superior material called

steel, nor shall I give any general description of the wholesale processes which have recently replaced those most elementary plans.

But there are some peculiarities of those new processes to which I
propose to call your attention, with a view of considering the chief
defect which is found in the working of them, and with the view of
analysing the circumstances which cause that defect. My practical
object is to give a key to the remedy for the defect which is now
encountered in the ordinary processes of steel-making.

It is of course known to all those who have followed the working of
steel-making that the most remarkable characteristic of the modern
changes in the process has been of this kind, that wholesale processes have been discovered and introduced in place of processes
which were only applicable to comparatively small quantities; and that
not only do the steel manufacturers work with larger quantities of
materials at a time than they did formerly, but they work more rapidly.
The modern processes to which I especially refer are two, which in
their general features I must presume to be known to you ; namely,
the Bessemer process and the so called Siemens-Martin process.   Of
course everybody knows that in the so called Bessemer process, air is
bubbled through the molten pig iron by a contrivance which has been
admirably adapted to that purpose, and one great merit of the process
has been the discovery of a simple and effective mode of doing that.
A great number of bubbles of air are forced in at the bottom of a
mass of molten pig iron, and they force their way upwards through
it whilst reacting on the molten mass through which they pass.   It
is known that the broad result of that action is to burn out the carbon
which was in the pig iron ; to burn out also the silicon, and at the
same time to burn a certain quantity of the iron itself.   It is
generally taken for granted, at all events, as far as my experience
goes, by those who refer to those processes, that the oxygen of the air
which is blown into the molten pig iron reacts directly upon the
carbon, inasmuch as the final result is that the oxygen which is
forced in and the carbon which was there, pass out in combination as
carbonic oxide, the lowest oxide of carbon.   It is indisputable that
carbonic oxide is formed from the carbon, but it does not follow that
the process is merely a direct combination of the oxygen with the carbon
and I think that only a very brief consideration of the facts will show

that such cannot be the case. I venture to affirm that the carbonic oxide which is the essential product of the process is not for the most part formed by the direct combination of the oxygen which is forced in with, the carbon which was present in the pig iron ; and the actual process by which carbonic oxide is formed, is of so much importance to the understanding of what takes place in later stages of the steel manufacture that I must claim your indulgence for dwelling a short time upon it. It deserves our most careful consideration.

If you consider that the iron which is used at first contains, say, four per cent. of carbon, and that the process of de-carbonisation is usually carried so far that the carbon is completely burnt out, we may say that the average percentage of carbon during the conversion of the molten mass is about two per cent.; for the first period it is four, at the end it is nothing, so that we may put down at two as the average. When a bubble of air escapes into this molten mass, you have a little spheroidal mass, dissolved in molten iron containing two parts by weight of carbon in one hundred, and there is no doubt if those two parts of carbon were dissolved in ninety-eight parts of inert matter the oxygen would combine with the carbon directly, and with nothing else ; but inasmuch as the whole surface of the molten mass which surrounds the bubble consists of a material which is pre-eminently combustible ; inasmuch as iron even at a lower temperature than that at which it is employed takes up oxygen with enormous rapidity the moment it comes in contact with it, it is impossible that the particles of oxygen which are in contact with the iron should fail to combine with it. If anybody held a contrary opinion I think you would agree with me that the onus would be upon him to show grounds for maintaining it. It is well known that where iron and oxygen come in contact at a high temperature they combine very rapidly and with intense evolution of heat ; so that when a bubble enters the molten mass there is some carbonic oxide formed at first, but there is much more oxide of iron formed by the first action. The film of oxide of iron lines the spheroidal cavity in which the bubble at first is, and when the bubble rises upwards and is displaced by molten iron containing some combined carbon and silicon, the oxygen in the thin film of oxide of iron is transferred to that carbon and silicon. I say therefore that the greater part of the combustion of the carbon takes place not

I

by the direct combination of carbon with oxygen, but by the action of the oxide of iron upon the carbon dissolved in the remaining iron. It is a well known fact that at that temperature and even at temperatures considerably below that of the Bessemer process, oxides of iron such as are formed do react very vigorously upon carbon ; in fact that carbon cannot exist in contact with them for any perceptible time ; the action is almost instantaneous.

With regard to the remainder of the process in Bessemerizing iron I will simply refer to the fact that when the carbon and the silicon have been burned out, together with some of the iron, by the process which I have briefly referred to, it is customary to pour into the molten mass a suitable quantity of the manganiferous pig, so called *spiegel.* The carbon which is to be introduced in order to form steel, can be most conveniently introduced in the form of that particular compound of carbon with iron and manganese.

The other or so-called Siemens-Martin process, is in its general features different from the Bessemer process. It consists in melting in a great hearth some five to twelve tons of pig iron, either with bar iron or with a pure ore of iron.

The workmen then apply a high temperature to this molten mass, upon the surface of which the oxide of iron is floating, and a reaction gradually sets in analogous to that which takes place in the old process of pig boiling ; bubbles of carbonic oxide come off while the oxide of iron gives up its oxygen to the carbon, of course oxydising at the same time the silicon which may have been present in the original pig.

Now I think it is of some importance to consider that if the statements I have made represent well-known facts, and I am not aware of any possibility of doubt on that point—that is if the process in the Bessemer reaction is mainly an action of oxide of iron upon carburet, the two processes are, in their chemical principles, the same. It does not follow from this, that the one process would yield results identical with those obtained from the other, for the mechanical conditions under which the process takes place in the Bessemer system are very different from those under which it takes place in the Siemens-Martin process. If you merely consider the exceeding tenuity which the films of oxide of iron must have when they are formed under the

conditions which I described to you just now, and if you compare the condition of these exceedingly minute films of oxide of
iron with the condition of oxide of iron shovelled in at the top of
the molten mass in the Siemens-Martin process, you will at once
perceive that there is this one important mechanical difference
between the two, that in the Bessemer process there is a far
more intimate mixture established at once between the oxide of
iron and the carburet upon which it reacts, than in the Siemens-
Martin process. The reaction therefore takes place far more rapidly
in the Bessemer process than in the Siemens-Martin process. In
fact the difference between the two, roughly speaking, is about
the difference between minutes and hours. Ten minutes in
the Bessemer process serve to do as much as about ten hours in
the other. Now there is one rather important circumstance which
results from this intimate mixture of the oxide of iron in the Bessemer and the comparatively less intimate mixture in the other,
namely, that in the one case—the Bessemer process—we get more of
the oxide of iron left in our product. Of course if you over-burn
the metal in either process there is a good deal of oxide of iron
diffused through the mass. It is found that if the oxidation of the
molten mass by the ore be allowed to continue too long, that
peculiar properties are found in the mass which no doubt correspond to the presence in it of oxide. The last operation of the
Siemens-Martin process, as you are no doubt aware, is the same
as that in the Bessemer process, viz. :—The addition to the de-
carbonized iron of the proper proportion of carbon which is required
to make steel of the particular quality which the manufacturer aims at ;
to add it in the form of an alloy of iron with manganese, containing
some five per cent. or so of carbon.

I now come to the point to which I am desirous more particularly
to draw your attention—namely, the casting of the steel when it is
made. It is of course well known that whether steel be made by
the one or by the other process it is customary to cast it in the first
instance into ingots ; there are cast iron moulds in which it is allowed
to cool, and when cooled it is taken out, heated again, hammered,
rolled, and beaten into the form in which it is required for use—
that is the same in both cases. Now there is during the cooling

of the steel in these moulds a remarkable process which has given considerable trouble to the manufacturers who work the one and who work the other process—a process which is frequently called by the workmen " boiling." Once the mass has stood for a short time in the ingot mould, bubbles rise to the surface of it, and if the mass is left to itself, if no means are taken to arrest the process, the gas comes off from it with such violence and in such quantity that a considerable proportion of the contents of the mould are thrown out and the portion which is left in the mould is honey-combed and is in a form which is very ill-suited for further treatment. This occurs in a greater or less degree with steel which is run into ingots either from the Bessemer converter or from the Siemens-Martin hearth. The usual practice in order to prevent this inconvenient occurrence, consists in closing moulds by a lid fixed down so strongly as to cork it all in, if I may use such an expression. Some sand is thrown over the top of the molten mass so as to cool the top of it, and some of it forms a natural lid, and then a strong plate of iron is put on and wedged in strongly so that it is capable of resisting the very considerable force with which these gases tend to come off. Now this particular process of boiling in steel has, I believe, always been attributed, for I have never heard any other opinion with regard to its nature, to the evolution of carbonic oxide gas. I purposely state that with the necessary reserve, because in scientific matters although there may be a strong presumption in favour of a particular conclusion, we are bound to distinguish and are in the habit of distinguishing even the strongest presumption from direct proof. It is supposed, and I think reasonably supposed, that the boiling of the mass is due to carbonic oxide being given off from it. Now various remedies have been proposed for the purpose of preventing either, or of obviating the inconvenience of this circumstance, and I think a brief analysis of the nature of those remedies, and of the results which have been obtained from them may be worthy of our consideration. In the first place it is familiar to those who are in the habit of seeing the making of steel, that it is much more easy to get a solid casting, in other words, that there is much less liability to the occurrence of this frothing, if a considerable percentage of carbon be left in the steel, say one per cent., than it is to get a solid casting under similar conditions in other respects if only one-half or three-tenths of a per cent. of carbon is present. In fact the chief difficulty in getting solid castings is, I think, well known to occur in those cases in which a soft variety of steel has to be made. Of course it is not allowable to put in any quantity of carbon that might be convenient for making the ingot. You must put in that quantity which is suitable to make the steel of the particular quality required for the use to which it is to be devoted. For some purposes you must have hard steel, and for others you must have soft, so that you have not a choice of carbon, and that remedy is not generally available for the cure of the evil.

Another remedy which has been not unfrequently adopted and which is worth careful consideration, consists in adding a greater proportion of spiegeleisen or of that richer alloy of manganese with iron now known as ferro-manganese, an alloy containing always more than ten per cent. of manganese, sometimes even more than fifty per cent.—in fact, putting a greater portion of manganese into the molten mass at the end of the operation. I think I may say that it is clearly established as the result of many careful observations that when a rich alloy of manganese is put in, not only to supply the carbon you want, but also to leave about one per cent. of metallic manganese in the molten mass, that then this frothing does not occur. To render this intelligible I ought to mention a fact which is well known to those who have witnessed the operation frequently, namely, that if at the end of the blowing in the Bessemer converter or at the end of the decarbonising process in the Siemens-Martin hearth, a given quantity of manganese alloy is put into the molten mass, and if the molten mass is then allowed to run out into one of those huge pots, and then run into the ingots, as a matter of fact a great deal of manganese which had been been put in is not found in the steel ; the manganese which is put in gets oxydised with great rapidity, and if there is not much of it, it usually becomes completely oxydised during the contact which there is between a molten mass and the air through which it is poured. That circumstance in itself is I think one well worthy of notice and of consideration in relation to the meaning of this frothing. It is beyond doubt that manganese which is present in the molten mass whilst it is in the hearth or in the converter, passes out of it by oxydisation ; I mean much of the manganese passes out of it whilst the molten mass is being poured through the air as it always is more or less—that there is in fact an oxidation. I stated therefore, just now, that it has been found that if not only the usual quantity of manganese alloy is put into the molten mass, but so much more manganese alloy that there is left in the ingot about one per cent. of the metal, a remedy is thereby obtained for the frothing and there is no carbonic oxide evolved.

With regard to the other processes for dealing with this evil I ought especially to mention one which is not so much in the nature of prevention as of cure. I mean that which has of late been introduced by Sir Joseph Whitworth in which the ingots of steel, made as carefully as possible, but still liable as they are to be slightly honey-combed, are subjected while still soft, to very great mechanical pressure from an hydraulic ram. It is found that by this treatment any little cavities which would otherwise have been left in the mass are not found in it. All the particles of the mass are in contact with one another, and very great advantages are indisputably obtained in steel, other things being equal, which has been treated in this way, over steel, which has not been so treated.

I think if we were to put aside for the moment the general results which are known, and if we were to consider, merely from the data which we possess, what must happen in this process, it would be easy with the knowledge which we have of the properties of the materials in question to say with confidence, not only that, under the usual conditions, carbonic oxide must be evolved, but also to say that the excess of carbon which is occasionally introduced would probably tend to diminish the amount of it ; and to say also that the introduction of manganese alloy in greater quantity than usual would very probably tend to prevent it ;—in fact, to predict the phenomenon and to assert that those remedies must be more or less effectual in preventing the frothing. But the same considerations will, I think, lead us to see that by observing similar conditions which can easily be established it is not difficult to prevent the formation of this gas—in fact, to resort to prevention instead of attempting cure.

I may mention that the usual remedies which I alluded to, viz., the introduction of a larger percentage of carbon, or of manganese, are often objectionable, and in some cases inapplicable ; for the presence of manganese in any perceptible quantity in steel although it doubtless has some advantages, has also a particular disadvantage, and that is, that steel containing an appreciable quantity of manganese does not weld in the way that it does when free from it; it is deprived to a very great extent of the power of welding, and therefore it is a matter of considerable importance to dispense with the presence of manganese in the steel when finally completed.

Now it is assumed that in the Siemens-Martin process the rest which is frequently allowed to the materials at the end of the process is sufficient to allow a complete subsidence of the metallic iron from the oxide. When we make that assumption we really assume something that is almost impossible; for the utmost care which I have known to be applied to the process, or which I could conceive to be applied to it can only, at the best, approach to that result. We can never get complete subsidence. I believe there always must be, when the spiègeleisen is added, some oxide mixed with it during the agitation that is taking place. But even supposing that such complete sub-sidence had taken place; there is no doubt that the conditions to which the molten mass is exposed while it passes from the hearth into the bucket and into the moulds must cause an oxidation. If any one sees the stream of molten steel running through the air, and notices the action of the air upon it, he has at once a very manifest indication of that. There is a column of molten metal pouring down, and there is no doubt a rapid current of air passing up by the sides of it, and burning the sides of it as fast as it can, burning the carbon with which it comes in contact, but mainly burning the iron. It cannot help being so, and there is, therefore, a thin film of oxide of iron being forced down into the molten mass in the bucket or ingot mould. Of course, if there is much manganese in the substance the action would mainly bear directly or indirectly upon it. We have therefore oxide of iron introduced into the mass in the form of this film which is formed on the surface of the flowing mass of molten steel, and we mix this oxide in the bucket or ingot mould with the carburet of iron.

Now, if we keep this mass quiet, the particles of oxide of iron will no doubt come in contact with some very small quantity of carbon in the adjacent mass; but if each little particle of oxide of iron be made to move about in the molten mass, of course, its action on

the carbon will in the unit of time be much greater. Now that is
exactly the difference between allowing your steel to cool in an
open, or in a closed ingot. In any case, we must expect that some
few bubbles will form; but if they are pent up and not allowed to
escape, they are prevented from doing what they otherwise would do
with such energy—prevented from stirring the mass about. The agitation which is caused by the bubbles when they are allowed to escape
freely is no doubt a beautiful sight physically, but it is an exceedingly
distressing sight to the metallurgist who wants a good casting. If you
wanted to establish the conditions most favourable to the formation of
carbonic oxide from the materials which are in your substances, you
could not do anything better than stir it about vigorously; and it is
therefore quite intelligible that if this carbonic oxide be formed by a
reaction between oxide of iron and carburet of iron, there will be less
of it formed, less of the reaction, if you keep the mass forcibly at rest
than if you allow it to undergo that violent agitation that sets in if you
do not close the mould. Again, if you have a great deal of carbon
present, say one or more per cent. instead of one-third per cent.,
there cannot be much oxide present in the mass and much carbonic
oxide formed in the moulds. Before the metal is settled in the mould
there is so much carbon in contact with oxide of iron that it is rapidly
reduced, and the carbonic oxide passes out of the mass before it
is in the mould. And there is a still further reason why the presence
of more carbon must tend to produce the same result. It is perfectly
obvious that if you are running down through air some steel, rich in
carbon, there is not so much oxide of iron formed as if you were
similarly to run down through air a steel containing little carbon.
If there is a half per cent. of carbon, or one part by weight in 199
parts of iron, then, of course, the chance of oxygen coming in contact with the carbon is smaller than if you have one per cent., or one
part to 99 of iron, and the oxygen combines in the former case to a
greater extent with iron.

With reference to the action of manganese, it is perfectly notorious
to every metallurgist and almost to every chemist, that if you have
oxide of iron in contact with metallic manganese at that high temperature, an interchange of metals instantly takes place; forming
metallic iron and oxide of manganese. The process takes place with

very great energy, because, as is well known, manganese combines with oxygen more powerfully than iron does under those circumstances, and holds it more firmly; but if the manganese, which is added to the molten steel, does decompose the oxide of iron and form oxide of manganese while decomposing oxide of iron, what is the effect of that? You have, instead of the oxide of iron which could easily be reduced by carbon, oxide of manganese, which is extremely difficult of reduction. And the oxygen is so firmly combined with the manganese that it does not combine with the carbon to form carbonic oxide.

I ought perhaps to allude to the idea which has been suggested, and which I have no doubt is familiar to most metallurgists, that the carbonic oxide is dissolved as such in molten steel, and given off by the cooling of the metal. Instances are known of a molten metal dissolving a gas which it cannot keep dissolved on solidifying; and, as far as I am aware, the only idea which has hitherto been brought forward to explain it is this : that steel heated to a temperature considerably above its melting point possesses the property of dissolving and holding in solution carbonic oxide as such, whereas steel at a lower temperature has not got that property. Now, in the first place, there is one objection to an explanation of that kind and to many other similar explanations, that it consists in giving a word instead of a fact. It is no doubt very convenient and very useful to name things, but it is still better and more convenient, although sometimes more difficult, to explain them ; and instead of merely being told that carbonic oxide is dissolved, we want to understand the condition in which its elements are dissolved. Now, as far as we know accurately, the condition of substances when they are dissolved, it appears to be some chemical process of combination ; every process of solution that we have been able to analyse carefully, has been explained in that way, and we have no proof whatever that there is any such thing as a process of solution in which the materials are dissolved in such a way that the two substances remain chemically the same when dissolved as they were when separate. I believe that the process of solution is in all cases due to some more or less powerful chemical combination, and distribution of the elements in some other form. So that my statement of the condition of the elements of the carbonic oxide consists rather in analyzing a fact, which is generally named without analysis.

Besides those considerations which I have laid before you, there are various others which I think tend very clearly to the conclusion that the carbon and oxygen are not combined with one another in the mass, but have to come in contact and to combine in order to form carbonic oxide. For instance, I have never yet been able to get any indication whatever that carbonic oxide is driven out by the addition of manganese, to a charge which was in such a condition that it certainly would froth without the manganese alloy. If the carbonic oxide were dissolved as such, and were not there when manganese is added, then we must see it go out when the manganese is added, and the opportunities for seeing such a process, if it occurred, are so numerous that I venture to think it could not have escaped notice, especially when it has been looked for. I think that circumstance is in itself entitled to some weight in support of the considerations I have laid before you, that the elements of carbonic oxide are not combined with one another in the steel to which manganese has to be added. The oxygen remains, but when you add sufficient manganese it is taken up by that metal, and remains as oxide of manganese, a compound which carbon cannot decompose under these conditions, at any rate can only decompose slightly.

Now I conceive that these conditions derive their chief interest and importance from having some definite practical application, and I think that in this meeting it can hardly be necessary that I should state in express terms what must be done in order to obtain solid castings of steel. If the key which I have given be the true one, the thing is so simple that I believe any metallurgist must at once know what he must do, and would not have the least difficulty in knowing how to do it. Of course you must have your charge made carefully and well subsided. You must take care that when it is poured from the " convertor" or " hearth," that it does not get oxidized. Nothing is easier than to surround it by a current of combustible gas, which is in most steel works present in considerable quantities; and it is necessary also to insist on other conditions which are well known.

If in the Bessemer process the oxide remains to any considerable etxent diffused through the mass, then of course you cannot get a good charge out of it, and there is no doubt that one difficulty of getting perfectly uniform products in the Bessemer process is due to the circumstance that it is so rapid that to stop it precisely at the required point is a matter of no small difficulty.

In speaking of steel, I ought perhaps to mention a substance which has frequently been designated of late by that name, though it might have been better to give it a name which would more naturally and properly represent what it is, a substance which is substantially pure soft iron, not steel. It does not temper. It is just now being made with considerable advantage by the Siemens-Martin process. It is also made by the Bessemer process. It is really pig iron which has been decarbonised almost completely, and then instead of being so far re-carbonised as to make it into steel by the addition of spiegeleisen, it is supplied with the requisite quantity of manganese with a much smaller proportion of carbon in the form of ferro-manganese, in fact, it can be made with two tenths per cent. of carbon when the process is conducted with care and regularity. This new substance is in fact perfectly soft iron. I conceive that it is likely to come into very extensive use, and to supersede to a very great extent the comparatively heterogenous and uncertain product obtained by puddling iron. The conditions which I have laid before you are likely to be found even more important because of the small quantity of carbon in it than for the manufacture of steel with a higher percentage of carbon.

The PRESIDENT: I am sure you will all agree with me that our thanks are due to Professor Williamson for this exceedingly lucid and instructive exposition of the steel manufacture. Professor Williamson has explained the source of the evolution of carbonic oxide, and the importance of its proportion in the manufacture of homogeneous steel. He has also touched upon some points of chemical theory, which those of you who were here this morning heard treated by another scientific investigator of eminence, Professor Guthrie, and you will no doubt have noticed that these two gentlemen have arrived at theoretical conclusions very dissimilar indeed as to what is to be understood under the name of a solution. These differences in theoretical opinion, I need not say, are continually occuring amongst scientific men actively engaged in investigation. But you will have noticed from Professor Williamson's remarks on the importance of a knowledge of what really is the case with this evolution of carbonic oxide, how greatly it is to be desired that the distinction, if there be one, between solution and chemical combination should be settled as soon as possible. It is only one of the very numerous cases in which this difference crops up, and which no doubt prevents the successful application of theoretical generalizations to manufactures like that of steel. There are no doubt many gentlemen present who have a special interest in the subject, and I trust they will favour us with their observations upon the paper. If there are no remarks from any gentlemen present I will ask Mr. Chandler Roberts to give us his paper on the apparatus used by the late master of the mint in his researches. We have in the exhibition here many of the instruments employed with so much ingenuity and success by the late Professor Graham, the master of the mint, and I feel sure we shall have an interesting communication from Mr. Roberts, who is so well acquainted with this apparatus, and who worked with Professor Graham in the use of it throughout the entire investigation.

On the Apparatus Employed by the late Mr. Graham, F.R.S., in his Researches.

W. Chandler Roberts, F.R.S. : Mr. Graham will probably be best remembered as a chemist, although the most important of his researches were either purely physical, or were devoted to the elucidation of questions which occupy an intermediate position between physics and chemistry. It is specially interesting, therefore, to observe what was the nature of the apparatus he employed in obtaining results of such importance as those with which his name is associated.

From the fact that the instruments on the table are those with which he arrived at all his more important conclusions, it will at once be evident that the appliances he used were both few and simple. Before I proceed to describe them, I should, as the time at my disposal is very limited, briefly state that Graham's labours were mainly devoted to ascertaining the nature of molecular movement in cases in which he was satisfied that no *mass* movement could take place, and, as Dr. Angus Smith has pointed out, while Dalton showed the relative weights of the combining quantities, Graham showed the relative magnitude of groups into which they resolve themselves. It is interesting to note that, as Prof. J. P. Cooke has observed, while Faraday was so successfully developing the principles of electrical action, Graham, with equal success, was investigating the laws of molecular motion. Each followed with wonderful constancy, as well as skill, a single line of study from first to last, and to this concentration of power their great discoveries are largely due.

The Royal Society's Catalogue of papers shows that his earliest paper was on the absorption of gases by liquids. It was published in 1826 in Thomson's "Annals of Philosophy;" in it he considers that gases owe their absorption in liquids to their capability of being liquefied, and therefore that solutions of gases in liquids are mixtures of a more volatile with a less volatile liquid. He concludes the paper by saying, that "All that is insisted on in the foregoing sketch is, that when gases appear to be absorbed by liquids they are simply reduced to that liquid in elastic form which otherwise, by cold or pressure, they might be made to assume, and their detention in the absorbing liquid is owing to that mutual affinity between liquids which is so common." It was a theoretical paper only, and no apparatus was even described ; I quote it merely because, in his last paper in the *Phil. Trans.*, forty years afterwards, he speaks of the liquefaction of gas in colloids in much the same terms.

In 1829, the *Quarterly Journal of Science** contains his first paper on the diffusion of gases ; he found that the lighter a gas is the more quickly it diffuses away from an open cylinder. The cylinders he employed were nine inches long, and 0·9 inches interior diameter ; they were placed in a horizontal position, and the gas under examination was allowed to diffuse outwards through a narrow tube directed either upwards or downwards according as the gas was heavier or lighter than air. It was therefore by the aid of a simple cylinder that he was led to believe, as he states in this his first paper, "that the diffusiveness of gases is inversely as some function of their density, apparently the square root of their density." He subsequently found that so great is the tendency of gases to diffuse into one another, that

* Quart. Journ. Sci., ii., 1829, p. 74.

this mixture of inter-diffusion will take place through apertures of insensible magnitude.   And in his paper in 1834,* he treats in detail of diffusion through porous septa, his object being " to establish with numerical exactness the following law of diffusion of gases :—The diffusion or spontaneous intermixture of two gases in contact is effected by an interchange in position of indefinitely minute volumes of the gases, which volumes are not necessarily of equal magnitude, being, in the case of each gas, inversely proportional to the square-root of the density of that gas."   He started from the well-known experiment of Döbereiner, who found, in 1825, that hydrogen kept in a glass receiver standing over water, escaped by degrees through the fissure into the surrounding air, the water in the receiver rising to the height of about 2¾ inches above the outer level.   In repeating Döbereiner's experiments and varying the circumstances, Mr. Graham discovered that hydrogen never escaped outwards by the fissure without a certain portion of air penetrating inwards, but with this essential difference, for every volume of air which penetrated into the vessel 3·8 volumes of hydrogen escaped.

The apparatus consisted of a graduated glass tube nearly an inch in diameter, having one end closed by a porous diaphragm of plaster of Paris.   This tube was filled with the gas to be examined, and the rise of the mercury indicated the rate at which the interchange of gas and external air took place.   He also interposed a bulb two or three inches in diameter between the diaphragm and the graduated tube with a view of increasing the capacity of the instrument, and of avoiding the interference of vapour.   In this paper he traced the relation which diffusion bears to the mechanism of respiration, but time will not permit me to consider this question.

These early results were repeated and greatly extended in a paper " On the Molecular Mobility of Gases,"† but in the experiments there described, thin plates of compressed graphite were principally used. The paper is chiefly remarkable for the clear enunciation of the fact that diffusion is a molecular, and not a *mass* movement, for Mr. Graham observes : " The pores of artificial graphite appear to be so minute that gas in mass cannot penetrate the plate at all.   It seems

---

* Edin, Roy. Soc. Trans., xii., 1834, p. 222.   † Phil. Trans., 1863.

that molecules only can pass, and they may be supposed to pass wholly unimpeded by friction, for the smallest pores that can be imagined to exist in graphite must be tunnels in magnitude to the ultimate atoms of a gaseous body. The sole motive agency appears to be that intestine movement of molecules which is now generally recognised as an essential property of the gaseous condition of matter.

"According to the physical hypothesis now generally received, a gas is represented as consisting of solid and perfectly elastic spherical particles or atoms, which move in all directions and are animated with different degrees of velocity in different gases." . . . If the vessel containing the gas "be porous like a diffusiometer, then gas is projected through the open channels by the atomic motion described, and escapes. Simultaneously, the external air is carried inward in the same manner, and takes the place of the gas which leaves the vessel. To this atomic or molecular movement is due the elastic force, with the power to resist compression, possessed by gases."

In order to demonstrate the diffusion of gases it is necessary to exaggerate the conditions of Mr. Graham's experiments. Instead of employing a tube closed with a disc of plaster of Paris, it is better to fix a glass tube into a battery cell and to employ it as the septum through which the gas is diffused. The following experiment was also shown :— A porous battery cell was attached to the short tube of a wash-bottle, both tubes being previously turned upwards ; when a jar of hydrogen was placed over the battery cell, the gas diffused through the cell, and the change of pressure caused the water to issue like a fountain several feet in height. I believe this arrangement was devised by Prof. Bloxam.

Now I must ask you to follow me a step further. In 1846 Mr. Graham read a paper before the Royal Society, " On the Motion of Gases." He showed that the *effusion* of gases through a minute hole in a platinum plug left no doubt of the truth of a general law that different gases pass through minute apertures in times which are as the square roots of their respective specific gravities, or with velocities which are inversely as the square roots of their specific gravities ; or in other words, he experimentally verified the mechanical law that the velocity with which a gas rushes into a vacuum through such an aperture, is the same as that which a heavy body would acquire in

falling from the height of an atmosphere, composed of the gas in question, of uniform density throughout. The relative rates of effusion and
diffusion are alike, but Mr. Graham is careful to observe that the
phenomena are essentially different in their nature. The former
affects masses of gas, the latter only affects molecules.

The apparatus Mr. Graham employed consisted of two glass jars ;
the one containing the gas to be examined was placed in a pneumatic
trough, and the other stood on the plate of an air-pump. They were
in connection, a series of tubes containing the usual reagents for puri-
fying and drying the gas being interposed between them. The jar on
the air-pump was exhausted, and the gas entered it through a minute
orifice in a platinum disc, the rate of passage being observed by the
aid of a mercurial column.

Three years later Mr. Graham published a paper giving the results
of an investigation on what he considered to be a fundamental property of the gaseous form of matter, which he termed transpiration. He
employed capillary tubes, and found that effusion and transpiration differed widely;* "for if the length of the tube is progressively increased,
and the passage for all gases becomes greatly slower, the velocities of
the different gases are found to diverge rapidly from their effusion
rates." The velocities at last, however, attain a particular ratio with a
given length of tube and resistance, and preserve the same relation to
each other for greater lengths and resistances, the most simple result
probably being that of hydrogen, which has exactly double the transpiration rate of nitrogen, the relation of these gases as to density being as 1 : 14.

	Diffusion Velocities.	Effusion Velocities.	Transpiration Velocities.	Rates of Passage through Caoutchouc.
Hydrogen	3·83	3·613	2·066	4·73
Oxygen	0·9487	0·950	0·903	2·224
Nitrogen	1·0143	1·0164	1·030	0·870
Carbonic acid	0·812	0·821	1·237	11·819
Carbonic oxide	1·0149	1·0123	1·034	0·968
Marsh gas	1·344	1·322	1·639	1·869
Air	1·0	1·0	1·0	1·0
	About 9600 c. c. of air pass per minute through 1 sq. metre of stucco 2·5 mm. thick.	78·3 c. c. of air pass per minute through a cer- tain small aper- ture in a brass plate.	62·9 c. c. of air pass per minute through a glass tube 6·6 metres long and 0·55 mm. in diameter.	16·9 c. c. of air pass per minute through one square metre of caoutchouc 0·02 mm. thick.

Note.—It is impossible to make the four columns strictly comparable on account of the
difference of the conditions under which the experiments were made.
* Phil. Trans., 1849, p. 349.

Thus, in what are very nearly Mr. Graham's own words, a gas may pass into a vacuum in three different modes ; that is, by effusion, transpiration, or diffusion, and I hope you will bear with me while I recapitulate them.

1. The gas may enter the vacuum by effusion, that is, by passing through a minute aperture in a thin plate, such as a puncture in platinum-foil made by a fine steel point. The relative times of the effusion of gases *in mass* are similar to those of the *molecular* diffusion, but a gas is usually carried by the former kind of impulse with a velocity many thousand times as great as is demonstrable by the latter.

2. If the aperture of efflux becomes a tube, the effusion rates are disturbed. The rates of flow of different gases, however, assume again a constant ratio to each other when the capillary tube is so elongated that the length exceeds the diameter by at least 4,000 times. The transpiration rates appear to be independent of the material of the capillary ; they are not governed by specific gravity, and are indeed singularly unlike the rates of effusion. The ratios appear to be in direct relation with no other known property of the same gases, and they form a class of phenomena remarkably isolated from all that is at present known of gases.

For instance it will be seen by the table already given that the rate of carbonic acid which is low for effusion and diffusion, becomes comparatively rapid when the gas passes by transpiration.

3. A plate of compressed graphite, although it appears to be practically impermeable to gas by either of the two modes of passage just described, is readily penetrated by the agency of the molecular or diffusive movement. The times of passage through a graphite plate into a vacuum have no relation to the capillary transpiration times of the gases, but they show a close relation to the square roots of the densities of the respective gases, and agree with the theoretical *times of diffusion* usually ascribed to the same gases.

These latter results were obtained by the graphite diffusiometer. It stood over mercury, and was raised or lowered by an arrangement introduced by Prof. Bunsen.

Mr. Graham subsequently employed the barometrical diffusiometer. It consisted of a tube in which a Torricellian vacuum could be produced. The upper end was closed by the porous septum, and a slow stream of the gas under examination was allowed to pass over the plate through the india-rubber hood by which it was covered.

I might mention that the very exact and illustrious experimenter, Prof. Bunsen, was led to doubt the accuracy of Graham's law of the diffusion of gases, but he employed plugs of plaster of Paris which impaired the results by introducing the phenomenon of transpiration; and probably also as Mr. Graham observed to me, by an actual retention of hydrogen in the pores of the plaster. It is interesting from our point of view, because it shows that the simple apparatus employed by Mr. Graham really gave the only trustworthy results.

The results of the later experiments led him to prove that mixed gases might be separated from each other by diffusion. Stems of tobacco-pipes were employed, arranged inside a glass tube, which could be rendered vacuous, the mixed gases being passed through the tobacco-pipe. For example, when this explosive mixture of 66 per cent. of hydrogen, and 33 per cent. of oxygen is passed through this tube, a mixture is obtained containing only 9·3 per cent. hydrogen, and is therefore non-explosive. With air it was found possible to concentrate the oxygen by 3·5 per cent.

With the apparatus now before us, Mr. Graham subsequently worked on liquid transpiration in relation to chemical composition. He started from the discovery of M. Poiseuille, that a definite hydrate of one equivalent of alcohol with six equivalents of water is more retarded than alcohol, containing either a greater or a smaller proportion of water. The rate of transpiration depending upon chemical composition and affording an indication of it, it thus appeared probable that a new physical property might become available for the determination of the chemical constitution of substances, and the experiments appeared to establish "the existence of a relation between the transpirability of liquids and their chemical composition. It is a relation analogous in character to that subsisting between the boiling point and composition so well defined by Hermann Kopp."* The apparatus consists of a strong glass jar closed at the top by a brass plate into which a condensing syringe is screwed. This plate also had

---

* Phil. Trans., 1861, p. 373.

a tube screwed into it, and into the tube the glass bulb with a long capillary tube was fixed. The fluid under examination was placed in the bulb, which communicated freely with the interior of the jar, containing compressed air.

To revert to the chronological order. His paper in December, 1849, formed the Bakerian lecture of the Royal Society. It was on the *Diffusion* of liquids, and the only apparatus employed was very similar to that adopted in his earliest paper on the diffusion of gases ; it consisted of a bottle and glass jar, the fluid under examination being placed in the bottle, which was immersed in the water with which the jar was filled. With this simple apparatus he found that when two liquids of different densities, and capable of mixing, are placed in contact, diffusion takes place between them much in the same manner as between gases, except that the rate of diffusion, which varies with the nature of liquids, the temperature and the degree of concentration is slower. Common salt when placed in the inner vessel will diffuse twice as rapidly as sulphate of magnesia, and this salt will diffuse twice as rapidly as gum arabic. Subsequently Mr. Graham modified the disposition of the apparatus and simply introduced the salt to be diffused by means of a pipette to the bottom of a jar filled with water. There experiments led to the very remarkable and important discovery that different compounds might be separated from each other by diffusion, and this was not all, for it was proved that a partial decomposition of chemical compounds was effected by diffusion. Thus ordinary alum was partially decomposed into sulphate of potassium and sulphate of aluminium, the latter being less diffusible than the first-named salt. Mr. Graham considered this research to be very important, and he remarks, " in liquid diffusion we appear to deal no longer with chemical equivalents or Daltonian atoms, but with masses even more simply related to each other in weight." We may suppose that the chemical atoms " can group together in weights which appear to have a simple relation to each other. It is this new class of molecules which appear to play a part in solubility and liquid diffusion, and not the atoms of chemical combination."

Continuing the investigation he described in a paper of singular beauty, his well-known experiments on the varying rates of liquid diffusion of various soluble substances, which led him to divide them into crystalloids and colloids, the former having a rapid diffusion rate, the latter being marked by low diffusibility. He placed the substance under experiment in a tambourine of parchment paper, which was floated on the surface of a comparatively large volume of water, the highly diffusive crystalloid passed through the membrane and the colloid remained behind, for "the diffusion of a crystalloid appears to proceed even through a firm jelly with little or no abatement of velocity."

I have here the very interesting series of colloids prepared by Mr. Graham, and of these perhaps the most interesting is the soluble silicic acid. If silicate of soda is poured into diluted hydrochloric acid, the acid being maintained in large excess, a solution of silicic acid is obtained. But this solution also contains, in addition to the silicic acid, chloride of sodium, from which it may be freed by the action of dialysis, and by this means a solution, which is not in the least viscous, is obtained, contaning 14 per cent. of silicic acid. The coagulation of the silicic acid is effected, however, by the addition of a solution containing the one-ten-thousandth part of any alkaline or earthy carbonate. Mr. Graham therefore described this gelatinous state as the "pectous," as distinguished from the "peptous" or dissolved form.

By a similar process Mr. Graham obtained specimens of soluble alumina, peroxide of iron, chromic oxide, and stannic acid, all of which have their pectous and peptous states. And he showed that in most cases alcohol, sulphuric acid, and glycerine can replace part of the water of these colloids. I cannot describe these interesting substances now, nor can I do more than remind you of the use of dialysis in medico-legal inquiries. I must content myself with summing up a few of Mr. Graham's conclusions with reference to crystalloids and colloids. Although chemically inert, in the ordinary sense, colloids possess a compensating activity of their own, arising out of their physical properties. While the rigidity of the crystalline structure shuts out external impressions, the softness of the gelatinous colloid partakes of fluidity, and enables the colloid to become a medium for liquid diffusion like water itself. Another and eminently characteristic quality of colloids is their mutability, as fluid colloids often pass from the fluid to the pectous or gelatinous condition under the slightest influences. The colloid is, in fact, the dynamic state of matter, the crystalloid being the statical condition. The colloid possesses *energy*, and it may be looked upon as the primary source of the force appearing in the phenomena of vitality.

The next instruments to be considered are those with which Mr. Graham studied osmotic force. When a solution of a salt, or a liquid, is separated by a membrane or porous diaphragm from a mass of water, a flow of liquid takes place from one side of the septum to the other. This action was discovered by Dutrochet, and is known as osmose. Dutrochet and Mr. Graham both used a narrow glass tube, having a funnel shaped expansion at the bottom, covered at that end by a piece of bladder. Mr. Graham also used porous earthenware and albuminated calico.

In some cases the flow of liquid through the septum is sufficiently powerful to sustain a column of water many inches high in the glass tube. Dutrochet inferred from his experiments that the velocity of the osmotic current is proportional to the quantity of salt or other substance originally contained in the solution. He attributed the action of the septum to capillarity, but Mr. Graham ultimately considered that the water movement in osmose is " an affair of hydration and dehydration of the substance of the membrane or other colloid septum," and that the diffusion of a saline solution only acts by affecting the hydration of the septum. The outer surface of the membrane being in contact with pure water, tends to hydrate itself in a higher degree than the inner surface does, the latter surface being supposed to be in contact with a saline solution. When the full hydration of the outer surface extends through the thickness of the membrane and reaches the inner surface, it there receives a check. . . . The contact of the saline fluid is thus attended by a continuous catalysis of the gelatinous hydrate, by which it is resolved into a lower gelatinous hydrate and free water. Now this question of hydration is perhaps the most remarkable instance of the persistent continuity of Mr. Graham's work, as Dr. Odling has pointed out,*—" it is noteworthy that for him (Mr. Graham) osmosis became a mechanical effect of the hydration of the septum ; that the interest attaching to liquid transpiration was the alteration in rate of passage consequent on an altered hydration of the liquid, and that the dialytic difference

---

* Lecture on " Prof. Graham's Scientific Work," Royal Institution, January, 1870.

between crystalloids and colloids depended on the dehydration of the dialytic membrane of the former class of bodies only."

I must now direct your attention to a section of Mr. Graham's work, which, although it was the last, was a reversion to some of his very earliest experiments. In 1829, under the title, " Notice of the Singular Inflation of a Bladder," he described the following experiment :—A bladder two-thirds filled with carbonic acid was introduced into a bell jar filled with carbonic acid gas ; after the lapse of some hours the bladder was found to contain 35 per cent. of carbonic acid, and to have become distended. Mr. Graham observes :—" M. Dutrochet will probably view in these experiments the discovery of endosmose acting upon aëriform matter as he observed it to act on bodies in a liquid state. Unaware of the speculations of that philosopher at the time the experiment was made, I fabricated the following theory to account for them :—The jar of carbonic acid standing over water, the bladder was moist, and we knew it to be porous. Between the air in the bladder and the carbonic acid without there existed capillary canals through the substance of the bladder, filled with water. The surface of the water at the outer extremities of these canals being exposed to carbonic acid, a gas soluble in water would necessarily absorb it. But the gas in solution . . . permeated the canal, and passed into the bladder and expanded it."*

You will remember that in the concluding experiments on the diffusion of gases Mr. Graham employed a tube, closed with a graphite disc, in which a Torricellian vacuum could be produced. In his experiments on the penetration of different gases through membranes the same apparatus was employed, only the disc of graphite was replaced by a film of india-rubber. He found that gases penetrated to the vacuous space at the rates given in the last column of the table. You will observe that the gas which penetrates most rapidly is carbonic acid, and you will also see that the rates of passage are in no way connected either with those of diffusion or transpiration.

A comparison of the relative rates of passage of oxygen and nitrogen led to a most remarkable experiment.　Oxygen penetrates $2\frac{1}{2}$ times as fast as nitrogen, therefore by dialysing air Mr. Graham actually

* Quart. Journ. Sci., 1829, p. 88.

increased the quantity of oxygen from 20·8 to 41 per cent., just as he had effected, by the aid of a tobacco-pipe, a partial separation of oxygen from air by the slightly greater diffusion velocity of nitrogen. The Torricellian vacuum was ill adapted for the experiments, and Mr. Graham gladly availed himself of the mercurial exhauster devised by Dr. Hermann Sprengel, and he considered that without the aid of this instrument it would have been impossible to conduct certain portions of the research. He was thus able to use larger septa of india-rubber, bags of waterproof silk being found to be most convenient. The vacuum was not even absolutely necessary, for the penetration of the nitrogen and oxygen of air through rubber into a space containing carbonic acid could be readily effected, the gas being absorbed by potash at a certain stage of the operations.

Mr. Graham considered this penetration to be due to an actual dissolution of the gas in the substance of the india-rubber, for, as he observes, "gases undergo liquefaction when absorbed by liquids and by soft colloids like india-rubber," words I think of interest, when we remember that the sentence only marks a slight extension of the view he expressed in his first paper in 1829.

These discoveries led Mr. Graham to inquire whether it was probable that the discovery of MM. Troost and Deville of the penetration of red-hot platinum and iron tubes by hydrogen, could be due to an actual absorption and liquefaction of the gas in the pores of the metal, and by submitting the question to the test of experiment it was proved that such an absorption did take place.

For instance, palladium was found to act as platinum, only in a more marked manner. A tube of palladium when attached to the mercurial exhauster did not allow hydrogen to pass in the cold, but when heated to redness in an atmosphere of hydrogen the gas passed through the walls of the tube at the rate of 4,000 cubic centimetres per square metre in an hour. This led to the remarkable discovery of the absorption or occlusion of gases by metals. It was found that nearly all metals appear to select one or more gases. Silver, for instance, absorbs many times its volume of oxygen, and under certain circumstances gives it out again on cooling. Iron is specially characterised by its absorption of carbonic oxide, but it also retains hydrogen, and this fact led Mr. Graham to extract from meteoric iron, the gas that probably affected its reduction to the metallic state, and which certainly exists in the atmosphere of certain stars.

The most remarkable results were obtained with palladium. I called your attention at the beginning of the lecture to the index which you will obseve has moved six inches.

I will now describe the apparatus ; it consists of a tall jar filled with acidulated water ; at the bottom of the jar two wires are fixed, and these wires are parallel throughout the entire length of the jar. Each is attached to the short arm of a lever, the longer arms of which are about five feet long. One wire is of palladium, the other of platinum, and they form the electrodes of a small battery capable of decomposing the water. The palladium now forms the negative electrode, and is freely absorbing hydroden, the excess of which is escaping from its surface. The absorption of hydrogen has been attended by a considerable expansion, as is shown by the fall of the index. The index attached to the platinum wire has of course remained stationary.

This expansion enabled Mr. Graham to calculate the density of the gas in its condensed form, and for reasons which I cannot give you now he was led to believe that hydrogen gas is the vaponr of a white magnetic metal of specific gravity 0·7.

Now by taking palladium which has been charged in the manner you have seen, and heating it *in vacuo*, I can actually extract and show you the hydrogen it contained. This little medal of palladium contains an amount of gas condensed into it which would be equivalent to a column of gas more than a yard high, and of the diameter of the medal.

The story of Mr. Graham's work has been much better told by Odling, Williamson, Hofmann, and Angus Smith, but what does it teach us from the point of view of a collection of scientific apparatus? Surely that, although in certain researches or for accurate observation and measurement, delicate and complicated instruments may be necessary, the simplest appliances in the hands of a man of genius may give the most important results. Thus we have seen that with a glass tube and plug of plaster of Paris, Mr. Graham discovered and verified the law of diffusion of gases. With a tobacco-pipe he proved indisputably that air is a mechanical mixture of its constituent gases. With a tamborine and a basin of water he divided bodies into crystalloids and colloids ; and obtained rock crystal and red oxide of iron soluble in water. With a child's india-rubber balloon filled with carbonic acid he separated oxygen from atmospheric air, and established points, the importance of which, from a physiological point of view, it is impossible to overrate. And finally, by the expansion of a palladium wire, he did much to prove that hydrogen is a white metal.

The CHAIRMAN : I am sure, ladies and gentlemen, you will return Mr. Chandler Roberts your thanks for this explanation he has afforded us of the exquisitely simple apparatus used by the late Professor Graham in his most important researches. I think we must all have been struck by the Daltonian simplicity, I may say, of these instruments before us, when we reflect upon the great results achieved by Professor Graham in that borderland, as it were, between chemistry and physics, in these researches on the laws of the diffusion of gases and liquids, and of the occlusion of gases and solids, we must be struck by the ability and ingenuity of the observer who could achieve so much by such simple means. We frequently have to remark that some of the most important discoveries in chemical science, at all events (it is not so much so in physics), have been made by instruments of the most simple construction. I dare say that some further explanations may perhaps be desired by some gentlemen present, and Mr. Roberts, I have no doubt, will be happy to afford any further explanation which may be required. I may point out perhaps one of the important applications of this method of diffusion of gases which was pointed out by Professor Graham was that it enables to separate the constituents of gaseous mixture, and also to prove the homogeneity or otherwise of gases which have to be submitted to investigation. For instance, you may have gas which yields certain results upon analysis which would be yielded by a mixture of equal volumes of ethyl gas and hydrogen, but those results would also be obtained from the compound known as ethylic hydride, and the problem is to ascertain whether it is really a homogeneous gas or a mixture of those two constituents. It is not easy to ascertain this by analytical experiment, inasmuch as the mixture yields in endiometric processes the same results as the chemical compound ethylic hydride ; but by employing diffusion you at once distinguish between a gaseous mixture and the homogeneous gas. The latter will diffuse equally through the porous septum whilst the mixture diffuses irregularly, and after you have submitted this gas for some time to diffusion if you analyse the residue remaining in the tube, in the one case it will remain unaltered, though mixed of course with atmospheric air, whilst if the gas has been a mixture, the hydrogen being by far the lightest of the two gases, will be diffused much more rapidly than the ethyl with which it is mixed, and consequently your residual gas will have an entirely different composition from that which it had before the commencement of the experiment. Thus this simple operation of diffusion enables us to decide whether a gas is a true chemical compound or merely an admixture. Numerous resistances of this occur in the investigation of gases belonging to organic chemistry. I have myself on one or two occasions made use of this beautiful process which was made known to us by Professor Graham to solve that problem of mixture or true chemical combination.

I will now call on Mr. Hartley to favour us with a communication

ON LIQUID CARBONIC ACID CONTAINED IN THE CAVITIES OF CRYSTALS.

Mr. W. N. HARTLEY : In the year 1822 Sir Humphrey Davey investigated the cavities which are found in rock crystal and the liquid contained in them. This he found to be in almost every case pure water or some weak saline solution. In one case he found a kind of mineral naphtha present, and there were also gaseous contents which were not particularised in his paper except in one case where the gas was nitrogen. In the following year Sir David Brewster announced the discovery of a remarkable liquid, or as he stated two liquids, in the cavities of topazes, in rock crystals and in chryso-beryls, and one of the most remarkable properties of this liquid was the enormous expansion it attained. The expansion is such that if they are gently warmed the cavities become entirely filled with liquid, and the bubble which is under ordinary circumstances seen in the liquid disappears. Sir David Brewster states that a Mr. Sanderson, a jeweller in Edinburgh, placed a topaz in his mouth and it exploded with great violence and hurt him. Mr. Alexander Bryson, of Edinburgh, in 1860 recognized a similar fluid in cavities in porphyry from Dun Dhu in Arran, in the schorl of Aberdeen granite, and also in the trap tufa of the Calton Hill, the basalt of Samson's Ribs, and green-stone from the Crags in the Queen's Park, Edinburgh.

There is a particular line of research which bears on these liquids. First of all the researches of Cagnaiard de Latour who showed that the effect of high temperature on the liquid in a sealed tube was to endue it with remarkable mobility; that the liquid suddenly disappeared in a state of vapour. This of course in the case of water, and ether, and such liquids, requires a considerable temperature,—in the case of water, that of melting zinc. Then Faraday showed that gases could be liquefied by the combined action of pressure and low temperature. Thilorier and Mitchell, independently, prepared liquefied carbonic acid in large quantities and examined its properties. Thilorier mentions the specific gravity of the liquid as being much less than water, it being for instance at 0° Centigrade ·83, water being 1, whilst at 30° Centigrade the specific gravity is ·6, the specific gravity of water not differing much from that at ordinary temperatures. The application of this work of Thilorier's to the examination of the liquid which was noticed in topazes is evident, and in 1869 Mr. Sorby and Mr. Butler examined the liquid which they found in a sapphire. Sapphires and rubies both contain this remarkable liquid, and they were fortunate in getting a cavity which was somewhat tubelike in shape and not irregular, and which was of such a size that by observing the liquid with a lens, its rapid expansion when put in warm water was easily noticed. Mr. Sorby measured this very carefully and found that 100 volumes of the liquid in expanding from 0° Centigrade to 30° became 150 volumes. Thilorier showed that between 0° and 30° Centigrade, 100 volumes of liquid carbonic acid became 145, so that the difference is very slight when the experiment is made under such circumstances.

In the same year, 1869, Vogelsang and Geissler examined the liquid in the cavities of crystals, in the following manner :—Taking a small retort containing a mineral in fragments, it was attached to a tube commonly known as a Geissler vacuum tube. There were platinum wires sealed at each end of the tube and to one end was affixed an ordinary Sprengel pump. After a vacuum had been produced so that an electric spark would no longer pass across from wire to wire, the fragments of the mineral were heated, the mineral decrepitated, because the cavity exploded, and the spark passed through, because the pre-

viously vacuous space then contained a gas. This gas had the spectrum of carbonic acid.

About four years ago I came across a specimen of quartz which
had been cut and polished for microscopical examination, and in
this I noticed a large cavity which contained a fluid which I at once
took to be water, and thinking that I would repeat an experiment
of Cagniard de Latour's with this cavity, I heated it strongly when
looking at it under the microscope. To my astonishment the liquid
disappeared, although the temperature was far too low to convert
water into vapour in such a cavity. I thought at first that the liquid
had escaped, but on carefully watching it for some time I found that
it returned. Professor Andrews has told us in his researches on the
continuity of gaseous and liquid states of matter that at a temperature
of 30·9 Centigrade liquefied carbonic acid becomes gas, and that a pressure of even 500 atmospheres will fail to liquefy it again until the temperature is reduced. Now these specimens with cavities were kept by
me some time for examination, being at first at a loss to know how I
should take the temperature with accuracy, but the plan which I adopted
I have since found perfectly satisfactory. I took water which was just
above 30 Centigrade, say 30.9, placed a specimen in it, took it out,
wiped it rapidly, and if the liquid still remained there I knew that it
was not liquid carbonic acid, or at any rate, I knew that it was not
pure carbonic acid. If at a temperature of 31 under similar circumstances I find that the liquid carbonic acid is no longer there, that
the bubble has disappeared, then I have two temperatures of which,
if I take the mean they will give me 30·95, and this is a temperature
very nearly that which Professor Andrews has found to be the critical
point, that is, the point at which liquid carbonic acid becomes gas.
Now, I have improved the method in a certain way, that is, I have
introduced a modification in the means of taking the critical point of
the liquid in minerals and crystals in cut specimens which certainly
gives me confidence in the results. I take a little glass trough and
place the mineral within this. I have small thermometers about two
and a half inches long which give me a temperature to the tenth part
of a degree Centigrade just above and below the critical point of liquid
carbonic acid, and by stirring these troughs containing the minerals
and thermometers in water at the right temperature, taking the trough

out, just wiping it, and placing it under the microscope, I can very accurately get the disappearing point of the liquid. In this way I have detected liquid in many minerals, for instance in quartz, topaz, and tourmaline, and taken the critical point. I found that in the quartz it is almost as exact an approach as anything can be to the critical point of carbonic acid. However, in other cases it varies somewhat. There is a method which gives one the means of detecting these volatile liquids in smaller quantities with very great ease—in fact it would be impossible without such means I think to detect small quantities of liquid carbonic acid, or some volatile fluid of that kind in cavities without this arrangement. I have put the instrument up at the end of the table. It consists simply of a tube which is heated by the flame of a spirit lamp, and attached to this tube is a ball syringe by which air can be blown through the tube. The tube is drawn out to a fine point, and projects a stream of warm air on to the object just at that part where the cavity occurs. By just blowing a current of warm air in that way, the liquid is instantly converted into gas, and the appearance which may be doubtful in certain small cavities, as for instance, cavities in granite, is at once shown to be that of a volatile liquid floating on water. On blowing with the ball syringe, the carbonic acid portion disappears, leaving the water, and the liquid state returns as the specimen afterwards cools, and in this way I have detected liquid carbonic acid in a cavity which I could just discern easily with a $\frac{1}{8}$ of an inch power, so that I do not think the carbonic acid could have been more than the 50,000th part of an inch in diameter. There are one or two very curious appearances seen in cavities containing this volatile liquid, and one of them is this: when the proportion of liquid carbonic acid to gaseous carbonic acid is less than one half, heat causes the liquid carbonic acid to get gradually less and less until it disappears altogether, but if the liquid carbonic acid is in the proportion of one-half of the whole contents, that is to say, if the liquid and gaseous carbonic acid at ordinary temperatures are equal, then the increase of heat causes the liquid to expand until it fills the entire cavity, and then it generally becomes gas. This would be very curious if one did not know what takes place with liquid carbonic acid, as fortunately we do. Thilorier says that if he takes liquid carbonic acid and seals it in a glass tube, if it occupies only one-third of the tube, it forms what he calls a retrograde thermometer, that is, one which is the reverse of a mercurial one, which does not rise but falls by increase of temperature ; in other words the liquid becomes gas as the temperature increases.   But if the liquid occupies two-thirds of the tube, then it is a normal thermometer ; the increase of temperature causes the liquid to expand, and the great pressure more than compensates for the effect of rise of temperature.   Then there is another appearance which is curious. In certain cavities, those namely in which the liquid expands by rise of temperature after the critical point has been passed, there is a curious boiling motion seen.   Mr. Bryson noticed that in the cavities he examined.   This curious boiling motion is due to the fact that first of all liquid spherules are formed in the cavity independent of the surrounding gas ; these spherules form a sort of mist, presently they begin to grow in size until they get so large that they touch.   As soon as they touch they begin to gravitate, and as soon as they gravitate and collect at the lower part of the cavity, gas is also entangled in them, and gas rises out of them.   The same thing is seen I believe in the tubes which Professor Andrews has sealed up containing liquid sulphurous acid ; these when heated and cooled shew the same sort of action.   In some cases this appearance in the cavity is very remarkable, because it seems as though there was a whirling action in the cavities in which the spherules float.   It is impossible to tell by simple inspection which are gas and which are liquid spherules, both having the same appearance, and they seem to chase each other round and round, the reason being that there is a descent of the liquid in one direction, and an ascent of the gas in the other. Of course, under the microscope it appears the reverse ; the liquid appears at the top, and the gas at the bottom.   Then another fact connected with these cavities is that in some of them, which I have had the pleasure of looking at and which belong to Mr. Butler, (large specimens of sapphire), there is absolutely no water present.   This we know by the fact that the liquid does not adhere to the sides of the cavity ; it has a convex surface just as mercury has in a tube, but whenever the liquid is in contact with a wet cavity, then it has a concave surface.   In this way one can very readily tell in any specimen whether there is water there or not, because although sometimes the water is not seen, you see that the liquid has this curve.   In some specimens water is distinctly seen in the cavities. The evidence of liquid carbonic acid throws light on the chemical reactions which must have taken place in the bowels of the earth. This subject has yet to be worked out, in great measure, but there is strong evidence of the enormous pressure under which these reactions must have taken place, because it is not at all likely that the carbonic acid would have been liberated at low or ordinary temperatures, and even at a temperature so low as 30° Centigrade, that of a warm summer's day, liquid carbonic acid would exert a pressure of 109 atmospheres, and this of course would be very much increased by any rise of temperature. The presence of liquid carbonic acid in minerals, is now I think, absolutely certain, taking together the recognition of it by means of the spectroscope, by means of Geissler and Vogelsang's experiments, by the co-efficient of expansion as determined by Messrs. Sorby and Butler, and the determination of the critical point which my own experiments deal with.

The PRESIDENT : Is it your pleasure, ladies and gentlemen, to return a vote of thanks to Mr. Hartley for this interesting account of the occurrence of carbonic acid in the cavities of crystals? Mr. Hartley has drawn our attention to the undoubted circumstance that the formation of this carbonic acid must have occurred under very considerable pressure. Whether it is necessary that the temperature should at the same time be high may perhaps be considered as a doubtful point, but I would venture to suggest that this reaction with the generation of carbonic acid can take place at the bottom of the sea, for instance. There appears to me to be no difficulty in comprehending how a slight accumulation of liquefied carbonic acid might occur in positions of this kind on the supposition that these minerals have been crystallised beneath water pressure. At a very moderate ocean depth carbonic acid would very naturally exist as a liquid, especially when we consider the low temperature that recent research has shown to be maintained pretty uniformly over the deep sea bottom, namely, one approaching the freezing point of water. At that temperature it would only be necessary for the mineral to be formed at a depth of something like forty times thirty feet, in order to have a pressure which would cause carbonic acid generated in the form of the mineral to assume the liquid instead of the gaseous condition. It may be that these curious instances of these occurrence of liquid carbonic acid in these positions may have been brought about by water pressure, but I would venture to hope as Professor Andrews is here, that he will be able to throw some light on the probability of the introduction of carbonic acid into these crystalline bodies.

Professor ANDREWS : I regret very much that I am unable to throw any light on the particular question, which is a very remarkable one as to how these cavities become filled. It is a problem of extreme difficulty, but one at the same time which we feel is likely hereafter, when solved, to throw some light on the condition of the earth when those liquids were formed. I wish, however, as he has kindly mentioned my name to say that Mr. Hartley has had the goodness to shew me his very interesting experiments. I wish at the same time to direct his attention to a circumstance which I was not aware of till Professor Bosscha, two days ago, drew my attention to it, that there is in this exhibition an apparatus employed by Vogelsang, (an able physicist, since dead,) and Geissler in a remarkable research, which was published in 1869 in Poggendorff's Annalen, on these very fluids to which Mr. Hartley has referred. They examined the nature of the fluid when converted into a gas by spectrum analysis, and also determined the temperature at which it disappeared when the cavity was heated. They found this temperature to be 32°C. Here we have an independent observation fully conceiving what Mr. Hartley said, and fixing nearly the same temperature as that at which these changes occur in these liquids. I thought it was right as Professor Bosscha had drawn my attention to this apparatus being sent here, to refer to it. I cannot avoid saying, in conclusion, that I think Mr. Hartley has a very rich subject before him for investigation.

The PRESIDENT : If there are no further remarks on this paper, I will ask Dr. Gladstone to make a very short communication with reference to some new applications of electrolysis to the decomposition, more particularly of organic compounds, which are of great interest. Dr. Gladstone is kind enough to give us this very shortly, quite at the end of the Conference, so as to put us in possession of the most important points connected with this decomposition.

The Copper Zinc Couple.

Dr. GLADSTONE : I have just brought from the adjoining cabinets this old bottle and tube—a remnant of the first copper zinc couple. It originally consisted of a piece of zinc foil wound round in various coils, with muslin to separate one from the other; that was washed over with a solution of sulphate of copper, so that metallic copper was deposited upon it in a very finely divided condition, the crystals being formed upon its surface and touching it at myriads of points. These rolls of zinc covered with copper were placed in water, and it was found that the water was gradually decomposed, and hydrogen gas was given off. This went on for days, for weeks, and for months, and it was found that the zinc was gradually oxidated, and that less hydrogen appeared, but it varied somewhat according to the state of the weather, coming off more freely on warm days than on cold days. We found we could produce the same result just as well otherwise. This is the next bottle that was put up. Here the zinc foil was crumpled, and it was covered in the same way with copper, but there was no muslin or anything of that kind employed. Gradually this also oxidated. We generally employ the same form now, but sometimes by heating the zinc at a particular temperature we make it all break down in pieces in a sort of powder; at other times we take granulated zinc, and cover it in the same way with copper. There are many other couples of metals which may be similarly employed—aluminium with copper, or with platinum, and so on. This copper zinc couple has lately been employed by Mr. Tribe and myself for the decomposition of many bodies, and also for the synthesis of many others. The way it acts appears to be this :—You are aware that zinc itself will not decompose water. Some metals will, such as potassium and sodium ; and I would direct your attention to a book which is in the adjoining room in which you will be able to read Davy's description of the experiment in which he first obtained potassium by electrolysis. There is his own record of his experiment, and you will also see the apparatus with which he performed it. Zinc, however, acts otherwise. You are aware that if we take zinc and copper and place them in contact in many liquids, pure water for instance, there is a certain amount

L

of action; the galvanometer will show that an electric current is flowing, and if we heat the water we find it flowing still more freely. Also the nearer we bring the copper and zinc plates together the greater is the power. Here in this copper zinc couple we have all those conditions at work. We have chemical action though not strong. Zinc, it is true, will not decompose the water, but still it is acting there exerting its pull on the elements; at the same time we have the voltaic action between the zinc and the copper, and that acting at an insensible distance, at a myriad of different points, so that there is practically no resistance in spite of the bad conductivity of the water. Then we can employ heat if we please, for we find generally these effects much augmented by heating the solution. What the couple has already effected is about this. We can obtain as I have already stated in this way the decomposition of water. That may seem to be a simple thing, but it is perhaps better than it appears at first sight, because we get in that way pure hydrogen. Even though the zinc be impregnated strongly with arsenic, the hydrogen will come off pure. This is a proof of it. We made some zinc about as bad as we could possibly make it, and when we prepared with it a copper zinc couple and put water upon it, the hydrogen that came off was passed through tubes heated in the ordinary way to show if there were any arseniuretted hydrogen, but there was no trace of it. Similar zinc which we treated with sulphuric acid gave an immense amount of arseniuretted hydrogen, which was reduced as you see in this tube ; so that there we have a proof of the purity of the hydrogen given off by means of the couple. Then it was employed by us very early to see if we could not produce in a simpler manner some of those beautiful results with which Professor Frankland had enriched chemistry in producing zinc ethyl and a great many other compounds. We found that instead of taking as Professor Frankland did pieces of zinc, and placing them with iodide of ethyl in closed vessels, and subjecting the whole to a very high degree of temperature, we had only to take a copper zinc couple and to warm that along with iodide of ethyl, and we obtain the same reaction, and we get his zinc ethiodide and other compounds. If we employ alcohol we get other products. If we took instead of iodide of ethyl, iodide of propyl, we obtained zinc propyl, a liquid very similar to Professor Frankland's zinc ethyl, and which is also instan taneously inflammable directly it is poured into the air. In this way a number of new compounds have been produced all of them originally, I think, zinc compounds ; and there are one or two of these upon the table which have been produced by the re-distribution of the elements of organic substances under the action of this chemical affinity aided by galvanic force. There are several other substances, for instance, acetyline from chloroform, iodoform, or bromoform, and several other hydro-carbons, which have been produced in this way more readily than by the ordinary modes given in the books. We may expect, I think, when a good many observers are turning their attention to this, and are making experiments with copper zinc couples—and Mr. Tribe as well as myself hopes that a great many will enter the field and work upon it—that many different substances may be broken up into products of decomposition, and that various changes may take place enriching our knowledge of organic compounds. The copper zinc couple is of service also by its taking to pieces organic compounds in such a quiet way, that we may be enabled to know something more of the rational composition as well as of the structure of these built up bodies.

The PRESIDENT : I have to propose that we give our thanks to Dr. Gladstone for this interesting communication with which he has just favoured us. The shortness of time at his disposal has not allowed him to do justice to the important subject which promises to furnish us with a means of considerable power in synthetical chemistry.

With this communication our chemical conferences come to a close, and I think you will all agree with me that we have not spent our time in vain during the two days we have been discussing the chemical exhibits here. I think many of us will be able to look with much more interest and intelligence upon many of the things, more especially those of historical interest that are exhibited here. This we should not have been able to do had we not been favoured by the many interesting communications that have been brought before us in this room. I have now only to thank you for your attention to these communications that have been given us to day.

Professor ANDREWS : Before the meeting closes, I think it is only due to our chairman that this meeting should record its high sense of the invaluable services he has conferred on this section, by passing a vote of thanks to him. I need not say that the success of such a Conference as we have had to-day is due very largely—mainly indeed—to the ability and urbanity of the chairman, and when you have those qualities combined in so remarkable a manner in one individual, the success of a Conference of this kind we may say is secured. I should further add that it is not merely in presiding here, but in the great labour which Dr. Frankland has had to undergo in the preparations that were necessary, and in inducing scientific men all over the world to send collections here, and to attend these meetings, that Dr. Frankland has conferred a favour upon us. His labour is not to be measured merely by the few hours he has spent in this room, but by the great toil which he has undergone before. I beg, therefore, to propose a cordial vote of thanks to him.

Professor ROSCOE : I have much pleasure in seconding this vote of thanks which has been so ably proposed by Professor Andrews. I agree with every word he says, and I am sure you will all thank Professor Frankland very heartily for his kindness and conduct in the chair.

The PRESIDENT : Ladies and gentlemen, it is very gratifying to have this expression of your appreciation of the services I have been able to render. Had more of my time been at your disposal I should, I trust, have done more to merit those thanks, but I have had very great pleasure in contributing the little quota I have been able to do towards increasing the interest of this exhibition, and I trust that the objects for which it has been founded may in the end be achieved, and that we may possess a national collection both of instruments and of chemicals, which I feel convinced will be of the greatest service to future investigators.

SECTION—BIOLOGY.

President: Professor J. BURDON SANDERSON, M.D., LL.D., F.R.S.
Vice-Presidents :

Dr. G. J. ALLMAN, F.R.S.

M. le Professeur VAN BENEDEN, F.R.S.

Herr Professor COHN.

Herr Professor Dr. DONDERS, F.R.S.

Professor MICHAEL FOSTER, M.A., M.D., F.R.S.

Colonel LANE FOX.

Herr Professor EWALD HERING.

Dr. HOOKER, C.B., P.R.S.

M. le Professeur MAREY.

Professor ROLLESTON, M.D., F.R.S.

PROF. J. BURDON SANDERSON'S OPENING ADDRESS.

IT having been made a part of the duty of the chairman of each of the sections into which this Exhibition is divided to deliver an opening address, I had no difficulty in selecting a subject. I propose to place before you a short and very elementary account, addressed rather to those who are not specially acquainted with biology than to those who are devoted to the science, in which I shall give you a description of a few of the methods which are used in biological investigation, particularly with reference to the measurement and illustration of vital phenomena. You are aware that the committee, in order to render these conferences as useful as possible, have thought it desirable that we should devote our attention chiefly to those subjects of which the instruments in the collection contribute the best examples.

Now these subjects are, first, the methods of registering and measuring the movements of plants and animals ; secondly, the methods of investigating the eye as a physical instrument ; and thirdly, the methods of preparing the tissues of plants and animals for microscopical examination. Of these several subjects it is proposed we should to-day concern ourselves chiefly with the first. I will therefore begin by endeavouring to illustrate to you some of the simplest methods of physiological measurement, particularly with reference to the *time* occupied in the phenomena of life, leaving the description of more complicated apparatus to Professor Donders, who will address you on Monday, and to my friend, Professor Marey, who is with you now, and who will give you an account of some of the beautiful instruments which he has contrived for this purpose.

The study of the life of plants and animals is in a very large measure an affair of measurement. To begin, let me observe that the *scientific* study of nature, as contrasted with that contemplation of natural objects which many people associate with the meaning of the word "naturalist," consists in comparing what is unknown with what is known. Whatever may be the object of our study—whether it be a country, a race, a plant, or an animal, it makes no difference in this respect, that the process in each of these cases is a process of comparison, a process in which we compare the object studied in respect of such of its features as interest us, with some known standard, and the completeness of our knowledge is to be judged of in the first place by the certainty of the standard which we use ; and secondly, the accuracy of the modes of comparison which we employ. Now, when you think of it, comparison with a standard is simply another expression for measurement ; and what I wish to impress is, that in biology, comparison with standards is quite as essential as it is in physics and in chemistry. Those of you who have attended the conferences on those subjects will have seen that a very large proportion of the work of the physical investigator consists in comparison with standards. From his work, our work, however, differs in this respect, that whereas he is very much engaged in establishing his own standards and in establishing the relations between one standard and another, we accept his standards as already established, and are content to use them as our starting-point in the investigation of the phenomena which concern us.

Now I wish to illustrate this by examples. The first objects which strike the eye on entering this collection—the collection in the next room—are certainly the microscopes. But you will say, surely the microscope cannot be regarded as an instrument of measurement. In so far as it is an instrument of research and not merely a pastime, it is emphatically an instrument of measurement, and I will endeavour to illustrate this by referring to one of the commonest objects of microscopic study, namely, the blood of a mammalian animal. Now as regards the blood I will assume that everybody knows that the blood is a fluid mass, in which solid particles float. With reference to the form of those particles, all that we see under the microscope is merely a circular outline. If we wish to find out what form that represents we must use methods which are really methods of measurement. By the successive application of such methods we learn that this apparently circular form really corresponds to a disc of peculiar bi-concave shape. But I will not dwell more upon the application of measurement to the form of the corpuscles, but proceed at once to a subject that can be illustrated by an instrument before you for ascertaining the *number* of the corpuscles. It will be obvious to you—even to those who are not acquainted with physiology and pathology—that the question of the proportion of corpuscles which are contained in the blood must be a matter of very great importance to determine. It has been long known that the colouring matter which is contained in the corpuscles is the most important agent in the most important vital processes of the body, because it is by means of it that oxygen, which is necessary to the life of every tissue is conveyed from the respiratory organs to the tissues. This being the case, it is evidently of very great importance both to he pathologist and to the man who interests himself in investigating the processes of nature, to be able to determine accurately what proportion of corpuscles the blood contains. Well, there are chemical methods of doing this. We can do it by determining how much iron the blood contains, because we know that the proportion of iron in the corpuscles is always nearly the same, and by determining the quantity of iron chemically, we can find out how many corpuscles there are in a certain amount of blood. But this is a long process, requiring first the employment of a considerable quantity of blood, and secondly, difficult chemical manipulations and a long time. Now by a method which has been very recently introduced, we have the means of applying the microscope even to a single drop of blood, to a drop such as one could obtain by pricking one's finger at any moment, or could take, in this way, from any patient in whom it might be desirable to ascertain the condition of the blood as regards the number of its solid particles.

The method consists in this. In order that you may understand it I will ask you to fix your attention upon this cube which I draw on the board. Suppose this cube is not of the size actually represented, but that it is a cube of one millimetre, *i.e.,* the one twenty-fifth part of an inch. How many blood corpuscles do you suppose are contained in a cube of that size? Such a cube we know to contain in normal blood about 5,000,000 corpuscles. Supposing we had a method by which we could count those 5,000,000 particles, it is obvious that the task would be endless, and even if we were to take a cube the 125th part of that size, namely, a cube of one-fifth of a millimetre in measurement the enumeration would be somewhat easier, but still impossible, for the number contained in such a cube would be enormous; and therefore, it is necessary to diminish the bulk of the blood in which you make your counting very much further. This you can only effect by a process of dilution. In order to get at your result you have not only to diminish the bulk of the quantity which you contemplate and in which you count, as much as possible, but also to dilute the blood so that your liquid may contain a very much smaller proportion of blood corpuscles. You dilute it then 250 times, and in this way you divide the cube of a millimetre from which you started, into about 31,000 parts, and count the blood corpuscles in the 31,000th part of a cubic millimetre. Supposing you find it contains about 160 corpuscles you will find by calculation that they amount to about 5,000,000 in the whole cube from which you started. This being the case, the question is how we effect the division. We do it in this way : You first dilute your blood in the exact proportion required, and for this purpose one uses the apparatus which is on the table. You take a capillary pipette which will only take an extremely small quantity, in fact, a cubic millimetre of blood. Then having filled your pipette you discharge it into a little eprouvette, into which has been introduced 250 times, or rather 249 times, the bulk of some liquid with which blood can be diluted without its corpuscles being destroyed. Having thus got this diluted liquid which contains blood in the proportion I have mentioned, all that you have to do is to place under the microscope a layer of a definite thickness—one-fifth of a millimetre— and count the number of corpuscles in a square of the same measurement. That is effected by this very ingenious arrangement, which was introduced by M. Potain, and has been finally perfected by Messrs Hayem and Nachet. The way it is done is this : An object-glass is covered by a perforated plate ; the perforated plate is of the thickness

I mentioned, namely, exactly one-fifth of a millimetre. Consequently if a very small drop of the mixture of the blood with serum (the diluting liquid) is placed within this space, you have a layer of the thickness I have mentioned which you can contemplate. You can cut off a cubic millimetre of that stratum of blood perfectly easily by means of a micrometer eyepiece, and in that way accomplish the required enumeration. You have in short before you a quantity of liquid which contains about the thirty-one thousandth of a cubic millimetre of blood, and consequently would obtain, if the blood were normal, 160 corpuscles. These can be very readily counted, and the whole process can be done in a very few minutes—in a much shorter time, in fact, than I have taken to describe it to you, and you get results which are not only equal to those obtained by chemical investigation, but more accurate. This, I think, is a good example of the application of the microscope as an instrument of measurement to an important question.

The next subject that I wish to draw your attention to is a different one. It is a question of measuring the time occupied in certain simple processes in which the nervous system is concerned. The examples I am going to give you are entirely derived from the physiology of man, and relate to the phenomena which we observe in ourselves. The measurement to which I wish to draw your attention is the measurement of the time occupied in what we call in physiology a "reflex" process. You may reasonably ask that I should endeavour to explain what a reflex process is, and the only way, or at any rate the readiest way in which I can do this is by giving you an example. Supposing this blank card, which has written on it previously some word, say the word "reflex," were suddenly turned over by a second person. It is agreed that at the moment I see the word upon it, I say the word "reflex." In that act it is obvious that there are three stages. First, the reception of the impression by my eye produced by seeing the word; secondly, the process which goes on in my brain in consequence of seeing it; and thirdly, a message sent out from my brain to the muscles which are concerned in articulation, by means of which certain movements are produced which give rise to the sound which you recognise as the word "reflex." That is one example. Let us now take another which is simpler. We cannot take one better than the act of sneezing. Some snuff finds its way into the nose ; an impression is received, a change is produced in one's nervous centres, and in consequence of that central change, a certain number of muscles are thrown into the action recognised as sneezing. These are different examples of reflex action. The brain, the highest part of the nervous system, has to do with the first ; whilst the other is one in which the nervous centres lower down have to do, and consequently it is simpler. The methods which I am going to illustrate to you are methods intended for the measurement of the time occupied in this process. First, let me draw your attention to the circumstance that you have here three stages. You have the stage of reception ; the stage corresponding to the changes which take place in the brain in consequence of the reception of an impression from outside ; and thirdly, the process by which you convey the effect to the muscles which act. Now let us agree, in speaking of this, to call the impression the " signal," and to call the muscular effect the " event." In that case the question before us is to measure how much time takes place between the reception of this signal by a certain person and the occurrence of the event, namely, the completion of the muscular action. There are a great many questions involved in this : thus you may measure either the whole process or one of its stages. You may measure, for example, either the time occupied by the reception, the time occupied by the discharge, or, on the other hand, the time which is occupied by the changes which take place in the centre itself. In the first instance I gave you just now—the example of reading a word aloud—the time occupied in the reception is extremely short, and the time occupied in the discharge is also extremely short. Popularly the whole thing is done as quick as thought, but, comparatively, the time the brain takes in going through these changes which connect the reception of the impression with the discharge is a very considerable one. All this we can make out with absolute accuracy by methods of measurement. Most of these methods are founded on this principle, that we measure the duration of a voltaic current which is closed at the moment the signal is given, and opened or broken at the moment that the act takes place. There are a great many instruments constructed on this principle, of which you will find illustrations in the next room. The general principle involved in all of them is shown on this diagram. In the simplest form you can give to such an apparatus you must have a surface of paper so placed that it shall pass horizontally by the point of the lever, and at an uniform rate ; thus, for example, it may pass at the rate of one metre in the second.   Supposing this to be the case, it is obvious that if you arrange the electro-magnet so that when you close the current a certain mark is made, and that at the moment of the break of the current when the magnet ceases to act, another mark is made, you will have a tracing on the surface of the paper which indicates the time.   So long as nothing is going on, the paper receives a horizontal mark, but at the moment the signal is given you have the point of the lever descending.   At the moment the act takes place the lever assumes its original situation, and you have again a horizontal line.   That is the general principle of the apparatus.   Now for its action.   We have here a voltaic circuit and a key by which we can give the signal.   I shall be the subject of the experiment, and you will see what the result is.   Here is the recording arrangement.   We have two electrical keys, one at the further end intended for making what is called the signal, and one here for breaking, which is placed close to the person who is to be experimented upon.   Mr. Page, at any moment he likes, will act upon me by sending an induction flash through my tongue.   I shall arrange the electrodes so that they shall be against the tip of my tongue, and at the moment I feel that flash I shall place my finger on the key.   Then the clockwork being in motion at the same time, we shall see by the length of the depression in the tracing the duration of the process.   If we take different sorts of signals, or if the person to be experimented upon is in different conditions, the time will be very different.   Thus we may compare the result which will be produced when I am attending and expecting the signal with the result which will be produced when I am not attending or expecting the signal ; or, on the other hand, I may compare those results with that which will be produced when I am expecting it, but Mr. Page, instead of giving it at the time I expect it gives it me at a different time ; in that case the time occupied would be longer than either of the other two cases.   A great variety of different cases can be investigated in this way in which we measure the total period occupied in the reflex. The arrangement is perfectly simple.   You see when Mr. Page presses on his key, which is the signal key, that a lever is set in vibration and makes a tracing, and at the same moment the voltaic current is made and the coil is acted upon inductively ; the result is that an induction flash passes through my tongue which I feel, and the moment I feel it I break the current.   Consequently the time between the moment at which Mr. Page makes the current by closing his key and the moment at which I break the current by placing my finger on my key, gives us precisely the time which is occupied by the reflex process.   We will make two experiments, first, with the signal expected, and then unexpected ; that is, in the one case I shall be on the *qui vive*, and on the other I shall not be so.   (The experiments were made accordingly.) We shall now repeat the process, so that instead of my receiving the information of the making of the current by means of the excitation of my tongue, the signal shall consist in my hearing the sound of an electrical bell.   In that case we shall find that, although the signa will come in exactly the same way, practically the time occupied will be very considerably longer, showing that a signal received by sound akes longer in producing its effect than one in which the signals is felt by the tongue.

In order to make all this perfectly plain I shall hand round this tracing.   You will see there several experiments made with expected and unexpected signals, which show the different results obtained in the two cases.

The next question which arises, and with that I must conclude what I have to say just now, is this :—You will readily see that the exact measurement of time depends upon the rate at which this clockwork happens to be going.   I happen to know that it makes twenty revolutions per second.   But suppose I do not know that.   In fact one would not trust to the accuracy of clockwork for such a purpose.   How should I then be able to measure the duration of time so exceedingly short as the one which now concerns us ?   In order to do this we always come back to a physical standard, to a standard of absolute invariability which we can depend upon as being true.   For this purpose we use a tuning-fork which produces vibrations, the rate of which we know, because we know the tone which the tuning-fork produces, and the arrangement which is always used for this purpose is the one shown here.   We have turned off the voltaic current we used for signalling, and turned it on the tuning-fork.   There are two electro-magnets on either side of the tuning-fork which react upon it, so that the moment you close the current the fork is thrown into vibration and produces its own characteristic note. All that we have to do is, during the time we are making our record, to bring this tuning-fork, which is now in vibration, into such a position that this little brass pointer shall make a tracing against the paper. If you look at the tracing I have sent round you will find there are tracings on it of a fork, which vibrates at the rate of 100 per second, consequently you have nothing to do but to translate the tracings which you have made and which correspond to the duration of the mental process which you have been investigating, into vibrations of the tuning-fork, and you get an exact measurement of the total duration of the process. While I have been doing this you hear the tuning-fork is in vibration, and Mr. Page has made the tracings. After it is varnished it will be sent round and you will see the tracing made by the fork over the traces corresponding to the different experiments we made just now.

I may observe that although the experiments made on that paper were made with myself, you find that the period occupied by the reflex is considerably longer than in the other which I sent round previously. But that one may very easily explain from the abnormal conditions under which the experiment has been made as regards myself.

I intended to go on from this subject to another mode of investigation, namely, to the very beautiful instruments which have been lately introduced for the purpose of measuring the finest differences of bulk in different organs, as for example, in the human arm, by which you can ascertain the condition of the circulation precisely by a very exact registering-measurement of the bulk of the arm ; but as there are several other gentlemen now ready to address you, I will defer that till this afternoon. I will now conclude what I have to say by asking you to listen to Dr. Hooker.

ON THE PLANS OF THE NEW LABORATORY FOR INVESTIGATIONS IN VEGETABLE PHYSIOLOGY AT KEW.

Dr. HOOKER : The subject which I have to bring before you this forenoon, and which will not detain you long, is the construction and equipment of a botanical laboratory for physiological purposes in the

Royal Gardens at Kew—a laboratory in which researches will be conducted into the phenomona that accompany the life-history of plants, during all stages of their growth, from their seedling to their seed-bearing state, and then to their decay and death—in life and in health.

Until very recently, vague answers were all that could be given to questions put to botanists with regard to the principle functions of plant-life, especially those which related to growth. There were many reasons for this, but one of the most effective was that it has only been quite recently that those physical phenomena on which the growth of plants depends have been themselves investigated. The phenomena of chemistry in particular, are of primary importance, because except we know the constitution of the air, the water, and the inorganic matters that plants absorb, we can never know the nature of their food or of the changes that take place when they pass into the substance of plants ; and so it is with heat, light, electricity, with attraction, repulsion, gravity, with endosmose, with capillary action—these are all physical phenomena explored in very recent years, and conversance with which is necessary before any progress can be made in physiological botany proper, especially in respect of those prcoesses that are concerned in the growth of the plant.

Another very important desideratum was delicate instruments. These were not invented, nor were the methods of constructing them known, nor, had they been, were artizans capable of turning out such instruments as are now required for conducting physiological experiments, whether in the animal or vegetable kingdom. You have had this morning from our chairman an illustration of how delicate and how beautiful those experiments are, and what extreme skill in manipulation is required in making instruments by which they shall be conducted ; and you have no better example of them than what this Exhibition shows. The marvellous skill of modern artizans is something that twenty-five or thirty years ago, when physiological botany was in its infancy, that we had no conception would ever be ealised.

I will now point out to you the nature of this laboratory, and say that it originated in the recommendation of the Royal Commission on Science, and that it is the first that will have been constructed in this country. There have been several constructed abroad, and it is not derogatory to the science of England to say that although physiological botany has from many points of view made less progress in England than upon the Continent, still botany has, on the whole, perhaps made greater advances in England than in any other country, and as I shall be able to show you, the great fundamental doctrines of physiology have all had their origin in this country. We may reasonably hope that in future the progress we shall make in those studies will be greater now that those opportunities will be given to students, which the construction and equipment of the laboratory at Kew will afford. The Kew Laboratory is now in process of building, and will in a few months be completed. It consists of a single separate building about forty feet by thirty. It has six principle apartments upon one floor below, and a small store-room for the storing of implements, and acids, and matters of that kind, and a little plant-house which will be heated in two compartments, in which plants can be put during the progress of the experiments. There will be a private sitting-room containing what books are required, and accommodation for instruments. There will be a north-room for microscopic observations, which will accommodate four persons comfortably, twenty feet by ten feet ; a chemical-room for chemical apparatuses, which will be lighted in the whole of the front, so as to give a very full amount of light ; a room for spectroscopic apparatus ; and behind, one for gas analysis. The spectroscopic-room will have a bay window, so that as much direct light can be got from the three divisions as possible, and the observer will be able, by the aid of a heliostat, to throw a beam of light into the gas analysis room through a slit in the wall. These are the principal features of the building ; it will accommodate a considerable number of students, and is placed in a part of the garden which is out of the way of the public, and where it is in close connection with the pits for the propagation of plants, and where young plants are reared. I need hardly say that at an institution like Kew, which is so very much used by the public, and where we have very frequently from 20,000 to 30,000 visitors in a single day, it is most important that the naturalists' work-room should be removed from the noise of these.

I have mentioned to you that most of the principal—indeed the fundamental researches connected with the phenomena of the growth of plants have originated in this country. I hold in my hand a brief statement which I will read to you of what has hitherto been done :— English discoveries in physiology proper, apart from histology or morphology, commenced in the seventeenth century, when a group of Englishmen discovered the sexuality of flowering plants. Nehemiah Grew, Secretary to the Royal Society, made out the anatomy of the stamens. Sir Thomas Millington, Savilian Professor at Oxford, suggested the function of pollen in fertilising the ovary. Robart, Curator of the Oxford Botanic Garden, verified it experimentally. Hales, an English Clergyman in the beginning of the eighteenth century, made classical researches still of great value on the movement of water in plants. He also first suggested that the function of leaves was to obtain some kind of nutrient matter from the air. Priestly found that carbonic acid was removed from the air by plants, and oxygen exhaled. Ingenhousz, a foreign physician, in 1779, published in England an investigation in which he showed that plants give off carbonic acid in darkness, and that the sun's light rather than its heat is effective in causing the evolution of oxygen. Knight published at the commencement of the present century a series of papers, one of the most important of which was the experiment proving that the downward direction of roots was due to gravity. Daubeney first experimented on the influence of differently coloured light on vegetative processes, and proved that the red end of the spectrum is most effective. Lawes, Gilbert, and Pugh, gentlemen whose names are known to you all, were the first who showed that uncombined nitrogen is not absorbed by plants. When we come down to the present day, there are many names I should further mention. I need hardly allude to Robert Brown, the greatest botanist perhaps who ever lived, whose discoveries—the fertilisation of the ovule—are classical, nor to Mr. Darwin, who is amongst us still, and whose observations are perhaps of more acuteness and importance than those of any philosopher who has ever made investigations on the physiology of plants. His works on " Climbing Plants," on " The Absorption of Nutritive Matter," and on " The Fertilisation of Orchids," are masterpieces of physiological research that have never been rivalled.

I have only a few words more to say, and that is to inform you that we are indebted to a private individual for this laboratory; my friend Mr. Thomas Phillips Jodrell, who endowed the chair of physiology in University College, so ably held by your president of to-day, is the one to whom the government and the public are indebted for both the building and equipment of this laboratory.

The PRESIDENT : I will now call on Professor Thiselton Dyer.

ON VARIOUS APPARATUS FOR INVESTIGATIONS IN VEGETABLE PHYSIOLOGY.

PROFESSOR THISELTON DYER, M.A., B.Sc. : I have been asked to explain to you this morning, in a few brief words, several pieces of apparatus which have been contributed by some distinguished botanists on the Continent, who are unhappily not able to be present and to give you the explanations, which would have come with much more force from themselves. I shall call your attention first to an apparatus in two parts on the table in front of me, the purpose of which is to register the growth of plants. Several pieces of apparatus have been devised and are in use in different laboratories for this purpose. The general principle of them however is essentially the same. The plant, the growth in length of which is to be studied, is placed under a stand of this kind, and a pin is passed through the apex of the stem ; then a thread is fastened to the pin and is carried over a small pulley, while to the other end a small weight is attached. You will easily understand that as the plant grows, the pin ascends and the weight decends. The result of that is that the pulley is turned round, and you therefore convert the upward growth of the plant into the revolving motion of the wheel of the pulley. Of course the growth is extremely small. It has often been said that under some conditions we can see how grass grows ; but as a matter of physiological nvestigation it is by no means easy to do so, and the great object is to magnify the rate of growth, so as to make it obvious and convenient for examining small alterations in its rate. One very simple plan of effecting this purpose is to attach to the axle of the pulley a very light and long index, such as a piece of very light wood, which sweeps through a large arc of a circle as the pulley turns round. The

M

simplest kind of instrument of this sort, which was at first used, was simply provided with a graduated arc, upon which the index moved. However, it was found that in order to get any interesting results from this class of investigation, it is necessary to make the observations over a very considerable period of time. The whole day of twenty-four hours is the least in which you can get any result which can be examined. It therefore became very desirable to have some method by which the plant would authentically record its own growth, and any variation of the rate of growth at any part of the day. Professor Sachs, of the University of Wurzberg, invented an extremely ingenious arrangement for this purpose. He employed a revolving drum, such as has been already used this morning, which revolved by clock-work once in an hour. The index of course in any one position will describe a trace on the drum. Professor Sachs used a drum, of which the spindle was not absolutely driven through the centre, but was eccentric, so that the surface only came into contact with the index during a part of the revolution, and in that way he got upon a piece of blackened paper, fastened upon the drum, a number of inclined lines, each of which was described by the index, and the inclination of which was due to the combination of the vertical movement of the index with the horizontal rotation of the drum. The curious result that came about was this, that when the traces that were made in the space of twenty-four hours were examined, while, if the rate of growth had been perfectly uniform, the space between each of these lines would have been the same, because the index would have moved through the equal spaces during equal times, it was found that an altogether unexpected cause interfered with the regularity, and that this cause was the variation in the amount of light. It has been found in point of fact that if the temperature remain constant the greater part of the growth of plants takes place during the night, and that as the day dawns, and the light intensifies, the rate of growth materially slackens, until towards the afternoon there is an actual minimum of growth. In the same way as daylight disappears and night comes on, the rate of growth increases, and towards early morning there is a maximum. That light should have this retarding effect upon the growth of vegetable tissues is one of the most remarkable results which has been made out in late years in vegetable physiology, and it is connected with a good many other points, on which time will not allow me to expatiate. One of the most curious is the explanation it gives of what is called heliotropism. Anyone knows that when plants are cultivated in a room, or are exposed to light from one side only, their stems and foliage incline towards the light. The explanation of that curious fact is not at first sight obvious, but it arises from this, that light arrests the growth of the tissues, and the consequence is that if a stem is evenly illuminated on all sides it grows upright ; if however it is only illuminated on one side the growth on that side is proportionally checked compared with the other, and the stem inclines over.

The arrangement of the present piece of apparatus, made by Stöhrer (of Leipzig), is somewhat complicated. The wheel of the pulley is provided with little teeth, each of which as it turns causes a small lever, to which a delicate spring is attached, to rise and fall. This spring makes and breaks electrical contact, which brings into play alternately one or the other of two little electric magnets, and the consequence is that a pencil which marks on a revolving disc is sometimes attracted to one and sometimes to the other, and you get marks on the disc corresponding to the teeth of the wheel. The disc is driven round by clock-work once in twenty-four hours, and the margin is divided into intervals of quarters of an hour. You can see by setting the machine to work how many teeth of the pulley have passed the lever at any portion of the day or night, during which the experiment has been going on.

Professor Cohn has sent us the apparatus he uses in his laboratory for demonstrating the very celebrated experiments to which the President of the Royal Society has already referred as those of the classical investigation of Thomas Andrew Knight. Everyone knows that when a seed is sown in the ground the root penetrates the soil and the stem ascends, but the causes of this divergency of direction are far from being properly understood. Mr. Knight advanced our knowledge to the point where it stands at present, by showing that the force which produced the downward tendency of the roots and the upward ascent of the stem was the action of gravity. The way in which he ascertained that was by growing the seed under conditions in which the force of gravity was eliminated. The seeds are transfixed by pins, the pins are stuck into the circumference of a little wheel, and a cover is put over the box in which it rotates which keeps the whole arrangement in a uniform state of moisture and excludes the light. At the top of the cover is a small pipe, to which a water pipe is attached, through which flows a stream of water which plays on a little water-wheel; this causes the wheel rapidly to revolve, and the spray keeps the seeds moist. The revolution of the seeds round the wheel eliminates the action of gravity, because the seeds are being continually reversed in position, and any tendency they might have to obey the action of gravity in one direction is promptly exercised in the other, so that they are practically removed from the action of gravity altogether. Under those conditions they are submitted to the action of a centrifugal force alone, and the consequence is that the roots are found to point outwards in the direction of the rays of the wheel, and the stems to point towards the axle. The experiment has been modified in various other forms so as to get various other results, all of which are of considerable interest.

Professor Cohn has also sent this large apparatus which perhaps does not require a lengthy explanation, though it is of considerable interest to investigators, because it has been employed by him in some very admirable work which he has accomplished on the natural history of the *Bacteria*. It is a kind of incubator containing a number of trays, in which the seeds which are to be germinated, or the matters which are to be acted upon by a constant temperature and humidity are placed; there is a glass cover which fits on the whole. The apparatus is warmed by a small gas jet placed underneath, and there is a gas regulator which, I am reminded by Mr. Page, is perhaps not of the best form, because it depends in part for its construction on the presence of a body of air, which of course, if the experiment is to be carried on for a considerable time, varies in volume with the atmospheric pressure, and therefore the temperature is not perhaps absolutely constant.

There are finally two large glass bottles, to which I will direct your attention, used by Professor Sachs of Wurzberg, in the important investigations as to the parts of the spectrum to which different kinds of physiological work in plants are to be attributed. The first experiments of this kind were made by Daubeney at the Botanical Gardens in Oxford. He used various coloured media and exposed the plants to the light which passed through them, and he found that the greater part of the vegetable processes which are characteristic of the life of plants, did not require the whole of sun-light for their efficient performance. In point of fact, the light belonging to the red end of the spectrum, including the red and yellow rays, does nearly the whole work, and the blue light is comparatively inactive as regards vegetative work. The interest of these two vessels is that they divide between them nearly the whole light of the spectrum. They are simply two large hollow bottles with double walls. A plant can be placed inside and the interval between the walls is filled in one case with a solution of bi-chromate of potash, and in the other with the well-known dark blue solution of ammoniacal oxide of copper. It is a remarkable thing, that it has been ascertained that the evolution of oxygen by plants will go on with perfect regularity under the red rays from which all the blue light has been sifted out ; on the other hand, the blue light is that which is effective in producing the very curious arrest of growth which I described just now. So that if you place a plant in the red vessel while it will perform all its normal processes of giving off oxygen and decomposing carbonic acid it is not heliotropic. If it is illuminated on one side with yellow light, it does not move towards it, whereas if it is placed under the blue bottle, it ceases to give off oxygen, but is strongly heliotropic. So that there is this curious partition in function of the different parts which compose sun-light, that one seems to arrest growth, and the other to stimulate the process of gaseous decomposition upon which the life of the plant depends.

APPARATUS FOR REGISTERING ANIMAL MOVEMENTS.

Prof. MAREY : The registering apparatus which have enabled us to carry so far the investigation of the functions of living animals are applicable to the analysis of movements of every kind in health and in disease. It is to this important application that I desire to draw your attention at the present time.

Most of the movements whose various phases we have to estimate must be transmitted to a distance, preserving at the same time all their characteristics. It is by the medium of the air that this transmission is effected, and its principle is as follows:—

Upon the organ (muscle, artery, heart) whose movements are to be investigated an apparatus called the exploriug drum is applied. It is a small metal basin closed by a caoutchouc membrane, and communicating by a longer or shorter tube with a similar drum, upon the membrane of which is supported a recording lever. The pen with which the extremity of this lever is provided inscribes the curve of the movement impressed on the membrane of the first drum on a cylinder covered with smoked paper and turning on a horizontal axis.

1. Let us at once apply the process of analysis to the muscular movements of man. For this purpose we may either grasp the muscles of the ball of the thumb between the flattened jaws of the pincers which I show you, or apply to the fleshy substance of any muscle an exploring drum, the knob of which rests upon the muscle. When by means of electricity we cause a contraction or tetanus of the muscle to be studied, the curve of the contraction or that of the tetanus is recorded at a distance upon the revolving cylinder.

This apparatus shows the thickening which a muscle undergoes during contraction, and it furnishes results identical to those obtained by investigating the shortening of the muscle during contraction in living animals. We are then quite authorised to interpret in the same way the curves obtained in both cases.

It is needless to insist on the numerous services which the myography of man may render to physiology and medicine. The study of the forms of movement, of the latent period, of muscle, and perhaps even the rate of transmission of impulses along motor nerves, by means of this apparatus may be as easily pursued in healthy or unhealthy men as in animals.

2. Without quitting the investigation of muscular movements, let us examine that of the respiratory movements, and we shall obtain valuable information as to the means by which the important function of respiration is effected. We apply to the chest this apparatus formed of an elastic plate and furnished with two lever arms, to the extremity of which is attached a band which surrounds the thorax. Each dilatation of the chest causes the spring to bend, and it resumes its position during respiration. This double movement is accompanied by a

rising and falling of the membrane of the drum which forms part of the apparatus, and which therefore becomes a regular bellows, causing the elevation and depression of the inscribing lever placed beside the cylinder.

The respiratory curves thus obtained present certain normal characteristics susceptible of being greatly modified when any obstruction interferes with the respiratory functions either by impeding the entrance or the exit of the air, or even by opposing its passage in both directions. In all these cases the curves have a special physiognomy, and their simple inspection enables us to recognise the seat of the obstacle to respiration. Clinical research will yet discover here many points for investigation.

3. But above all there are the phenomena of circulation, which have been minutely investigated both in man and animals. The apparatus by means of which we can completely analyse the movement of the heart, the arterial pulse, &c., have already rendered great service ; we are, however, right in expecting yet more from it, by making use of it in clinical investigations.

Of various cardiographs, that on which I wish to dwell differs little from the explorer of which I have already spoken. The knob with which it is provided is applied to the region of the apex of the heart, and each beat of the organ is transmitted to the recording lever. There is seen in this pulsation of the heart the same elements which the physiological cardiograph has revealed in the higher animals. This beating of the heart is then a complex act, and the numerous details which have been discovered by graphic analysis have each considerable importance from the point of view of functional investigation. One part of the tracing shows us how the ventricle is emptied into the artery ; another enables us to appreciate the play of the auricle, the beating of the sigmoid valves, &c. You will easily see that the precise diagnosis of affections of the heart, already carried so far, thanks to auscultation, will be greatly improved by the application to man of the cardiograph applied to the study of the pulsation of the heart.

The arterial pulse cannot be separated from the pulsation of the heart in the study of the phenomena of circulation in man. Already numerous researches have been undertaken by means of the direct sphygmograph ; but much more may be expected from the use of the air sphygmograph (*sphygmographe a transmission*).

I place this apparatus upon my wrist, and the artery raising a spring connected with the membrane of the exploring drum, transmits its movement to a distance by means of the tube filled with air, which enables this sphygmograph to communicate with the drum to which the recording lever is attached. By recording simultaneously the traces of the pulse and those of the heart much information may be obtained and many errors avoided.

4. I shall present to you, in conclusion, a new method of investigating the peripheral circulation. This method is based on the principle that the variation of the calibre of the blood-vessels in any part of the body is faithfully indicated by the variations of the volume of that part. Without dwelling on the history of these investigations, I may tell you that they originated many years ago, Dr. Piégu, of Paris, having pointed out in 1846 the alternate expansion and contraction of the tissues in connection with the dilatation and contraction of the blood-vessels. Since that time Chelius and Fick in Germany, Mosso in Italy, Franck at Paris, have carried on and extended these researches.

The recording of the movements of a column of water inclosed in a tube communicating with a receiver filled with water and into which the hand and forearm is plunged, was first effected by Fick by means of a float armed with a pen. Ch. Buisson hit on the happy idea of transmitting to a distance, by means of tubes filled with air, the oscillation of the column of water, and it is with his apparatus that M. Franck, in my laboratory, has executed a series of researches. You see the apparatus in action. The hand is plunged into this jar filled with water and hermetically closed. A vertical tube, furnished with a bulb to avoid the effects of the speed acquired by the liquid, serves to transmit to a recording lever the oscillation of the column of water. You will remark that these oscillations are rhythmical with the heart, and if we record them by the side of the cardiac pulse registered by the transmitting sphygmograph, we can establish the identity of the variations in size or, as we may term them, the pulsations of the hand and of the pulsations of a single artery. With this apparatus we may perform numerous experiments on the mechanical effects of compression of the arteries or veins, the action of the vaso-motor system of nerves, direct or inflex, &c.

I shall not explain to you by the side of this method of investigation, that which we owe to Mosso, of Turin. His plethysmograph, which ought soon to be presented to you, permits the estimation of changes of volume of the hand, and, assuredly, the combination of these two processes ought to lead to important results in the investigation of the phenomena of peripheral circulation.

I have sought to submit to you some of the points more immediately applicable to man, without dwelling on the investigation of the movements among animals. But these two orders of researches complement each other. We may say that most of the data furnished by experimentation on animals are now susceptible of rigorous verification on man, healthy or unhealthy. This verification we owe to investigation by means of precise apparatus and to the recording of the smallest movements, thanks to the registering instruments, the principal specimens of which are shown in this collection.

The PRESIDENT: Ladies and gentlemen, I do not know whether it is usual in these conferences to propose any direct expression of thanks to those who address you, but I am sure on the present occasion you will not think I am doing wrong in asking you to allow me to convey your special thanks to Mr. Marey for the very valuable exposition he has given us of these instruments.

(The Conference then adjourned until two o'clock).

The PRESIDENT : The first communication this afternoon will be a very short one from Mr. Schäfer, in which he will bring before you another kind of apparatus for recording and measuring certain variations in the internal pressure in the heart of a frog after it is removed from the animal.

ON SOME RECENT IMPROVEMENTS IN RECORDING APPARATUS.

Mr. E. A. SCHÄFER : We have here on the table two or three kinds of recording apparatus. The one in front of me, which looks rather a large instrument is what is called an "endless paper apparatus ;" by using which one can go on making an observation for a very long time without altering the apparatus at all. On one portion there is a

roll of paper rolled by machinery, some hundreds of yards long.
The whole object of the machine is to unroll this paper, moving
it along at a regular rate, and passing it over these brass cylinders,
and then as the paper is passing, any movement in the animal body
which we wish to record is marked on the paper as it passes along.
You must have noticed in the shops of mathematical instrument
makers, clocks provided with a rotating drum, on which a piece of
paper is carried round very slowly according to the movement of
certain wheels of the clock, and you have also noticed, marking upon
such a drum, a pencil which moves up and down, and this pencil
is connected perhaps with a barometer. The apparatus here is
founded on precisely the same principles. It may be moved either
by a gas engine, a steam engine, or by clockwork, and the rate
is capable of being varied by a screw underneath. A curve may be
traced on the moving paper by means of a bent U tube or manometer,
filled with mercury to a certain height, and open at each end. At the
top of the mercury in one of the legs of the tube, lies a float, which is
connected to a little pen and follows the movements of the mercury.
The pen is wetted with ink so as to write on the white paper and
trace the movement upon it. If we wish to take the pulse in the
arteries we have only to connect the other end of the tube by means
of an india-rubber pipe with an artery, and then any movement in
the artery will cause the mercury to rise and fall. In this way many
of the principal discoveries, with regard to the phenomena exhibited in
the circulation of the blood, have been made. This apparatus we
owe to Professor Ludwig, and it is the ordinary apparatus used in
physiological laboratories for obtaining a trace of the movement of the
blood and the variations in the pressure which is found within the
system of blood-vessels.

The recording apparatus on the right is the most recent one which
has been made, and embodies many improvements. In the first place
it works by clockwork ; it can be set at a certain rate, and it will
maintain that rate throughout ; the rate can be varied considerably,
so that the paper will travel along at the rate of three inches per
minute, or as quickly as six feet per minute. It has been devised by
Mr. Dew-Smith.

Lastly, as a form of recording apparatus much in use, I would draw your attention to this one. It is beautifully made, and will go with a very uniform velocity for a considerable time, but the rate can also be varied by turning one or two screws, and if necessary by shifting the position of a wheel in the clockwork. By setting the clock in motion, the vertical drum, which is now covered with blackened paper, revolves, and the pen, connected as before with a little float which is being moved up and down by the mercury in a U-tube, instead of making a mark straight up and down as at present makes a curve recording the alterations of level of the mercury. The apparatus is at present arranged to shew the movements of the heart. The heart may be obtained for this purpose from either a frog or a toad. This one was prepared about an hour ago. The animal was killed, and then the heart was removed and was fastened to this Y shaped tube at the bottom of the apparatus. The interior of the heart is in communication with the apparatus, which is filled with a little serum of blood obtained this morning from the butchers. Serum acts in the case of the heart of a frog or toad very much the same as the blood itself would, but is cleaner to work with. The serum is pumped by the heart itself through the apparatus, if necessary round and round again, so that a heart which is entirely removed from the body, over which there is no control by the will of the animal itself, which has been long since dead, this heart can be kept living and moving as you see it now, simply by being supplied with fresh blood, or, as in this case with fresh serum. The arrangement of the Y tube is this : the heart being attached to the bottom of the leg of the Y, the serum comes in by one fork, passes into the heart, and goes out by the other. The advantage of separating a heart in this way and causing it to make a tracing on the cylinder is this : that with a separated heart we can try what is the effect of drugs. We may perhaps discover a new drug, and we may wish to know whether it has any action on the body. It is easy to take a little of the drug oneself ; if it is poisonous one need not take enough to kill oneself ; but we may take a little and observe the effects. We may find that it has an action in slowing the pulse, the pulse being an indication of the number of the beats of the heart, and we may find that this particular drug has that action. But this action it may exert in various modes, directly on the substance of the heart or upon the nervous system. It may exert an action on the brain or upon another part of the nervous system, and through this upon the heart. But if we have removed the heart entirely from the body and placed it in an apparatus by itself, and then by mixing a little of the drug with the serum we employ, and allowing it to flow through the heart, we get a distinct action in the heart, of course it cannot be through the medium of the brain or the central nervous system, but must be a direct action on something contained in the heart itself; and as we know pretty well what the structure of a frog's heart is, we have very much simplified the problem in ascertaining what is the action of the drug on the system. This heart I have no doubt if kept supplied with fresh serum would go on beating many hours, and after the sitting any one who cares to do so is at liberty to come near and watch the working of the apparatus.

The PRESIDENT : I will now call on Dr. Gilbert for a communication—

ON SOME POINTS IN CONNECTION WTH THE NUTRITION OF ANIMALS.

Dr. GILBERT : A few days ago Professor M. Foster wrote to me to say that he intended to bring the subject of nutrition forward on this occasion, and asked me if I would take part in the discussion afterwards ; and as he and I had had a good deal of correspondence and conversation some little time ago about the important question of the sources of the fat of the animal body, I coucluded it was probably to that subject he wished me to devote my attention. At any rate, I looked up hurriedly the materials which Mr. Lawes and myself have collected in relation to that subject, and some allied points, and propose, with your permission, to lay the facts before you shortly, although Professor Foster has not given you his paper.

Thirty-five years ago, or more, I believe the view generally accepted was, that the carnivora found the fat which existed in their bodies ready-formed in the herbivorous animals they consumed, and that the herbivora in their turn found all the fat of their bodies ready stored up in the plants they consumed. About that time Liebig, in reviewing the composition of vegetable food, came to the conclusion that this was simply impossible, taking into consideration the amount of fat which was stored up by many animals in proportion to the known quantities in the food. He put forward the view that the carbohydrates of the food—starch, sugar, and so on—were important sources of the fat of the herbivora. For a short time this view was opposed, but only for a short time, by Dumas and Boussingault, and some other experimenters in France, though they afterwards accepted it.

The investigations of Mr. Lawes and myself, it must be borne in mind, have always had an agricultural object, so that if they were not conducted exactly in the way which the physiologist will say they might have been, it has been because we had not the same object before us, that is a purely physiological one. Very soon our own experiments led us to believe that Liebig was right in his conclusion on this point, but that he must be wrong on some other points in relation to the feeding of animals which he so ably discussed. We found it was pretty certain, from the consideration of the feeding experiments, that the fat must have the source which he assumed. On the other hand, he assumed that the value of food to the animal was measured by the amount of nitrogen which it contained ; that is to say, he maintained that in the formation of meat, in the formation of milk, and in the exercise of force, the measure of the value of the food required, for these purposes, was the amount of nitrogen it contained ; and in the case of the exercise of force, the amount of urea which was eliminated. We found, however, that we could give twice or three times the quantity of nitrogen within a given time to one animal as to another, both at rest, and that the amount of nitrogen eliminated in urea was almost proportional to the amount of nitrogen in the food, and had no direct connection with the amount of force exercised.

The question of the constituents in the food, which were of the most importance for the exercise of force, and for the making of fat, remained in this condition until the experiments instituted in Munich, about 16 or 17 years ago, with Pettenkofer's beautifully contrived respiration apparatus, a model and drawings of modifications of which are in the next room. I am glad that after very much trouble on my part to get such an apparatus brought to this Exhibition, and entirely failing, it has after all been sent by some one. It consists of a chamber in which an animal can be put, and by a water wheel, or by some other power, the air is gently aspirated through the apparatus, then it

passes through gauges, and through solutions, which absorb the carbonic acid, &c., and so the amount of air passing is gauged, and the
products of respiration are determined. It is not the apparatus itself,
but the results which it has brought out, which I wish to refer to
on this occasion. In 1860, Bischoff and Voit published their first
results. They kept a dog for many months without change as to
movement, without giving it any special exercise, but varied its food
immensely, and they found the urea eliminated was almost in proportion to the amount of nitrogen taken in the food. But inasmuch as
the then existing view required this to be connected in some way
with the exercise of force, they explained that so much more force
was exercised in the actions within the body in dealing with the increased amount of nitrogenous substance consumed ; so that after all
the amount of the urea eliminated was a measure of the exercise of
force, although it was in these internal actions, and not in the
voluntary exercise of muscular power. I happened to be in Germany
at the time that book first came out, and went to Munich to see
these gentlemen on the subject. I ventured to call in question the
conclusions at which they had arrived, and I think I was pitied for
my ignorance. But a few years afterwards it was found by others
also that the amount of urea eliminated had no direct connection with
the amount of force exercised, and that what is the most pronounced
when there is an increased exercise of force, is an increased elimination of carbonic acid by the lungs. I believe there is now no doubt
about that matter. Messrs. Fick and Wislicenus, Dr. Frankland, and
Dr. E. Smith, brought that prominently forward, and I believe it is
now accepted that the elimination of urea is no measure of the
muscular force exerted within the body.

After putting forward these views, Messrs. Bischoff and Voit put
their dog into a kind of tread-wheel, and they found that the amount
of urea eliminated was not in proportion to the exercise of force, but
the amount of carbonic acid was so, and eventually they themselves
admitted the truth of this.

Then came the question of the sources of animal fat. On this
point, again, Voit has worked almost exclusively with the dog, which
is a carnivorous, or, at most, an omnivorous animal. He has found,
which I do not wish to call in question, that in the case of the carnivora,

and in some cases of the herbivora, the fat may be formed from the nitrogenous substance of the food. But from the results obtained with this carnivorous animal he has come to the conclusion, that not only in such cases, but in all, the fat stored up in the animal is derived from the albuminous substance of the food or of the body. I have roughly noted a few of the experiments of Voit, which I believe are the strongest or most conclusive for his view of the question. He found that when a dog was fed on starch or sugar alone, or with albumin, or with fat and albumin, the carbon stored up, that is to say, the carbon which was not eliminated in any way from the body, was never more than that in the fat of the food, *plus* that in the albumin which was broken up as indicated by the amount of urea eliminated. He concluded that this was a proof that fat was not formed from the carbohydrates. In another case, which perhaps was stronger, he fed the dog with starch and a little fat, but no albumin whatever, and the carbon stored up was equal to that of the fat in the food, *plus* that due to the oxidation of albuminous tissue, and when he gave more starch to this food the amount of carbon stored up was reduced ; that is to say, he argues that the carbohydrates in this case protected the albumin of the body from disintegration, and did not in any way serve for the production of fat; and that there would have been a greater storing up of carbon if this additional starch which he gave to the animal had been the source of the fat. He also argued, from a number of experiments, that starch and sugar are quite oxidised in the body, yielding carbonic acid, &c., within twenty-four hours. He maintains that the same must occur with herbivora as with carnivora. The carnivora are found absolutely to digest vegetable food, and take it into their system as an herbivorous animal; and he argues that, to establish a different source of fat, it must be shewn by experiment that fat is formed in excess of that in the food, *plus* that which can be formed from the oxidated albumin. Now this, I think, I shall be able to show you we have done. We have not accepted the challenge in the way of making new experiments for the purpose, but I think we have old experiments which are perfectly conclusive, and do meet exactly the requirement which Voit says is essential to disprove the view which he maintains with regard to the herbivora.

But before entering on our own experiments, I will just say what has happened in answer to the challenge in Germany. Weiske and Wildt conceived, as I shall be able to show afterwards was a very right thing to conceive, that the pig was the very best animal to experiment on for this purpose. He is certainly *the* fat-maker of all the animals that we feed, and there are other reasons why he is the best of all others to experiment upon in this particular. They had, from a theoretical point of view, a very good conception of what was necessary. They took four pigs, slaughtered two of them, and determined the fat and other constituents in those animals. Then they fed one on food very poor in nitrogenous substance, and one on food exceedingly rich in nitrogenous substance. It happened that the pig fed on food very rich in nitrogen had so much that it became unwell, and that experiment failed entirely. With regard to the one fed on food poor in nitrogen, the food was so poor that the experiment took too long a time ; in fact, too much food was passed through the body in proportion to the increase produced; and when eventually they slaughtered that animal, and analysed it, they found so much nitrogen had passed through the body during the time, that the whole of the fat that had been formed might be derived from the nitrogenous substance consumed. Weiske and Wildt did not conclude therefrom that it was established that fat could only be produced from the nitrogenous substance, but they admit that the experiment was not conclusive.

In the experiments of Mr. Lawes and myself we have used a great many animals, and we have brought our results into calculation, although the experiments were not at the time arranged with the special view of determining this question. The table shows some results of experiments with sixteen oxen, 249 sheep, and fifty-nine pigs. You will see that the proportion of stomachs and contents to the body is 11·6 per cent. with the oxen, 7·5 with sheep, whilst it is only 1·3 in the pig. The intestines and contents, on the other hand, shew in oxen only 2·7, in sheep 3·6, and in the pig 6·2 per cent.; with it very much more therefore than with either of the ruminant animals. We know that the character of the food is such in the case of the ruminants that they must pass an enormous quantity of very crude stuff through their bodies, and must elaborate it first in one stomach and then in another, and the result is they have not only a very large capacity of stomach, but also a very large proportion of contents in relation to the whole body. In the case of the pig, on the other hand, the stomach is exceedingly small; the natural food of the pig is starchy seeds or roots (which are the food of man also), it contains exceedingly little necessarily effete matter, their stomachs have comparatively manageable stuff to deal with, and they have a very small stomach, while on the other hand their intestines are very large. It is known that the degradation of the starch goes on almost throughout the intestinal canal, so that we can easily understand how it is that with such starchy food these animals have an enormous amount of intestines compared with either oxen or sheep. If we look at the proportion in the live weight of the further elaborating organs of the body, the heart, the liver, the lungs, the pancreas, and so on, the percentage by weight in the bodies of the three descriptions of animals is almost identical. Now, for every 100 lbs. of live weight of animal the amount of dry substance consumed within a given time is 12·5 by oxen, 16 by sheep, and 27 by pigs; that is to say, 100 lbs. live weight of pig will consume much more dry substance of food, and as I have stated, that food is of a highly nutritive kind, and more easily digested than that of oxen or sheep. Again, the increase per week is only 1·13 per cent. on the live weight of oxen, 1·76 of sheep, and it is 6·43 of pigs. So that the proportion of the increase to the weight of the body is much the greatest with the pig. Then, if we take the facts in relation to the amount of the food, for 100 lbs. of dry substance of food, the ox will give in increase only 5·2 of fat, the sheep 7, and the pig 15·7. Suffice it to say, that there is less effete matter in the food of the pig, and therefore its live weight and its increase indicate more nearly the real increase of body, and not the fluctuating matters in the alimentary canal. Its food is of a higher character, so that a larger proportion of it is stored up. That which passes through the system is more completely used, and the amount of fat which is produced is also very much higher. Therefore, I say the pig is by far the best animal to experiment upon for this purpose.

Whilst on this subject I may refer to a portion of the table which vegetarians will perhaps not be much pleased to see. If we are to judge that the size of the stomach indicates to some extent the character of the food, its crudeness or concentration, as no doubt is the case with the other animals, and if we compare oxen, sheep, pigs, and man, we find the proportion of stomach by weight per cent. is, approximately, in oxen 3·2, in sheep 2·44, in pigs about 0·88, and in man only 0·38 ; so that going from one animal to the other you should have more concentrated and more digestible food in the case of man than of the pig; and you have animal food as well as starchy seeds, roots, &c.; and the indication is, I think, that man was not made to consume potatoes and cabbages by the bushel.

The next point is as to the indications of merely practical results. Without going into the chemistry of the subject, or discussing whether the food of the animal does contain enough or not enough of nitro-genous substances to yield all the fat produced, I will call attention to some results which will indicate the general relations of the food to the necessities of the body.   On the coloured diagram you have the results of twenty-six separate experiments on pigs.   The plan was this : we gave to a certain set a fixed amount of highly nitrogenous food and let them take whatever they liked of less nitrogenous food. To another set we gave a fixed amount of food low in nitrogen and rich in starch and such matters, and let them make up whatever they wanted with highly nitrogenous food.   So we rang the changes in a great many more cases than are here represented, but in this way it will be seen that the animal fixed its own diet according to the neces-sities of the case ; and the question is, was it the nitrogenous sub-stances, was it the non-nitrogenous substances, or was it the total dry substance, nitrogenous and non-nitrogenous together, which guided the amount consumed by a given live weight within a given time, or rather guided—for these were fattening animals—the amount of in-crease which was produced ?   The lowest amount of nitrogenous sub-stances consumed by 100 lbs. of live weight of pig per week in any one experiment being taken as 100, in some cases they took 300, and in most more than 200.   In the same way the lowest amount of non-nitrogenous substance being taken as 100, in no case was nearly as much as 200 consumed, and the average was about 140 parts.   When we come to the total dry substances, including both nitrogenous and non-nitro-genous, we find that the quantities ranged more closely together; that is to say, the gross organic substance seems to have been the measure of what was required, and that the nitrogenous might possibly act for the non-nitrogenous substances if there were not enough of them.   But

it is quite clear that the measure was either the non-nitrogenous substances or the total organic substances—certainly not the nitrogenous
substances. Then the question arose, whether the same thing would
hold in relation to the amount of increase in the weight of animal produced. It was always assumed, I think, until these experiments of Mr.
Lawes and myself, that when animals were not fed on highly nitrogenous
substances the amount they stored up was comparatively small. These
experiments show the amount of these three classes of constituents consumed in producing 100 pounds increase of live weight in the different
cases. 100 pounds being the lowest amount of nitrogenous substance
required, 282 was the highest, the animal fixing his own diet,
and in many cases it was over 200; that is to say, more than twice
as much as satisfied him when he had enough of other matters to
make up. At any rate it would seem that fat can be formed from nitrogenous substances, provided there is a deficiency of non-nitrogenous
substances in the food; and I may say that the nitrogenous substances
are of a higher food capacity, irrespective of the nitrogen, containing
more carbon, more hydrogen, and less oxygen ; they have more useful
matter in them than an equal weight of starch or any substance of
that kind.

Now the question arises, what is the state of affairs when we attempt
to calculate these results and to see whether or not the food did contain enough nitrogenous substances, or albuminous matter, to supply
the whole of the fat produced ? These experiments were not instituted
with a view to settle that question, but they were calculated afterwards.
It was about twenty-six years ago that we took two pigs of the same
litter, very carefully selected both by practical and scientific eyes as
being as nearly as possible exactly alike. One was slaughtered, and
the total amount of dry matter, fat, nitrogen, mineral matter, and so
on, determined; and then the other animal was fed. At that time we
had not arrived at such distinct conclusions as we did afterwards as to
the desirability of giving a greater proportion of starchy matters. We
gave the animal a great deal more than the proportion of nitrogenous
substances existing in what may be called the normal fattening food of
the pig—barley meal. In the first column of the table, the results of
that experiment are calculated out to show whether the food did contain enough nitrogenous matter to yield the fat produced. You will

see the proportion of non-nitrogenous matter to one of nitrogenous is 3·6. Now, the proportion in barley meal, which is the best fattening food for the pig, is between 5 and 6 to 1; so that we gave too much nitrogen according to what we now know is the best proportion. There was a considerable amount of increase in ten weeks, eighty-eight pounds, or 85·4 per cent. on the original weight of the body. The question is, how much fat was in the food? and that is shown in the second division of the table. It is calculated that for 100 pounds increase in the live-weight there were stored up 63·1 pounds of fat. There were of ready-formed fat in the food 15·6 pounds; leaving 47·5 pounds fat to be produced from some material or other. Out of 100 of nitrogenous substances consumed as food, there were stored up in increase 7·8, leaving 92·2 parts of nitrogenous substance which might be used for the production of fat or might not. If we calculate how much carbon there was in the produced fat, and how much there was left available in the nitrogenous substance for the production of fat, we find that there were 7·4 pounds more carbon possibly available from the nitrogenous substance than was necessary for the production of the fat ; or, put in another way, there were 120 of carbon available from the nitrogenous substances for 100 required. According to this mode of calculation, therefore, there was enough nitrogenous substance to justify the conclusion of Voit; or rather, the result does not in any way disprove his conclusion that fat has been produced from the disintegration of nitrogenous substances in the body. This table was calculated some years ago, and we have intentionally put the results in the worst aspect that we could for our own side of the case, that we might not exaggerate the conditions. For instance, we have assumed that the whole of the fat in the food would be taken up, which it certainly would not; and we have assumed that the whole of the nitrogenous substances of the food would be digested, and would come into play, which they certainly would not. If we assume in our own calculations the estimate adopted in Germany, that 100 pounds of nitrogenous substance cannot yield more than fifty-one of fat, even this experiment shows a little deficiency of nitrogenous substances, and in fact would be in favour of our side of the case. The two next experiments given in the table show a still higher proportion of nitrogenous substance in the food; and there was, accordingly, a great deal more carbon available from the nitrogenous substance than was necessary for the formation of the amount of fat produced. The next two experiments (four and five), were with more natural fattening food of the animal, one entirely Indian corn-meal, and the other entirely barley-meal. A pig requires for rapid fattening very little, if any, more nitrogenous substance than this represents. But here we have only 60 per cent. or a little over, 60·8 in one case, and 60·5 in the other, of the carbon of the fat produced in the animal, possibly derivable from the nitrogenous substance of the food. So that we have in those two cases nearly 40 per cent. of the carbon of the produced fat which could not possibly come from the nitrogenous substances, and must have come from the non-nitrogenous matter, in fact from the carbo-hydrates. But an objection may be raised to this calculation; the animals were larger to begin with; and the weights were heavier at the end; so that the composition of the lean animal, and of the fat animal, as derived from the direct analyses, does not absolutely apply; but we could not possibly thus get rid of this forty or more per cent. which the calculations would show to be derived from the non-nitrogenous substance of the food. The remaining four experiments are also entirely in favour of our view. The animals were about the same weights as those analysed: the food was more nearly the proper food for fattening, being rather lower in nitrogenous substances, but much higher than in experiments 4 and 5. But even here we found 18·9, 18·8, 25·2, and 14·1 per cent. of the total carbon of the produced fat could not possibly have been derived, and certainly a great deal more was not derived, from the nitrogenous substances of the food.

I need not trouble you further with these results. But I should say that the contrary view has been adopted not only by some physiologists, but in Germany in some text books on agricultural chemistry. I hold in my hand one of these text books in which the evidence of these experiments is discarded, and it is assumed that if you cannot experiment with the respiration apparatus the results are good for nothing. I would not wish to depreciate the importance of the results obtained by the respiration apparatus in any way. I have taken the greatest interest in them, and I think they lead to the most important conclusions; but I also think some observers have come to very

erroneous conclusions from the results of such experiments. I submit
that if you experiment with *the* fat-producer—the pig—and if you take
two carefully selected animals (or more if you like) kill and analyse one,
and feed the other as rapidly as possible, that is, let him take as much
of the most appropriate food as he will take, you may, without any
respiration apparatus, determine this point. It is most important that
it should be definitely settled. Since the recent publications on the subject, Mr. Lawes and myself have gone thoroughly into the question,
and re-calculated most of our results ; those relating to oxen and sheep
as well as pigs. They point to this, that the ruminant animals, which
have such elaborate machinery, and do so little productive work, do
pass so much nitrogenous substance through the body in relation
to the amount of increase, that they do not show that fat can be
derived from the non-nitrogenous substances of the food ; but in the
case of pigs the evidence is perfectly conclusive. Having re-calculated
our own experiments in this way, and the results being absolutely conclusive so far as the pig is concerned, Mr. Lawes is unwilling to be at
the trouble and expense of further experiments on the question ; but
it really is one of great importance, and one which public institutions
might well take up. It is of importance, not only agriculturally, with
reference to the proper way of feeding stock, but also in its bearings
on the nutrition of man.

[For the tables and diagrams referred to above, see—" On the
Sources of the Fat of the Animal Body," *Philosophical Magazine*,
December, 1866 ; and—" On the Formation of Fat in the Animal
Body," *Journal of Anatomy and Physiology*, Vol. xi., Part iv. ; and
for other points, and detail—" Food in its Relations to Various Exigen-
cies of the Animal Body," *Philosophical Magazine*, July, 1866 ; and
the papers therein referred to.]

The PRESIDENT : I do not know whether any physiologist present
would like to discuss the points noticed in this paper, for I find it
is not contrary to the precedents of previous conferences to do so;
but if not I will call upon Mr. Sclater to make his communication
on the subject of the drawings which are exhibited by the Zoological
Society of London.

ON DRAWINGS CONTRIBUTED BY THE ZOOLOGICAL SOCIETY
OF LONDON.

Mr. P. L. SCLATER, F.R.S. : After the many elaborate discussions you
have had before this assembly I am afraid you will think the drawings
in the next room concerning which I have to say a few words, a rather
trifling subject ; nevertheless I hope to show you that there are some facts
connected with them which are worth the attention of biologists. Most
people here, no doubt, are acquainted with the great national institution
of botany at Kew, under the direction of Dr. Hooker, who has been
addressing you to-day. That institution we are justly proud of in England, and consider it as one of the most perfect scientific institutions
perhaps in the world. You will recollect that in that great institution
there is in the first place a herbarium for the deposit of plants and dry
specimens from all parts of the world ; and in the second place
a collection of living plants in which they may be studied in their
life. Thus you will see that the botanists of this country have an
almost perfect institution for the study of their branch of biology. The
same high place, I might say, ought to be the case in zoology. As in
botany, so in zoology, in order to acquire perfect knowledge of an
animal we require to study the living form and the internal anatomy
The living forms are not generally well represented in museums. I
mean to say that those stuffed animals usually exhibited in museums
give by no means a good idea of the living form. The internal anatomy
may be in many instances very well studied at a museum, but still
there is a large number of animals which cannot be preserved entire in
a museum from their size and other difficulties, and therefore with the
anatomy of which we should be entirely unacquainted if we only had
our museums to rest upon. It being granted that a collection of living
animals is a proper supplement to a great national museum, let us
consider what are the facts in this country—how are we situated? In
Great Russell Street, as is known to you all, there is the National
Museum of dead animals. It is mixed up with a number of other
collections and a great public library with which it has really no
connection. When it was founded it was considered merely to
be a kind of appendage to the great public library which was

the original source of the institution in Great Russell Street, and even to the present day the director of that institution is called the " Principal Librarian." Happily it has now been determined for some years to move the national zoological collection into an independent institution, and a splendid building is now rising near us, which will accomodate the zoological collections ; and therefore I will assume that in a few years this country will be provided with an adequate museum of zoological sepcimens. As regards the living specimens, no sort of attempt has been made in this country by the government to supply that want. Such being the case, in this as in many other cases in this country, a private society has taken up the task which has been left unaccomplished by Her Majesty's government. About the year 1826 the Zoological Society of London was instituted for the general advancement of zoological science, and it seized upon this weak point in the national institution and constituted those gardens in the Regent's Park, with which many of you no doubt are acquainted. I am not going on to discuss how far has that society executed or accomplished the object before it. It may suffice to say that the living collection now in those gardens is admitted on all hands to be the most perfect of the sort in existence. The two principal points in which that collection of living animals is of advantage, are these : in the first place it affords opportunities for persons to become acquainted with the character and the external forms of many animals which they could not otherwise see ; and in the second place, when those animals die, there is an opportunity given for the anatomist to carefully dissect them and to learn their internal structure. It would be impossible of course for the hunter who had shot an elephant in the middle of South Africa to sit down and examine it muscle by muscle and write a memoir on its internal structure. In the same way with many other large animals, which hitherto have only been shot in far distant lands, and have never yet been properly described, when they are brought alive to this country, and are kept in such gardens as those in Regent's Park, and come under the scapel of the anatomist when they die, and form materials for numerous memoirs and papers.

But the point that I have to remark upon at present is rather in relation to the external form of animals. In the next room there is a collection of upwards of 100 water-colours. These are accurate drawings of the external form of the rarer and larger animals belonging to the Zoological Society's gardens, which have lived in those gardens for the last twelve or fourteen years. The object of the Zoological Society in making this collection was to improve our knowledge of the external form of animals. It is a great misfortune that not only in popular works on natural history, but even in most-text books, the same figures are copied over and over again. The artists employed to illustrate those books will not take the trouble as a general rule to go and draw from nature ; it is much less trouble to go to a figure already published and to copy it with a few alterations. The consequence is that errors in the external forms of animals are repeated over and over again, and it is only from the existence of such gardens as those of the Zoological Society and the use made of them that those errors can be avoided, I may state that in order to discountenance this practice as much as possible, the Zoological Society give free admission to drawing-students—to any person bringing a proper recommendation—on the morning of every day in the week ; and therefore there is no difficulty on the part of any art-student in obtaining an admirable series of figures and of copies to draw from. I may mention just one animal that is commonly very incorrectly drawn in books of natural history, and that is the rhinoceros. You generally see described, and you get a distorted figure of the best known species, which is the *Rhinocerous unicornis* of naturalists ; an inhabitant of Assam in northern India. You generally find it described and figured in popular books, covered with an impenetrable coat of mail, divided into plates something like this beautiful drawing shows. But so far from there being only one rhinoceros, we have in the Zoological Gardens examples of no less than five different species. Here is a drawing of a large African one which does not exhibit the coat of mail which is commonly supposed to be universal in these animals. I have brought these five beautiful drawings which have been executed by Mr. Wolff, just to show the object of the society in making this exhibition to the Loan Collection at South Kensington.

The PRESIDENT : I am sure, ladies and gentlemen, we all feel extremely grateful to Mr. Sclater for coming to give us so pleasant a relief after listening to the somewhat dry and tough, intellectual food we have had before us so far ; but we must now return to the subject of registering apparatus, and I must next call on Mr. Gaskell to give an account of another apparatus, having somewhat the same object as the one last described by Mr. Marey.

ON THE REGISTRATION OF THE VARYING BULK OF ORGANS.

Mr. GASKELL : I am sorry to say I have not a perfect specimen of the apparatus, because the inventor of that apparatus, Dr. Mosso, of Turin, was not able to send me one. He has only two ready made, one of which he is using at the present time, and the other he has unfortunately broken, so that the best he could do for us was to send over a diagram, by means of which I shall be able to explain to you the nature of the apparatus. He has given it the name of the Plethysmograph. It is a name which implies an apparatus for measuring the changes in volume in any organ, in this case that of the forearm. If the arm is kept perfectly steady, one finds that the volume of the arm is sometimes larger, and at other times smaller, and the only way in which this change can be accounted for, would be in the amount of blood supplied to it ; so that practically the apparatus in this case is one for registering the movements of the blood in the forearm. It consists of a glass cylinder filled with water, so arranged that the forearm will fit into it easily. At one end it is provided with a strong india-rubber band, which fits over the elbow so tightly that there is no possibility of any water escaping. The other end is drawn out into a small tube, to which is attached an india-rubber pipe. The arm and apparatus is suspended from the ceiling to prevent any slight movements of the body affecting the measurements that it is desired to register. There are two openings at the top of the glass cylinder, in one of which a thermometer can be introduced to measure any difference of temperature, and through the other you can apply electrical stimulation. The india-rubber tube is continued into a fine glass tube fixed rigidly in a holder. This is bent at right angles at the further end so that its longer arm stands absolutely vertical and dips into a small test tube which is attached to a thread passing over a pulley, and counterbalanced with a small weight to which is affixed a pen. A burette and clip is connected by means of a ⊥-piece with the india-rubber tube for the purpose of regulating the amount of fluid in the test-tube. The test-tube hangs in a large beaker filled with fluid so as to float and easily rise up and down. Every variation in the volume of the forearm communicates a motion to this test-tube and consequently to the pen, which makes a curve upon a revolving cylinder of paper. The great difficulty which Dr. Mosso found he had to contend with was to keep the pressure on the forearm constant during the whole of the experiment, i.e., to arrange, so that the level of the fluid in the large beaker and in the test-tube should always remain at the level of the exit opening in the glass cylinder, whatever might be the amount of fluid in the test-tube ; he succeeded in effecting this, by using a thin test-tube, and filling the beaker with a mixture of alcohol and water, of such specific gravity, in comparison to that of the water poured into the test tube, that the level of the fluid inside and outside the test-tube remained the same, under all positions of the tube. These two levels thus always remaining the same, it was only necessary that they should both be constantly at the level of the exit opening of the glass cylinder. This was attained by placing the fixed vertical small glass tube so that its free extremity just reached the level of the fluid in the beaker, and using a beaker with a diameter so much larger than that of the test-tube, that the rise of the fluid in it was not appreciable, even on full immersion of the test-tube. Dr. Mosso has sent over a couple of rolls of the curves produced by this instrument, which show accurately what tracings can be obtained by this means. On one there is a red line and a blue line, which represents the two curves produced by the two arms, the person operated upon being seated between two of these cylinders and having one arm in each. The object of this experiment was to see the effect produced upon both arms by the compression of one brachial artery. The normal line being horizontal, when the brachial artery was compressed the curve descends, and when the compression is removed it rises again. When the compression is applied, the volume at first diminishes slightly. Then it tends rather to increase, owing probably to a collateral circulation being set up, and then when the compression is removed it suddenly springs up again. On examining the curve of the other arm not acted upon, it appears that that varies in a converse direction to that of the arm operated upon, as if a larger portion of the blood were diverted to the arm not compressed. With regard to the electric stimulus, for some little time after it is applied, no effect is produced. Then the curve commences to fall and sinks considerably. On taking off the stimulus, the sinking continues for a short time afterwards, and then it gradually rises to the normal line. In this case it is noticeable that there is almost precisely the same curve in the two sides, although the stimulus was applied to one arm only. Amongst these curves there is one rather interesting. The instrument which applied the electric stimulus was supplied with two keys, one of which acted but the other did not. Dr. Mosso after several times putting down the key connected with the electrodes, which sometimes caused a certain amount of pain, sometimes a slight pricking, and sometimes no sensation at all according to the strength of the current, then without saying anything put down the key not connected with the electrodes, and instantly the curve sank just the same as before although nothing was done to the arm; the sinking of the curve being produced purely by the imagination. This apparatus therefore is very sensitive for showing the effect of mental action. So you see it might be used in an examination room to find out in addition to the amount of knowledge a man possesses, how much effort it caused him to produce any particular result of brain work—for instance, how much effort it caused him to multiply 267 by 8. I will show you on this curve how much effort it caused Dr. Pagliani to do that operation. The Doctor was told at a particular point to make this calculation, and to make a sign when he had finished, and the curve showed most distinctly how much more blood he required to do that little multiplication sum. It would therefore enable one to measure how much mental power it took, and certainly it would enable one to measure the different amount of mental power that different people would take to produce the same result in working out any mental problem. When Dr. Mosso was in Turin he had the apparatus fixed up in his room, and a classical man, who came to see him, looked very contemptuously on it, and asked what was the use of it—it couldn't do any body any good. He replied, " Well now, I can tell you by that whether you can read Greek as easily as you can Latin." He would not believe it, but he was put into the apparatus and given a Latin book to read. A very slight sinking of the curve was the result. Then the Latin book was taken away, and a Greek book was given him, and this immediately produced a much deeper curve. He had asserted before that it was quite as easy for him to read Greek as Latin, and that there was no difficulty at all; but Dr. Mosso was able to show him that he was labouring under a delusion. It is also a very sensitive apparatus for finding out what you are dreaming about—or, rather, how much you are dreaming. Dr. Mosso once succeeded in persuading Dr. Pagliani to go to sleep in the apparatus, and the effect was very marked indeed. He said afterwards that he was in a sound sleep, and did not remember a single word of what had been going on in the room—that he was absolutely unconscious; but yet every little movement in the room, the slamming of a door, the barking of a dog, and even the knocking down of a bit of glass, were all marked by these curves; therefore, you see, in every case, how the blood left the arm and went off to the brain; and one could see sometimes the lips muttering, and that he was dreaming, and that it was all marked on the curve. The amount of effort required for dreaming took away a certain amount of blood from the extremities. Again, it measures the emotions to a very great extent. It was found every time, whether it was Dr. V. Frey or Dr. Pagliani who was in the instrument, that when any of the students who were working in the laboratory came into the room, not much effect was produced—just a little sinking; but as soon as Professor Ludwig came into the room the arteries contracted quite as much as they did upon a very strong electrical stimulation. This is only the commencement of a series of experiments which Dr. Mosso intends to make with this instrument. But I think I have shown you sufficient to be able to say that this will, in future time, I hope, take its place amongst other physiological instruments, and be of very great service in the study of physiology.

The PRESIDENT: I am sure, ladies and gentlemen, it must be a great satisfaction to have heard so lucid and clear a description given of this apparatus, and I am very glad I did not do as I intended, give you an account of it in my opening address this morning. I was fortunate in persuading Mr. Gaskell to describe it, and I am sure we must all be very much gratified, inasmuch as he has given so vivid and complete an account of its mode of working, and of the extremely valuable results which are to be obtained from it. I just permit myself to observe, with reference to the other apparatus for the same sort of purpose, which was brought before you by Professor Marey this morning, which you were told was for the purpose of measuring the variation in the bulk of the human hand, and variations in the condition of the circulation, and which, therefore, is apparently to accomplish the same problem as that which is accomplished so beautifully by this plethysmograph, that in point of fact the purposes of the instruments are somewhat different. That which Mr. Gaskell has been describing to you has for its object to obtain measurements which extend over relatively long periods, whereas that of Professor Marey's gives you extremely delicate indications of changes which take place in the bulk of the hand in extremely short periods. For example, the contractions which take place in a minute, whereas the other changes take place over longer periods. This, in fact, is an apparatus which gives you an absolute measure of the changes of the form of the organ, whereas Professor Marey gives you delicate indications of the changes which occur within a short period, so that, adopting the phraseology which is used in physics, you might call the one a plethyscope, and this other one very correctly a plethysmograph. The next paper is a description, by Dr. Brunton, of several improved myographic apparatus.

Dr. LAUDER BRUNTON, F.R.S.: I believe the title of my paper is a description of new myographic apparatus, but I see so many here to whom a description of some of the old ones might be more interesting, that instead of confining myself entirely to the dry details of the apparatus I have here before me, I shall say a word or two on some of the old instruments which are exhibited. We all are naturally interested in the way in which we happen to move. We discover life in our neighbours, in plants, and in animals, only when we see movement, and we wish to know how those movements are brought about. Physiology is only advancing just now, but I think one can take no better illustration of the advance which has been made in the knowledge of physiology, than to quote a description which I once saw in one of Sir Walter Scott's novels. In describing one of his heroes, Scott said that he did not possess one ounce of superfluous flesh, he was all muscle. Every schoolboy knows now that muscle and flesh are the same; everybody knows that the large muscle which he feels in his arm is his biceps, and that it is absolutely similar in appearance and in properties to the flesh he sees in a butcher's shop. One great thing about muscle is that it is elastic, and I wish to show you some peculiarities in the elasticity of muscle. If we take a piece of indiarubber, and pull it, for every increase in the amount of force we pull with, or in the amount of weight we hang upon it, we find we get an equal amount of stretch; but this is not the case with muscle. Professor Donders has invented this apparatus (No. 3963) for showing that living muscles, —the muscle of the human arm, for instance,—are elastic. It is a very simple apparatus, but most ingenious. There is a rest for the shoulder and the elbow, and a weight is held in the hand. The arm is kept bent exactly at a right angle, and if you let the weight go the arm will spring up a little, which is due to the elasticity of the muscle. We might say, in fact, that the muscle really had been contracted to the position it assumed after the weight fell off, and that it had only been a little lower down on account of being stretched, just as a piece of indiarubber might have been stretched by a weight. Muscle has a second property; besides that of elasticity, it is able to contract. Here we have an apparatus, belonging to Professor Marey, intended to show the contraction of muscle (No. 3997,12). Here is a little telegraph, and here is the muscle of a frog, prepared in very much the same way as you would see it in France prepared for the table, except that it is not cooked. If I touch this muscle with these electrodes, and pass a galvanic current through it the muscle will contract, if it has not been too long in this chamber. This contraction might have been due to a number of causes. People were naturally interested in trying to find out how this contraction took place. They thought first of all that it was likely to be due to something analagous to magnetism. This was the theory propounded by Prevost and Dumas, that the contraction was due to the small particles of which the muscle is composed, being attracted together very much like a number of small magnets. If this were true, the more the muscles contracted the stronger ought the contraction to be. You remember that in the old story of the "Arabian Nights," when a ship went near the mountain of Loadstone, as it got nearer and nearer it went faster and faster, because the Loadstone mountain exerted a stronger and stronger influence upon it, until at last, when it got very near, it could not go fast enough, and the nails all started out of the ship and flew to the mountain. So if the particles of the muscle cause the contraction in one another like so many little magnets, then the closer they get together the more power would the muscles have to contract still further. In order to test if this were true, Professor Schwann, of Liege, invented this instrument (No. 3806), which is extremely interesting as being the first one which was applied to investigate the so-called vital forces in the same way that we are accustomed to do the so-called physical forces. The way in which he did it was this : He had a muscle fastened firmly to a plate of cork at one end, and at the other to the beam of a balance. Whenever the muscle contracted, the arm of the balance was drawn down, and the motion was made all the more evident by means of a long lever attached to the other arm. He found then that when he caused the muscle to assume the position of the greatest contraction by depressing the lever, instead of having an enormous power as it ought to have had on the theory of magnets, it had almost no power to contract at all, and that, on the contrary, the more the muscle became elongated the greater was its power of contraction. Here, then, we have two very interesting facts. The longer the muscle is, the less extensible is it, and the greater is its contracting power. When, on the contrary, the muscle is contracted and short, it will not readily contract further, but can be very readily stretched to a great extent. We can see how very useful that is for the movements of our body. We want, for example, to contract or extend our arm rapidly. Now, when the arm is contracted, the biceps muscle in front of it is very much shortened, and can, therefore, be very easily stretched, but the triceps at the back of it is already very much stretched, and, therefore, has great power of contracting at the very time when this power will be of most advantage. When I wish to straighten my arm, the triceps muscle at the back acts quickly and at a great advantage, and has no resistance to overcome on the part of the biceps. On bending the arm again after it has been extended, we find that the triceps being contracted offers little resistance while the biceps being stretched has then its greatest force, and is also least extensible.

A muscle when irritated once gives just one jerk ; but this is not the sort of motion we wish to have when we are going to perform any mechanical work ; we are obliged to have a continuous motion. Now the way in which we get a continuous motion is this. If we get a contraction we can see nothing more than a sudden rise and sudden fall, but if we were to transfer this motion to some of the recording apparatus, we should learn from the curve shown by these instruments, that there was something more which had escaped our eye, that the rise was not quite sudden, and that the fall was still less sudden. In order to have continuous motion we must have several of these curves, and putting them one over the other, instead of having a sudden rise or sudden fall, we get one almost continuous line. The number of contractions in a second which are requisite in order to get a muscle constantly contracted is about twenty. This has been determined by means of the so-called muscle-tone in an experiment which you can to a great extent repeat for yourselves. If you contract your thumb very strongly, and put it against your ear, you will hear a humming sound ; or if you clench your jaw very firmly, and lay your head upon the pillow, you will again hear this humming sound. This is due to the contraction of the muscle in somewhat the same way as the tone you get from a tuning fork is due to the alternate vibrations of the fork. You can test this for yourselves, because whenever you cease to contract the thumb, the sound will at once disappear, and will again re-appear as soon as the muscles are put in action again. This continuous contraction is called tetanus, and it was supposed this could not be kept up by very rapid stimulation—that when you applied a stimulus to the muscle very frequently it ceased to respond. Professor Schäfer, however, has shown you that instead of this being the case, as many as 22,000 stimuli in a second may be applied, and still you get tetanus of the muscle. The muscle can be set in action by a stimulus applied to it directly, but in order to get all the different muscles of the body working co-ordinately together, you must have some means of inter-communication. We have then a sort of telegraphic arrangement in the body, consisting of the nervous system, which makes one muscle act at one time, and another at another, just as they are required by the motions of the body. We can cause a muscle to contract by stimulating the nerve which is attached to it. A muscle does not begin to contract immediately after a stimulus has been applied to it, but takes some little time, and this is termed the latent period. In order to ascertain the latent period of a muscle we have such an

arrangement as that placed before you, which has been sent by
Professor Schwann.   We require, first, to apply a stimulus to a muscle
at a definite time ; second, to ascertain when the contraction of the
muscle begins ; and third, to measure the time which elapses between
the application of the stimulus and the commencement of contraction.
In the apparatus we have a plate of smoked glass which is drawn
rapidly along by means of a spring.   Against it rests the point of a
lever, which is drawn upwards when the muscle contracts.   The
moment of contraction is thus indicated by an upward stroke made
by the lever's point on the smoked surface.   Attached to the plate
is a piece of metal, which strikes against a spring beneath, and
closes the circuit of a battery, and thus irritates the muscle as the
plate is drawn along.   The point of attachment of the metal to
the plate indicates the time when the irritation is applied to the muscle.
The point when the upward stroke of the lever begins indicates the
time of contraction of the muscle.   The distance between these
points indicates the latent period of the muscle.   To convert the
distance between the two points on the plate into time we must
know the rate at which the plate is drawn along by the spring.
For this purpose a tuning fork is attached which vibrates a certain number of times in a second.   Its vibrations are recorded by
a point, which rests against the smoked plate.   As the point vibrates
vertically, and the plate is drawn horizontally along, each vibration
is recorded as a wave.   This corresponds to a definite fraction of
a second, and by counting the number of waves in the space between
the metal point which applies the stimulus to the muscle and the
upright stroke which indicates its contraction, the latent period is
exactly ascertained.

   There are just two points I may mention before concluding. I have
already spoken of the great use of the elasticity of muscle in allowing
us to make rapid movements.   Now we find under certain conditions, more especially under the action of poisons, that muscles no
longer extend again as they do in the normal condition after contracting. There is especially one poison, called veratria, which prevents
the muscle relaxing again after it has once contracted. The consequence of this is that an animal poisoned with veratria is able to
execute a rapid movement just for once.   It can extend its limbs, but

there they remain, and can only be drawn in very slowly and with great trouble. After a time it can again execute the same movement, but has again the same trouble in bringing the limbs back again to the proper position.

The other point I would mention is the influence of cold on the muscles. The curve you get in the normal condition is very much altered by cold. It is slight and long, and if the cold is sufficiently great the muscles will not respond at all. We had a very good example of this some time ago in a man who tried to swim across the Channel. After swimming for some time very bravely, he was obliged to be taken into the boat, not because he felt tired, but because he could not go on. When he was taken into the boat he could not move his limbs, neither his arms, nor his legs. He was quite well otherwise, but his limbs would not obey his will. The reason simply was that the cold had acted upon his muscles and chilled them down so much that they no longer obeyed the stimuli of his nerves, and so they were no use to him. Lately, Captain Webb succeeded in crossing the Channel, to the great astonishment of most people. I, for one, was inclined to predict beforehand that it was utterly impossible for Captain Webb, or for any one else, to cross the Channel, simply because in the process of crossing his limbs would be so chilled that his mucles would fail to obey his will, just as the muscles of the swimmer had formerly failed in the attempt made some months previously. It did not occur to me that there might be certain cases in which men might possess the same power as whales. Whales, you know, are warm-blooded animals ; yet notwithstanding the extreme cold to which they are subjected, they are able to swim comfortably about in the Polar seas. This is due simply to the immensely thick coating of fat, which protects the muscles from the influence of cold. We know the astonishment which has been expressed by some persons at the coating of fat which Captain Webb possesses. They seem to think that a coating of fat and great muscular power are not to be found together, but we now know that if it had not been for this he could never have succeeded in performing the feat which he acheived, and that if any one tries to follow his example, he must inevitably have either a natural coating of fat, or else must apply it in some way artificially.

The PRESIDENT : I am sorry to say the time for our meeting has expired, and although there are several other interesting communications, we must adjourn until Monday next, when the proceedings of this Section will be continued.

The Conference then adjourned.

SECTION—BIOLOGY.

Monday, May 29th, 1876.

The PRESIDENT : Ladies and Gentlemen, you will remember at our last meeting we were unable to get through the business of the day, and that we have as a remanet the account which Dr. Klein, who has been appointed as an official deputation for the Austrian department of this exhibition, will·give on the part of Professor Hering, of Prague, one of our Vice-Presidents, on the collection of apparatus sent by the University of Prague.

ON RECORDING APPARATUS EXHIBITED BY THE PHYSIOLOGICAL INSTITUTE OF THE UNIVERSITY OF PRAGUE.

Dr. KLEIN, F.R.S. : The apparatus I have to bring before your notice to-day have been sent by Professor Hering of Prague. One has been already brought before your notice by the President—the myoscope. I will trouble you now with giving a detailed account of two others. One is a complex kymographion, and the other an apparatus for artificial inspiration and expiration. The kymographion, as you see from these diagrams, is composed of a number of manometers, the object of which is to register the pressure of the blood—an apparatus belonging to the same class as that introduced by Professor Marey for registering movements of different organs. There are first of all two of Ludwig's kymographs—ordinary measuring manometers provided with writers, each of these having besides its own abscissa writer. Then there is one other kymograph, the so-called *feder* kymograph of Fick, one cardiograph, one time-marker, and two markers for other purposes. With reference to the two Ludwig kymographs there is this to be said, that whereas in Ludwig's kymographs the pressure is produced by a so-called pressure bottle, it is produced here by a syringe, which can be fixed to one arm of the kymographion and may be used for producing the over-pressure in the artery.   The kymographion of Fick, which is marked number three, is very delicate.   It is better than the ordinary instrument, and may be used on account of its delicacy for registering the pressure in the veins.   Then there is the ordinary cardiograph, number four, the same as that introduced to you last Friday by Professor Marey.   Then you have a time-marker in connection with a horizontal electro-magnet which may be introduced into an electric circuit, and a pendulum for marking seconds which may be made to close the circuit at every second.   There are two additional vertical magnets for marking other (nerve) proceedings.   There is only one other thing, with reference to the apparatus, which I may mention in addition to what I said just now—that is the very easy way in which it may be adjusted.   The registering cylinder may be brought into use very easily.   There is a small trough into which turpentine is poured, and by its flame the paper on the regulating cylinder may very easily be blackened.   This is an advantage over the ordinary way of preparing paper for the ordinary cylinder.   The other apparatus is here before you.   It is for artificial inspiration and expiration.   In the ordinary process of artificial expiration only inspiration is induced and the expiration is left to the elasticity of the lung and thorax itself. Under those circumstances we are working under abnormal conditions. An amount of air is constantly blown into the lungs, and the expiration is limited to the elasticity of the lungs and chest, and therefore the lung never gets rid of the surplus air, and becomes expanded more than in its normal condition; this produces first a lowering of the arterial pressure from mechanical causes, and secondly it produces a greater frequency of the pulse due to reflex action.   By the use of this apparatus these abnormal conditions are eliminated, and the lung is completely ventilated.   It is worked in this way : you have a working wheel which is brought in connection with our wheel.   The axis of one of these wheels is fixed to one cone, and another cone may be brought to revolve by putting a thick, strong leather strap round both.   It is quite clear, if you push the leather strap to the right the upper cone will be made to revolve much more quickly than if you pushed the strap towards the left.   The upper cone drives a horizontal bar which is in connection with a vertical bar, which I may call the lever; this vertical bar drives the piston rod of two cylinders, each of which acts like an inspiration and expiration pump ; consequently, by making the upper cone rotate you make the piston rod slide backwards and forwards and so inducing compression, first in one and then in the other cylinder, the air will be driven into the lung and again out of the lung. Each of the cylinders is provided with two tubes, one leading to the lung and one to the outside air. By expanding the air in one pump, you draw air from the outside and bring it into the cylinder, and at the next stroke it is driven into the lung ; the other cylinder acts in the reverse way, pumping air from the lung and driving it out into the external air. Instead of using the ordinary atmospheric air, you may introduce here an apparatus charged with any other gas, and in the same way you may introduce an apparatus for the quantitative or qualitative determination of the air expired. Professor Lovén drew my attention to one thing which appears not quite in favor of the apparatus, and that is that in analysing the air, you must be prepared to have a slight error on account of the air passing over metal, and being brought into contact with grease and oil which may slightly contaminate it. In the ordinary apparatus, for instance in Pettenkoffer's, you have the air passing only over mercury and glass ; but still the apparatus has advantages, for, as I have said, you have not only inspiration produced artificially, but also expiration, and you always work with a definite quantity of air and the lung is completely ventilated. The apparatus looks very large, but I have no doubt will in time prove useful.

The PRESIDENT : I am sure the Conference will be very glad to hear any remarks that anyone may like to make on the subject. The constructor of these beautiful pieces of apparatus—I refer particularly to the recording apparatus of Herr Rothe—will I am sure be glad to give any explanation to anyone who understands German, who may wish to have further details. The apparatus stands in the other room, and can be very readily understood, although it appears somewhat complicated in its aspect in comparison with this drawing.

Professor Rutherford is now ready to exhibit to you his freezing microtome, and I will ask him to proceed with his communication.

Professor RUTHERFORD : I have been asked by the President to show an apparatus invented by me some years ago for the purpose of making excessively thin sections of tissues and organs, for the purpose of examining them by means of a microscope. Of course I do not require to inform any microscopists here present, how extremely important it is to make an excessively thin slice of various organs, but for those who may not understand this matter, I may say that you may just as well attempt to see through your hand as endeavour to look through a section made for the microscope, unless it be excessively thin. If you take a piece of cartilage, the ordinary gristle that covers the end of a bone, one can without any difficulty make a very thin slice of it ; that tissue is just dense enough to resist the knife sufficiently, and moreover when you make a slice of this tissue, it being dense, does not go to pieces. If, however, we take a kidney, or lung, or liver, and endeavour to cut it, we find that it is not dense enough to resist a knife, and moreover when we make the section, the tissue being so soft falls to pieces and we really can make nothing of it, so that in all cases where we have to deal with soft tissues, and most tissues are soft, we require to harden them by some means. There are various ways by which they can be hardened ; we can put them in alcohol, in chromic acid solution, or in a solution of picric acid, or in a solution of osmic acid, and all these things produce a degree of hardness sufficient to enable us to make sections ; alcohol and osmic acid will harden tissues in a few hours ; picric acid will do it in a day or two ; chromic acid requires some weeks. But when we harden a tissue by these agents, there is the difficulty that we are apt to overstep the degree to which the hardening is carried, and often a coagulation of albumen takes place, which produces misleading appearances. Then again when we have the tissues hardened by these agents, if we desire to cut them they may be large enough to be held in the hand, and sliced with an ordinary knife, but often the tissue is very small, and it is difficult to hold it, so in that case we require to embed it in some material such as paraffin. If we embed it in paraffin, that is in ordinary block paraffin melted, we find that the paraffin adheres to the tissue. We make a slice of the paraffin as well as of tissue, but there is a difficulty in holding the embedded tissue and the knife with sufficient steadiness, to overcome which, various appliances have been invented. I may describe them generally by saying they consist of a plate of metal with an aperture in its centre, into which a tube is placed with a washer and a screw. When you desire to embed the tissue in paraffin, you place the tissue in the tube, pour melted paraffin over it and allow it to cool; you then simply glide a knife across the opening of the tube, and so make a slice of whatever projects above it, and, by having an indicator placed on the screw, we can get sections of any thickness we please. But we always find that paraffin has a tendency to adhere to the knife, and it also adheres to the tissue, and there is often great difficulty in getting rid of it, and we find we have a slice of paraffin as well as the tissue, and the particles of the paraffin are apt to be troublesome. Very often, in making sections of an embryo, it is very difficult to accomplish, and you often get it spoilt in getting it free from these things. There is thus a difficulty in rendering some tissues sufficiently hard by the ordinary methods, without spoiling them. Then there is the difficulty about getting rid of the paraffin without spoiling the section. These two difficulties may be overcome by the freezing method. We can rapidly harden tissue by freezing; that is a method which has long been had recourse to, but it has been employed under circumstances of such difficulty that really it has been very generally neglected. The ordinary process has been to place the tissue in a metal cup, and to set it in a freezing mixture consisting of powdered ice or snow and ordinary salt. No doubt the tissue freezes rapidly in that way, but you require to take it out of the metal cup and hold it in some way or other in order to cut it, and it usually begins to thaw again immediately after you take it out of the mixture; you have nothing to support your hand whilst endeavouring to make the slice, and although you may succeed once in ten times, that is about all. With a view to overcome these difficulties, I invented this instrument, which I have named the freezing microtome. It consists of one of the older instruments I have already alluded to, with a box for a freezing mixture placed around the tube that holds the tissue. You may pour ordinary water round the tissue, and then place in the box a freezing mixture of powdered ice or snow and salt. Around the freezing box there is a coat of gutta-percha to prevent the entrance of heat. My late assistant, Dr. Urban Pritchard, suggested an important improvement, namely, to use, instead of water as the fluid to be frozen around the tissue, a solution of gum arabic; and I think, next to the freezing microtome itself, that is the most important improvement. As you all know, water, when it is frozen becomes crystalline, and if you attempt to cut it, though you can make thin slices, it necessarily splinters; but with this solution of gum arabic, when frozen, it can be cut like a piece of cheese. You may take this razor and try it, and you will be satisfied that what I am saying is correct, that you can cut that frozen gum and the tissue which it surrounds as easily and conveniently as you can cut a piece of cheese or a piece of gristle, so that really I do not know what more can be done to render the process of making these slices in this way more perfect than it is. Since I suggested this freezing microtome there has been several others proposed. The majority, I am bound to say, are no improvement at all on this piece of apparatus. One gentleman stated in the *Journal of Anatomy and Physiology* that he had always heard it said, that everyone who had used this freezing microtome never could succeed with it. No doubt it was judicious for him to state that he "had heard it said," because a little experience would have shown the contrary. I have the original instrument, which has been in almost daily use since it was invented, and I have never found any difficulty. No doubt I have made a slight alteration of my original instrument; having made the freezing box a little bigger; and here it is, and you can see there is no difficulty in using it. Certainly, in a hot room like this it takes a little longer than usual to freeze, perhaps a quarter of an hour, whereas usually it can be frozen in ten minutes, but still I think for a quarter of an hour to elapse before the freezing process is completed is not very unsuccessful, or a very great disadvantage. With this instrument you can make a hundred sections of the retina in half an hour, and when you have cut the slices you may, in a freezing condition, transfer them to a slip of glass and readily spread them out. If you have to deal with such an organ as the ovary of a plant, or a minute animal, an embryo or something of that sort, you may take the slice in a frozen state, lay it out on the slip, and thus you have everything placed in just the proper position for the purpose of mounting. You have no difficulty in removing the embedding agent, because of course the melted mucilage immediately flows away, and you can, if you like, pour a stream of water or some other fluid over, to wash it perfectly clean. All that can be done in a few minutes with the greatest readiness, so that I have no doubt you will find, if you use such an apparatus, it presents great advantages.

The only form of this instrument that meets with my entire approval is made by Mr. Gardiner, South Bridge, Edinburgh.

PROFESSOR ALLEN THOMSON : This instrument is a very succesful adaptation of our ordinary microtomes. I am quite satisfied of its efficiency from a number of circumstances. I rather rose for the purpose of calling the attention of the meeting to the great value of such an apparatus in making sections of organs in which there is much disparity in the consistence of the tissues. Many years ago I made sections of the human eye by means of freezing. Unfortunately at that time I had no such apparatus as is now possessed by physiologists which makes this process so much more easy and perfect in its results, but I found the freezing process of the greatest advantage in making sections of the eye, because by the consolidation of the fluids, I was able to obtain a perfect view of the exact seat of the different parts. Every one who has attempted such sections must know that the moment you attempt to cut into the ball of the eye, air enters, or some other circumstance happens which interferes with the exact relative position of the parts. I was able in this way to see that the lens was in close contact with the iris, and that in the young subject the lens was placed very near the cornea, from the small quantity of aqueous humour, and many other circumstances, which could only be brought out by this process. I think the instrument which Dr. Rutherford has so well explained may serve many useful purposes.

The PRESIDENT : I am sorry to find that Dr. Royston Pigott, who is announced for the next communication, viz., a communication on the Microscope, with Complex Adjustments, Searcher and Oblique Condenser Apparatus, is not here. He lives at a distance, and I presume something has arisen to prevent his coming. As Professor Flower is present, I will call upon him to give his communication

ON THE OSTEOLOGICAL PREPARATIONS EXHIBITED BY THE ROYAL COLLEGE OF SURGEONS.

PROFESSOR FLOWER, F.R.S. : Ladies and Gentlemen, I have a few words to say in exposition of the specimens which the College of Surgeons has sent to the Exhibition, showing the method now used in that Institution, and which has been adopted in many other museums, which I think you will all see is a great improvement on the method of mounting articulating skeletons, which was in use till about ten or twelve years ago. At that time every skeleton, either human or of any vertebrate animal, so far as I am aware, both on the Continent and in England, was mounted on the principle of the one now shown, sent by M. Tramond, of Paris, to this Exhibition. It looks extremely well at a distance for showing the general form and proportions of the animal, but when you wish to go further than that, and as all osteologists do now, to compare each individual bone with another of the same skeleton, or with the corresponding bone of another skeleton, as is required for a proper scientific study of comparative anatomy, such a skeleton as this is of course of no use, because the bones are all fixed immoveably together. They cannot be separated without destroying the articulation ; the limbs cannot be removed ; even the head cannot be taken off, so that if you want to examine the under surface of the head, which is a most important part of the structure, or even the teeth, you are entirely baffled. That was the difficulty I found at the College of Surgeons, and other museums. All the skeletons were mounted on that principle, and consequently there was a great difficulty in the study of comparative osteology, and also palaeontology, because when you had a fossil bone, by which all extinct animals are known, and wish to compare it with the skeleton, you often found that the particular part you wished to examine in the skeleton, was covered up, being in contact with the next bone, so that you could not possibly compare your fossil bone with the corresponding one in the skeleton. The consequence was, it became necessary to have duplicate specimens of the bones kept loose in boxes. That was an impossible thing to do in the case of a great many rare animals, and very few museums could afford a double set of every skeleton, and even when we had all the bones of the animal loose in a box it took a great deal of time and trouble to find the particular bone required for examination and comparison. I happened, most fortunately, when I took charge of the museum of the College of Surgeons, to have a most excellent assistant in the articulating department, who by a curious coincidence has the same name as myself, Mr. James Flower. He is a very ingenious man, and I at once pointed out to him the difficulty about these skeletons, and suggested to him that with his great mechanical knowledge he should try to apply some other method of articulating and fastening the bones together. The first thing to be done was to be able to take off the bones of the limbs, so as to compare them with those of any other animal. That was done without much difficulty by the introduction of brass tubes, and wires fitting into those tubes. There was a great idea at one time, with all the large skeletons, that nothing should be shown in the way of wire-work, but that they should look as if they were living things walking about without any external support. The bones had holes made at each end, and then had strong irons running through them so as to fix them firmly together. That was sacrificing a great deal too much for the sake of appearance, so we have quite given it up, and have substituted external supports, as you see. We began with this brass tubing and wire, and that led us at once to the method of mounting skeletons, which is now adopted by Professor Huxley, at the museum which was formerly in Jermyn Street, but is now at South Kensington, and was carried out by his able assistant Mr. E. T. Newton, who exhibits this specimen of a dog's skeleton. By that method, as you see, every bone has a wire fastened in it, which is fixed into a brass tube, from which it can be taken and examined by itself. That is a most excellent plan for teaching or working purposes, but we want more than that in a museum of comparative anatomy. We want, at the same time, to give some general idea of the size, form and proportions of the animal, which you do not in the least by this method. We have, therefore, improved upon it at the College of Surgeons, as it is a method which admits of infinite variation of application, according to the size and nature of the skeleton. I hope that any one who is interested in the subject will visit the museum, and see the different modifications in which this plan is carried out.

The two skeletons which we have sent here will give you an idea of the system. You see they are capable of being taken to pieces. I will show you how that is done, and the facilities we have for examining the separate bones. The first thing is, the head must take off, and that is an extremely simple thing. In the next place the limbs must come off. They are easily taken off by the use of wires sliding in the brass tubes, and the joints between the bones are made of strands of twisted wire. By these means the bones may be twisted backwards and forwards as if they were on a hinge, so that you may move a bone in any direction, and see every articulating surface. Then in the feet and hands the bones are all fixed on the same principle by twisted wire, so that they can all be separated, and turned one on another, and still they are kept fastened together, and in their proper position. These little bones when they are loose, as in separated skeletons, took a long time to find in order to compare with any fossil you had to examine. Here you find each at once, and in its proper place. There was a difficulty for some time with the bones of the vertebral column. By Professor Huxley's method they are all mounted on separate pieces of wire ; but to do that in a mounted skeleton such as this would involve a rather complicated apparatus. A very simple method at last was thought of to overcome it. The different pieces are all fixed to a flexible piece of twisted wire, fastened along the under surface. It may be turned about in any way, so that each one of the vertebræ may be inspected. In the fixed skeleton you cannot look at the front and back of the vertebræ, but here you can compare one vertebra with another, or a vertebra of this animal with a corresponding vertebra of another animal in a similar way. The ribs and breast bone are put together as a single piece, because by that method you see everything required. You see the heads of the ribs where they join against the vertebræ, which cannot be seen by the old method, and you can take them off and turn them round, and look at the breast bone, which is a very difficult thing to do when the skeleton is fixed in this position, so that we have almost every part of the skeleton now capable of being thoroughly examined, just as if the bones were all separate and loose, whilst at the same time it gives you an opportunity of seeing the general form and character of the animal. Of course the method of mounting skeletons like this is more troublesome and a little more costly than the old method in which all the bones are fixed together, but that is really hardly a consideration in a museum devoted to educational purposes because one skeleton mounted on this plan is better than any number mounted on the old plan. Here you have skeletons that are really serviceable for advancing education, whereas when you have a skeleton which may look very pretty at a distance, but cannot be examined, it is really almost an impediment to the subject rather than an advantage. I think this is all I need say at present, but if any one feels disposed, while they are here, to pay a visit to the museum of the College of Surgeons, I shall be very glad to show them many modifications of this method applied to many other skeletons. We have already sent specimens to various parts of the world—to America, and to many of the museums in Europe, and I shall be very happy to give any further explanation, having found myself the great advantage of this method when well carried out.

The PRESIDENT : We shall be very happy to hear any remarks by any anatomist on this method of articulating skeletons which appears so important for advancing the study of osteology, but if not I will call on Professor Crum Brown.

ON ANATOMICAL INVESTIGATIONS OF THE SEMI-CIRCULAR
CANALS OF THE EAR.

Professor CRUM BROWN : Ladies and Gentlemen, I feel that I must apologize, not for my presence here, because I am not responsible for that,—the blame, if blame there be, rests with the committee,—but for the way in which I shall make use of it, because I am, to a certain extent, intruding upon a section of which I am not a member. The subject however has an important bearing upon biology, although the methods of investigation more strictly belong to physics and engineering ; it is as to the question of the anatomy and physiology of the semi-circular canals of the internal ear. The matter I have to bring before you is first a method of investigating anatomically the position of these structures, and, secondly, a theory as to their use. As to the theory of the use there is a point which I wish to mention now in case I should overlook it—that is, the historical question of priority. After I had communicated my views on the subject to the Royal Society of Edinburgh I observed in *Nature* a paper read by Professor Mach, of Prague, before the Academy at Vienna, upon the same subject, and from communication with him I find that his theory was very nearly identical with the one I had propounded, and shortly after that, Dr. Breuer, of Vienna, propounded a similar theory. Of course what I have to speak of is simply the views I take upon the matter, but I make this statement that I may not be supposed to be claiming more than is due to me. These questions of priority are not of very great importance, except perhaps to the individuals concerned, but it is right that they should be put in the proper point of view. Professor Mach was, as far as I know, the first to suggest the theory. Dr. Breuer was next, and I was third ; the curious coincidence is that it was about the same time we all fell upon the very similar explanations of the use of these structures.

In the first place, as to the anatomy ; and the mode in which I investigated their anatomical position—because of course any theory as to the use of them must be based upon the position. The method which I adopted was this : I obtained an exceedingly well macerated skull, and made upon it two saw-draughts through the outer table of the skull, forming a large angle with one another. It does not matter how deep you go so long as you do not go through, and it does not much matter whether they are at right angles or not. The saw draughts pass through the mastoid part of the temporal bone and are continued into the neighbouring parts of the skull. I then got two pieces of steel plate, cut so as to fit into the saw-draughts, the position in which they fit being indicated by marks on the plates and on the mastoid bone. The greater part of the mastoid and the whole of the petrous portions were sawn out and the steel plates fixed to the portion of bone removed. I then fixed the plates in firmly by means of wire, and the bone could be replaced in precisely the same position it occupied before, by placing the ends of the steel plates into the portions of the saw-draught remaining on the skull. The removed portion of bone was then plunged into a bath of fusible metal and placed under the receiver of an air pump. On exhausting, bubbles of air escaped from the cavities in the bone, and on re-admitting air these cavities were filled with fusible metal. By repeating this operation ten or twelve times in different positions all the air was pumped out, and the cavities completely filled with fusible metal. The bone was then removed from the bath of fusible metal, and the adherent metal cleaned off. It was then placed in melted paraffin so as to cover the steel plates and the greater part of the mastoid portion. When the paraffin had solidified, the whole was placed in a vessel containing dilute hydrochloric acid and allowed to remain there until the bone was completely softened, and then with a pair of forceps, the pieces of softened bone were taken away, leaving a cast of the cavities in fusible metal. Besides the internal ear which I wanted to observe, this cast included the casts of the mastoid cells. I do not know whether anatomists are aware how extensive these ramifications of the mastoid cells are, but I think it would surprise even an experienced anatomist to see what a large amount of metal, consisting of the casts of the mastoid cells, was laid bare when the bone was removed. Even in the space between the semi-circular canals there were cells so that only a very small shell of solid bones surrounds the canals. These were taken away with great care with the aid of a hot needle, and all the casts of vascular canals were also removed. The cast of the internal meatus was also removed, but I could not find it in my heart to remove the cast of the cochlea, it looked so pretty; and was obliged to leave it there. The next thing was to get it in position for observation. I soldered a brass pin to the mastoid portion, and then the pin was fitted on an apparatus which is here. The pin fits into a socket which has practically a universal joint, and is fixed to a table which rotates on a vertical axis, so that in fact it is a large goniometer and may be used for measuring by means of reflection, the angles between reflecting surfaces. There are arrangements here, about which I need not go into detail, by means of which the intersection of the different surfaces may be brought to coincide with the axis of rotation. I then placed the bone in position so that one of the semi-circular canals should be horizontal, and brought it as nearly as possible over the axis of rotation of the machine. Then I retired to a distance with a telescope so as to bring the canal in question upon the cross wires of the telescope to ascertain that it was really as horizontal as I could make it. We cannot make these canals truly horizontal, because they do not lie in one plane. There is a certain amount of tortuosity in the canals, and therefore we must get the average plane as it were and make that average plane horizontal. I get it as nearly horizontal as possible by this arrangement of the telescope, and then screw the joints tightly up that it shall not shift. Then I remove it again and fit it into its place upon the skull, which is done easily by means of

P

these planes and catches which are numbered. Of course the canal being horizontal before, and nothing having been done to change it, it must be horizontal now when it is replaced in the skull. Then I place upon the skull, and in a horizontal position, a plate of glass with sealing-wax, which holds it quite firmly. It does not change the position at any ordinary temperature, and if it yields at all it comes off. Then we repeat the same thing with the other canals, and then put in a fourth plate which I have here marked " M " parallel to the mesial plane. That is easily done by fixing the skull in such a position that the mesial plane is horizontal, and its being so can be ascertained in the same way by means of the telescope. Having ascertained this, this plate marked M is soldered on. We then easily obtain the angles between these different plates either by the rough method of an application goniometer, known as a bevelled square which carpenters use for taking angles, or by getting it into a position in which the intersection of two glass plates coincides with the axis of rotation of the table, and measuring it as one would two faces of a crystal in a reflecting goniometer.

Next, what are the facts brought out in this manner. In the first place this, which is ordinarily called the horizontal canal, is not, as I suppose was known before, strictly speaking horizontal in the ordinary position of the head ; but the head requires to be bent a little forward to bring it horizontal. In the infant, the normal position of the head is not its natural position in the adult head, but it is inclined forward, giving the infant the appearance of a large intellectual forehead. It would be worth while to examine whether this amount of rotation about a right and left axis, which takes place as we grow up, may not be the same as the angle between the plane of the horizontal canal and a truly horizontal plane in the natural position of the head. This plane, however, of the horizontal canal is at right angles to the mesial plane, and it is so, as nearly as I can make out, in all animals, so that the two horizontal canals are in the same plane.

The next point is with regard to the position of the superior and posterior canals which are often spoken of as the sagital and frontal, as being parallel to the sagital and frontal sutures, as if they were fore and aft and right and left planes, but that is not so. They form nearly equal angles with the mesial plane. This seems to be very nearly the case in all the animals I have examined anatomically, so that we have one canal at right angles to the mesial plane, and two others equally inclined to it, and therefore, of course, altogether we have three parallel pairs of canals ; the two horizontal canals which are in the same plane, the right superior and the left posterior which are in parallel planes, and the left superior and the right posterior which are also in parallel planes. This coincides with the view I was led to take before, from the physiological point of view, of the two internal ears, that the two sets of semi-circular canals form really one organ, the whole of which is necessary for the use which it serves. Each canal has an ampulla, but if you take the two parallel canals together the one has an ampulla at the one end and the other at the other end. If you take a pair of horizontal canals you can draw them both on the same plane. But the ampulla of one is at one end, and the ampulla of the other is at the other end, so that if you were to turn the whole thing round, one ampulla would proceed and the other follow its canal, and the same with the other pairs ; but it is obvious that one canal at right angles to the mesial plane, and two other canals making equal angles with it, is the only arrangement which would secure this along with the bilateral symmetry of the body.

Leaving, then, this subject of anatomy to turn to the physiology, the use of the semi-circular canals has been the subject of a great deal of discussion, and one theory, an old one, and one which still has adherents, is, that they have to do with our perception of the direction of sound. If we have to propound any other theory, we have first of all to see whether we have any other means of accounting for our knowledge of the direction of sound. The first experiment I made on this subject brought out, what was to me, an extremely curious fact, that we have no direct knowledge of the direction of sound. Such knowledge is made up from a great many different observations, and if you limit yourself to hearing alone you find that you do not possess that knowledge. Part of it arises from our having two ears, and although, considering the size of the waves of sound, and the smallness of the human head, the shadow, so to speak, which the head throws upon the ear, is a very faint one, still it is sufficient to give us an idea of whether the sound is to the right or to the left. We do hear a sound which is to the right of us louder in the right ear than we hear it in the left, and that gives us an idea that it is on the right, but I do not think we have any notion of the direction of sound whatever from the sense of hearing alone. If we are allowed to turn our head round, then we can find out the direction of noise, especially if the sound is continuous or repeated, because we turn our head until it arrives at that position in which we hear it most distinctly. Most persons hear better when facing the sound from the shape of the external ear, although some have more advantage in that respect than others. If we look at the lower animals we see the same thing. A cat pricks up its ears, and turns its head round until its head is directed towards the object from which the sound comes, and then it runs towards it. It is easy to make the experiment by a little apparatus, which most people carry in their pocket, namely, three coins, by means of which you can produce a very sharp click. A bell has been tried, but it is not a good sound; no musical sound is a good one, because it is reflected, or reverberated from objects which are in tune with it, but such an unmusical sharp sound as a click of three coins is quite distinctly heard, and can be varied in intensity, and can be repeated any number of times. If the person to be experimented upon has his eyes bandaged, and his head kept in a fixed position, and this sound is made in various positions about him, and he is asked to point out where it is, he will usually point in the wrong direction, particularly if you keep in the mesial plane, so as not to allow him the advantage of knowing whether it is on the right or left. If you make the sound in front of his nose, he will as probably point to the back of his neck as in any other direction. In fact, a friend of mine, with whom I made the experiment, pointed almost always to the back of his neck, which seemed to be his favourite position for hearing everything. It is a very curious thing, which I recommend to psychological physiologists, that though you may be perfectly wrong in your judgment, you are perfectly convinced, and have not a moment's hesitation. You hear the click, and knowing that the sound must come from some place, you instantly fix, on the most wretched ground of evidence, upon some direction, and make up your mind that that is the direction. It is very curious to see the surprise and incredulity of the person experimented upon when he is told where the sound really came from, and he will ask to have it repeated with his eyes open, and then thinks it is quite different, so incapable are we of abstracting from our judgment the facts which we know. Having then shown that this was not the use of the semi-circular canals, or that if it was, they performed it very ill, the next point was to find what is the use of them, when what occured to me was this. We have here a set of canals, and let us consider them as containing liquid and various light organisms, floating more or less loosely in the liquid. Suppose we turn the whole of this round on an axis at right angles to the plane in which they are situated. The liquid contained in these canals is continuous from one end to the other through the vestibule ; if the head is quickly turned round the liquid will lag behind; as anyone may see by pouring water into a tumbler, putting a few float-ing bodies into it, and then quickly turning the tumbler about a vertical axis, the floating bodies remain practically at rest, and you can turn the glass round without moving them. We should expect then that the liquid would remain steady while the canal moves round, or what would come to the same thing from the subjective view, that there would be a flow of liquid along the canal, and as we have two canals parallel to one another, the flow in the one case would be from the ampulla into the canal and thence into the vestibule, and in the other from the canal into the ampulla, and thence into the vestibule. If this rotation is continued in the tumbler, after a time, from the viscosity of the liquid and its friction against the glass, the whole of the liquid in the tumbler begins to go round steadily with the tumbler, and there is no more any relative motion. Accordingly if you turn the arrangement of the canals round for some time, after a time depend-ing on the friction, there will be no more any current flowing through. If then we stop the tumbler, the water in it continues to go round as before, and the bodies in it revolve and after some time that secondary motion also comes to an end. Now supposing, we stop the rotation o the head of an animal when relative rest has been attained, then the liquid will continue to flow round, and we have a real flow through the canals in the direction opposite to that in which the original flow took place. In the first place the bones moved round, while the liquid remained behind ; but when we stop it, the flow takes place in the other direction. because the liquid has now taken up the motion which the bones originally had. In the one we have the liquid **at rest,** and n the other the bones at rest, but as far as the animal is concerned it comes to the same thing, the relative motion being all we have to look to.   Now there are some phenomena which almost all of us have experienced to a slight degree, which we can also experience to a greater degree by means of proper apparatus.   Suppose we turn ourselves round, the simplest way of doing so is to rotate by means of our muscles.   But here we have another means of judging how much we turn ; we know how much we move our feet in turning round.   A better plan is to stand or sit on a revolving table.   The first one I used was an ordinary anatomy demonstration table, and an assistant turned me round.

It did not go very smoothly, and there were certain jolts from time to time which informed me of certain places, and I knew when I had gone once round by means of those particular jolts, but it answered pretty well.   After being turned about once round, sitting on a stool in the middle of the table with my eyes bandaged, I lost the distinct sense of rotation ; and afterwards in trying the experiment again on a much more perfect apparatus, the revolving table of a lighthouse, kindly put at my disposal by the Commissioners of Northern Lights, which went very smoothly, after about one or one-and-a-half turns, all sense of motion was abolished, and I felt as if I were at rest ; and after three or more turns I felt perfectly steady, so that I could scarcely believe that the motion was still going on.   A slight rolling noise of the rollers on which the table turned was all that I heard, and that might have been unconnected with my motion for anything I knew.   Then upon the stopping of the table I felt a violent shock, and felt as if I were turned round as rapidly as at first I had really been in the opposite direction.   To make a quantitative experiment is a little difficult.   The one we attempted first was that the person experimented on sat on the table, and when it was turned round he was told to call out when he had turned through a right angle, and again through another right angle and so on.   We found that the action was quite uniform, but the reason was that he fell into the habit of counting one, two, three, four, five, six, &c., and on asking him " Do you really feel it turning round," he said, " Oh, no ; I was going on counting from habit."   Thus knowing it was a uniform motion he was using his reason to assist him, not his sense.   The better mode was to say 'estimate the first right angle and then say when you feel at rest.' We got better results in that way than by trying to count the other right angles. The first right angle was almost always a little over the right angle, and after perhaps one-and-a-half turns, differing in different individuals, the sense of rotation vanished, and then on stopping the rotation I asked him to say when he had turned through a right angle, and compared the time he supposed it took with that occupied in the actual performance of a right angle in the first instance.

The non-horizontal canals can be brought into play by lying down on the table, and it was possible to compare the time requisite for the abolition of the sensation of rotation, the head being in one position, with that requisite when the head was in another position. It was found that this time was longer when the axis of rotation was at right angles to the plane of the horizontal canals than in any other position.

Now if you look at the semicircular canals you see that the horizontal canals are wider and shorter than any of the others, and therefore one would expect it to take a longer time for the friction and viscosity to abolish the relative motion of the liquid in these canals than in any of the others : the others being longer and narrower there would be more friction, and the relative motion would soon come to a stop.

Another point with reference to these experiments which I may mention is this : That if you change your position while you are rotating, a very curious sensation is felt, which I have tried upon myself, having previously however calculated out, which I should not have done, what I should feel on the theory, and I did feel it, but as we are so apt to be deceived if we know what ought to happen according to our theory, that we are almost certain to feel it if it is possible to do so. I therefore asked the engineer who had made the table, and who was watching the experiments, to lie down upon it, and make the observation. He did so, and he felt the same thing that I had felt which was this : If you change the direction of the head while the rotation is going on, turning round say from having one pair of canals horizontal till you get another pair of canals horizontal, you experience a combination of two things—first, the new real rotation, because of course the head is now being rotated about another axis ; you feel that real rotation, and you feel the subjective rotation resulting from the old real rotation. So that you have a resultant effect which varies from moment to moment, because the real rotation is going on, and the sensation of it is dying away. The sensation of the imaginary rotation is also dying away, but not necessarily at the same rate on account of the difference in the dimensions of the canals. If, for instance, from lying on the side you move over to the back, you feel the feet tilt up or tilt down, and the rotation about a right and left axis is perceived which dies away, and then you feel at rest again.

The next question is, how is it that if the rush of liquid through the canal excites the nerves in the ampulla, we are able to perceive any difference between turning in one way and turning in the other. If the hairs in the ampulla are agitated by the flow of the liquid, it is difficult to see how it should make any difference to them whether you stroke them one way or the other. A nerve current is started along the nerve to the nerve centre, and one can scarcely see how it should be that irritation by one method or another method should convey opposite impressions. I understand Professor Mach's original view was that there is such a difference; that there are in the ampulla two sets of nerves, and that one set is irritated by motion in one direction and the other set by motion in the other direction. My idea (which had also occured as a possible one to Professor Mach, and which I understand he now thinks most probable), was that as we have two canals for each axis one answers the one purpose and one the other, that the nerves of one are irritated by the motion in one direction and those of the other by motion in the other direction, so that, supposing the motion with the ampulla first is the one which causes sensation, if we turn the head round in one direction then one ampulla is irritated, and if we turn the head round in the other direction then the other ampulla goes first and we have it irritated in a similar way. We have thus in the six semi-circular canals, six different organs, each of which appreciates a special rotation, the whole six being necessary for the full appreciation of rotation in general.

Is there anything we can suggest as to how this should be? My first idea was, from my ignorance of the minute anatomy of these organs, and supposing that was right which we see in ordinary text books, namely, that the membranous canal hung loose in the bony canal, and that there was a quantity of perilymph surrounding it. My idae was that the membranous canal would lag behind along with the liquid, and if the ampulla went first there would be a pull upon the ampulla through the membranous canal. But that must be given up, because I understand it is now settled that in every animal the membranous semi-circular canal is tacked down to one side of the bony canal pretty firmly, so that there cannot be a great amount of relative motion of the membranous canal, and we must therefore look to the motion of the liquid in the interior of the membranous canal.

Can we suggest any explanation why the liquid rushing in one direction should act differently from the liquid rushing in the other direction? It may be so, but it would be necessary to make special observations for this purpose which I am not qualified to make, to see whether the liquid rushing through the small opening from the canal into the ampulla might not wash through without disturbing the hair cells, whereas coming in through the wider opening it might wash round and so affect those hair cells. No doubt they project to some extent into the ampulla, and it is not difficult to see how there may be a very considerable difference in the mode in which the hairs may be affected by washing in the one direction or in the other. Take the case of a tidal harbour; when the tide runs out we have quite different eddies round the quays than when the tide is coming in. However this may be, it seems to be shown that we could not do with one labyrinth as we can do with one cochlea, by the fact that the destruction of the whole of the internal ear on one side does produce disturbances in the equilibrium of the animal.

The anatomists and physiologists present know far better than I do the experiments which have been made on this subject by the destruction of various portions of the canals ; I understand such experiments are exceedingly difficult to make from the smallness and deep seated position of the organs, but all such experiments seem, as far as I have been able to follow them, to confirm this view. Pathological observations also seem to do so.

One curious disease is connected, as a rule, with one side, one ear is affected, and it is difficult to suppose how it could occur in such a small organ as the internal ear, that changes should took place in it, confined to one semi-circular canal and one ampulla. In this disease there is a very marked sense of rotation of the precise kind which one would expect to find if the whole of one ear were affected, supposing the labyrinth to have the function above suggested.

I show here a preparation of the skulls of several birds in which the semi-circular canals have been dissected out, leaving the hard bones which surround the canals. These have been painted so as to show the three pairs of parallel canals, and by turning them in different positions it will be seen that they are sensibly parallel to one another in various birds. Other specimens I have here are of considerable anatomical interest. They are casts in fusible metal of the internal ears of various cetacea. Most of the bones were given to me by Dr. John Anderson from cetacea which he obtained in the first expedition to Yunnan from Burmah, and, these specimens having been figured by Dr. Anderson in his work on the zoology of that expedition, he asked that they might be presented to some museum where they might be properly preserved, and they were accordingly presented by him and by me to the museum of the Royal College of Surgeons.

The PRESIDENT : Ladies and Gentlemen, the subject of the relation between the planes in which the simi-circular canals of the ear lie, and the function of those canals, is one which has been the subject of very important researches recently in Germany, and in particular the theory which has now been brought before you, which is often spoken of as the mechanical theory, has been very warmly discussed, and as we have before us no German physiologist but a Dutch physiologist who is *par excellence* the representative of the physical side of physiology, I of course allude to Professor Donders, we shall be disappointed if he does not address a few observations in respect to the subject which is before us. I am sorry to see he has left the room. We have also another gentlemen present who, to English readers, is necessarily associated with the subject of physiological acoustics, Mr. Alex. J. Ellis, who I hope will say something on this occasion.

Mr. ALEX. J. ELLIS, F.R.S., drew attention to the fact that Lord Rayleigh's experiments on the perception of the direction of the source of any sound (communicated to the Musical Association on 3rd April this year), seemed to show that the quality of the tone, and, consequently, if Helmholtz be correct, the cochlea was concerned in the result obtained, and not merely the semi-circular canals.

Dr. ALLEN THOMSON made some remarks on the functions of the semi-circular canals of the ear as connected with the appreciation, either of the direction of sounds or of the position of the head. He admitted that no satisfactory proof could be adduced of these canals being the seat of sensations which give a knowledge of the direction of sounds.

With respect to the second view of their function, as illustrated by Dr. Crum Brown's apparatus and experiments, while acknowledging the value of these experiments, and the ingenuity of the theory connected with them, as bearing upon the possible relation of the semi-circular canals and their ampullae to the appreciation of the position of the head and the sense of rotation, Dr. Thomson stated that he was still inclined on anatomical grounds to doubt the probability of such a variation of pressure, or the existence of such currents or tides of the perilymph within the membranous semi-circular canals, as would cause the sensations necessary to account for the phenomena. The whole subject he regarded as still involved in great difficulty and requiring farther investigation.

Dr. CRUM BROWN : I should just like to say one word as to what Dr. Allen Thomson has said. Professor Maisch states that there is no real current, but there is pressure which would produce a current if it were not brought to a stop instantly by the extremely small size of the tube, and it is the pressure and not the current which produces the sensations.

The PRESIDENT : I will now call on Dr. Pritchard to give us his communication

ON MICROTOMES.

Dr. URBAN PRITCHARD : Dr. Rutherford has said a great deal in favour of cutting microscopial sections by means of freezing when we have to do with soft tissues, and, therefore, I need not say anything more about that. But there are two points which he might have insisted upon ; first of all, that when you freeze the tissue, you render it of such a nice consistency, that it is very much easier to cut a thin section by hand, and as a great many persons prefer cutting the sections by hand—this mode of freezing helps them. Then again, sections cut by freezing are thinner than the sections cut in the ordinary way, for this reason ; in freezing ordinary soft tissues, which must contain water, the size increases, and therefore in thawing, the sections become thinner than when first cut.   There are three other microtomes in the collection here. First of all we have one by Dr. Mulder. This consists of an ordinary hand microtome, with a screw below, and a chamber outside for the ice and salt.   The tissue frozen outside is put into this glass tube, and then the ice and salt is put around it, and the slices made.   I can hardly say what are the advantages and disadvantages of this, because one has to use it before one can tell, but I should think there is not sufficient rest for the razor.

The next one is a very simple one, which I devised six months ago, from using Dr. Rutherford's machine.   I noticed that the gum water always froze on the top of the metal, very much quicker than inside, and I dare say many of you are aware of the practical joke which German students are apt to play on another in cold weather.   They ask somebody to touch their tongue against the handle of the door, and it freezes there.   It occurred to me to freeze a piece of metal and put the tissue on the top. I tried it and it succeeded very well. The advantage is that it can be used much more quickly and readily, and two or three can be used at the same time.   Again the expense is a mere nothing. I have here an ordinary child's pail, in which I put some ice and salt, and I have immersed all of these microtomes, if they are to be so-called.   I then moisten the tissue, which may be as soft as you like, with a little gum water, and placing it on the top it immediately freezes to it, and then putting over a felt cap, the whole is frozen in about two or three minutes.   When expense is a great object, or when many microtomes are required, little pieces of brass, turned by an ordinary workman, with a plug of firewood for a handle, will do very well, and then you can have a dozen or thirty persons using them at once with a very small expense. This apparatus may also be frozen by means of ether spray. This ether spray takes at the ordinary temperature about two-and-a-half minutes. I have put a little gum here on the top which will indicate when the freezing commences.   In cutting frozen sections it is always well to cool down the razor or else you are apt to thaw the tissue too much. A little ether spray will do that, or a little ice and water will do as well. This is frozen, and you see it can be cut with the greatest of ease. As regards the thickness I think nearly every one who has invented a microtome will allow that they can cut thinner sections by hand than can be cut by means of any arrangement for judging the thickness, simply because you get a more wedge shaped cut. This will remain frozen long enough to cut 100 sections or more—which is quite sufficient time.

I have here a more complicated microtome, which is a modification of the same. This is modified by Mr. Williams, and it consists of a thick wooden box to prevent the radiation of heat, and inside that is ice and salt. Any tissues can be frozen to the plate in the top and kept frozen. Here we have the arrangement which is used in America of a triangle and a knife which slides over the top of the section, and by means of three screws the knife can be raised or lowered, and by that means the thickness of the section regulated. This, no doubt, is a more complicated form, but it is exceedingly useful for many things. Under ordinary circumstances, as it is natural I should, I prefer my own instrument, because it is so quick and can be used so readily. You see it is just beginning to freeze. The temperature of this room is very much against freezing, and warm weather is one of the great objections to freezing microtomes of all kinds. Of course the ether is very much more expensive than ordinary ice and salt, but there are times and places when it is impossible to get the ice. It is scarcely worth while going any further with this experiment, because you will see it is sufficiently frozen for ordinary purposes.

The PRESIDENT : I now call upon Professor Gerald Yeo to give his communication

ON MICROTOMES.

Professor GERALD YEO : You have just heard from Professor Rutherford and Mr. Pritchard that the requirements of the rapidly growing science of histology demand several kinds of instruments for cutting thin sections of various textures. Among the many difficulties which have to be overcome is that of cutting sufficiently thin slices of very soft substances, such as the majority of organised tissue. These must first be hardened in one or other of several ways.

The process of freezing them until they are hard you have just seen demonstrated. There are, however, many tissues which cannot be frozen without their structure being spoiled, and, therefore, other methods must be used to make them tough enough to be easily cut in thin sections. The piece of tissue being held in the hand may then be cut with a razor. When the piece is very small, however, it is advisable to *imbed* it in some material of like consistence, in which it can be cut as part of the entire mass. These imbedding mixtures, of which there are several here, are commonly made of varying proportions of parafin, wax, lard, &c., which melt when warmed, and then can be poured around the piece of tissue. Thus imbedded they are sometimes cut in the free hand. Most histologists prefer one of these instruments—microtomes—of which you see a series here. In general they consist of a flat plate, upon which the razor can rest. In the centre of the plate there is a cavity to hold the tissue and imbedding material. This mass can be moved slowly above the level of the plate by means of a screw working from below. The thickness of the section can thus be accurately determined. With these instruments more extensive thin sections can be cut than with the free hand. The thin slices are so very easily torn that the razor and the tissue must be kept moist; otherwise they would stick to the surface of the knife and be dragged to pieces. The moistening may be attained by plunging the razor in some fluid before each cut, or what is better, the tissue may be kept moist by means of this simple dropping bottle, the outlet of which is to be placed over the surface to be cut. If the sections are very large, or the tissue brittle, they must be cut completely under water, and then floated off the knife. Dr. Rutherford said he could not imagine any improvement upon his instrument; here is a microtome which certainly has the advantage in size. It enables you to make enormous sections under water, of large organs such as the brain. The entire human brain can be embedded in this cavity, and then split into a series of microscopic sections, showing the relative structure of the entire organ. Here are some others for cutting smaller objects also under water. With it a small trowel-shaped knife is used, off from which the thin slice may be easily washed and separated from the imbedding material. Besides the many more simple microtomes which are to be seen in the Exhibition, there are some worthy of more special notice. Here is one which enables the histologist to cut sections in various directions. The object is embedded in a kind of wedge-shaped cavity; one side of the mass can be pushed up to the action of a lateral screw; the other—thin end of the wedge—remaining fixed. By this instrument, fibres or vessels which pass in a radiating direction may be followed by the plane of section which continues to be parellel to them, and thus a much greater number of longitudinal sections of the elements is given than one could otherwise obtain. Moreover, the sections cut by this instrument taper off to one edge in such a way that a part, at least, of them must be extremely thin and fitted for examination with the highest powers. The microtomes you have seen so far only fix the object to be cut, and steady the knife, which is held in the hand. Here is another set of instruments, in which the knife, as well as the preparation, is fixed in a supporting frame. The most practical of these are the so-called *inclined plane* microtomes. The first of these was devised by M. Rivet. On one side of a metal stand is a horizontal groove, in which runs a frame carrying a knife. On the other side of the stand is a groove on an inclined plane, upon which slides the clamp, holding the preparation to be cut. By moving the clamp up the incline the preparation is raised to the level of the edge of the knife, so that the thickness of the section may thus be regulated as accurately as with a screw. The knife which is moved on the horizontal plane cuts the object at a fixed level. In this instru_ ment, by Fritsch, there is the improvement that the object is imbedded in a case for the purpose, and not held in a clamp, which of course would tend to crush a delicate structure. All the microtomes, in which the knife or razor is fixed, have a great disadvantage, as they do not allow the edge of the knife to be drawn along the substance while cutting. No matter how obliquely the knife is set it must, more or less, push through the tissue and not cut cleanly.

The Conference then adjourned for luncheon.

Mr. BOWMAN took the chair, in the temporary absence of the President.

The CHAIRMAN: I will call first on Professor Donders to describe the instruments he has constructed to explain the movements of the eye.

Professor DONDERS showed some apparatus illustrating the movements of the eye, and used in researches on the laws of these movements (phaenophthalmotropes, isoscope, cycloscope, horoptero-scope, controller of the laws of Donders and Listing, and other apparatus). He described the four types of movement of the eyes, with parallel visual lines, with convergent visual lines, symmetrical rotation and unilateral rotation ; and explained their origin in connection with the function of monocular and binocular vision. (The subject needs demonstration, and a report without the apparatus would not be intelligible, even with numerous illustrations. Full details are to be found in "Proeve eener genetische verklaring der oogbewegingen, in *Onderzoekingen physiologisch laboratorium der Utrechtsche Hoogeschool,* 1876, iv., 1 ;" "*Archiv f. die gesammte Physiologie von Pflüger,* B. xiii., 1876 ;" "Essai d'une explication génétique des mouvements oculaires, in *Archives Nederlandaises des sciences exactes et naturelles,* xi., 1876 ;" and "*Annales d'Oculistique,* T. lxxvi., 1877.")

The PRESIDENT : I am sure, Ladies and Gentlemen, that we are all much obliged to Professor Donders. I hope he will also give us a demonstration of his apparatus and instrument for the purpose of determining what time is taken in the process of thought ; in fact his apparatus for investigating the velocity of thought.

Professor DONDERS : Ladies and Gentlemen, you know that if we have the will to move the hand the muscles contract, and if we irritate the skin we feel it. But the will is connected with the brain, and so is the feeling. Now the nerves convey the action from the brain to the muscles, and from the skin and other organs to the brain. Thirty years ago, Johannes Muller predicted it would never be possible to measure the velocity of the currents in the nerve. As du Bois-Reymond found that the action of the nerve is not a propagation of a simple mechanical movement, but a propagation of a process of chemical change, he thought it very likely that it would not be so very rapid as was supposed. Indeed, it cannot be compared with the waves of light, nor with electricity, but it may be compared to a train of gunpowder ; when lighted at one end the action runs along. It goes rather quickly, but not so very quickly. Such an action has to take place at every point, from section to section. It is a new development of chemical change, of physical (electro-motor) action, of heat, at every point. On this ground du Bois-Reymond thought the velocity might be measured, and even described a method; but Helmholtz was the man who first did it. The method is very simple. You kill a frog and take a muscle with its nerve. You may irritate the nerve at a certain point A, and determine the moments of irritation and contraction in the way which has been demonstrated here by your honoured President in opening these sittings. Immediately afterwards you may irritate it in B, nearer the point where it is attached to the muscle, and you find that the contraction comes a little sooner. The difference was the time necessary to conduct the action from point A to point B. If that has the length of one centimetre it would be a difference of the three-thousandth of a second, because it has been found that the velocity of these conductions is about thirty metres in one second.

There have been many different physiologists, and especially the astronomer Hirsch, who determined what we call physiological time— that is, the time required for reacting by a movement on a nervous irritation. If there is an irritation on the skin near to the head and you have to give a signal with the hand, it follows in the seventh of a second. If a sound acts on the ear, it follows after one-sixth of a second; if light irritates the eye the signal follows after one-fifth of a second: and these times are the physiological times. Now in making some experiments on acoustics with a phonautograph, I saw that this was a very good instrument for determining the physiological time. Here you see a simplified phonautograph. This is an elastic membrane which may vibrate by resonance, and on it you may find a very small feather which vibrates just in the same way. If I utter any vowel sound "A" or "O," it vibrates. If I ask our President to sound "A," and I follow him as soon as possible, the feather register-ing on a rotating cylinder, and at the same time the vibrations of a tuning-fork being recorded upon it, whose vibrations are 250 in one second, we can tell by the number of vibrations between the mark registered by the first sound and that registered by the second, how long a time has elapsed from one to the other. While determining the physiological time with this instrument, it occurred to me : would it not be possible to superadd to it some psychical action? Of course you know nothing about the time of the psychical phenomena in the experiments described. You hear the sound by the ear ; it is translated after some latent period, to certain nervous cells, and hence translated to the brain, and from the brain itself to other cells, the ganglion cells, and to other nerves, and then to the different muscles of the chest and larynx, and then follows the signal "A" as you have heard it. Now, how much of that time, which is one-sixth of a second, belongs to the psychical process? You do not know. You only know it is less than one-sixth of a second. It is clear that some time is wanted for the conduction in the nerves ; perhaps a little more for that in the cells ; but at all events you do not know how much. Then how can you introduce a psychical process? I will ask the President again to sound some vowel, but not to tell me which he is going to sound, and as soon as I hear it I repeat the same sound. It is quite certain if the experiment has been well done, that it takes a longer time than the first, because I have not only to catch the sound, but to be certain that it is the true sound, e.g., the "O," and that requires more time. Calling the times in the first experiment a, and in this b, the difference between a and b is the time necessary for judging of the *timbre* which is given by the vowel, and for giving the will to the muscles to shout out the same vowel. But here there are still two psychical actions, the one for distinguishing the *timbre*, the other for giving a corresponding will to the muscles. Is it possible to separate them? Yes. I have only in repeating the last experiment, to answer exclusively to the sound of "A." If our President sounds any other vowel, I do not answer. In this case I have only to distinguish the character of the sound, the muscles and larynx being already prepared for shouting out the sound agreed upon, and I have nothing to do but to emit the air so as to produce the sound just as in the first experiment. Calling "c" the time found in this experiment, a—c is the time required for simply knowing A, and b—c is the time necessary for volition.

You only require a cylinder and a tuning fork for making the experiments ; but I have tried to extend the field as much as possible, the rather complicated instrument you see here being the result. I have called it a "noëmotachograph." At one end of the cylinder are two rings, partly covered with copper, partly with ebonite. On each ring two pairs of electrodes rest; the lower two break and close a constant current, and the upper ones an induction current. The induction currents pass through the tuning fork, and when the tuning fork is struck and the cylinder moved half a turn, the lower electrodes leave the copper, and the induction current makes a small hole in the chronoscopical line of the paper on the cylinder. In this way, using one of the rings, several kinds of experiments may be made. We can have the spark visible without being audible, or audible without being visible; or we can have a shock produced, the spark being neither audible nor visible; and in this way we can determine alternately the physiological time for the different senses. But we can do much more. We can irritate different points of the body, or we can produce different sounds, or show different colours, or different letters, all by the spark, and, by subtracting the simple physiological time, determine the time necessary for recognising, either reacting by the voice on the phonautrograph or with a movement of the hand. In this way we can vary the experiment enormously, and even more complicated psychical processes may be introduced.

The second ring has only a small conducting plate, upon which, at the rotation of the cylinder, the constant current is closed and broken, after contact with the electrodes of the induction current has followed. Thus a second spark is made to pass, perforating the chronoscopic line, and at the same time either visible near to the first, audible, or sensible at another place of the skin, sooner or later after the first spark: by sliding the second ring, this interval of time can be modified. Hence we learn how much time is required between two impressions on the same sense, or on two different senses, for enabling us to judge about the priority of the two. This time seems to represent the time for a simple thought.

More accurate notion about this time is obtained by another instrument, which I called *noematachometer*. You see it is a high vertical board with a long metallic prism suspended from a sliding electromagnet. The prism has in front and behind a pair of sliding arms upon which a rod can rest, at its back part two small sliding metal wire frames, which can contain a very small piece of burning coal. On closing a current the prism falls, and is received in a box with asbestos. Now, the eye is before a slit in the board. During the fall, the rods, at meeting respectively a wooden and a metallic arm, each produces its characteristic sound ; the frames with incandescent coal pass the slit, one to the left, the other to the right. Thus either two sounds, two lights, or a light and a sound can be made perceptible with *changeable* difference of time (by sliding the prism, rods and frames). If we now increase the difference, we learn accurately how much time is required between two impressions, for enabling us to judge with increasing chances to perfect certainty, about the priority of one of these.

As to the time required, I found in the first experiments with the noëmatachograph, that, in a simple dilemma, $b—a$ equals seventy-five thousandths of a second, which is the time required for recognition and corresponding volition ; $c—a$ only requires forty thousandths of a second, which is the time required for simple recognition, thus leaving thirty-five thousandths for volition. Now, I think it is very remarkable, that the same time, the twenty-fifth of a second, is found necessary for enabling us to judge about the priority between two irritants, acting on the same sense, in the noëmatachometer.— So we obtain the same result for the time of a simple thought, in following two quite different methods. If the two irritants act on different senses (sound and light), more time is found necessary for enabling to judge about the priority. Also the recognition of a letter by seeing its form requires more time than by hearing its sound, and indeed the first is a more complicated psycho-physical process.

The time of the twenty-fifth of a second for a simple thought is found in a man of middle age, and who perhaps thinks not so very quickly. Some of my younger hearers might want less ; but the difference would not be very great. In all my experiments I never found less than a fortieth of a second.

I think, Ladies and Gentlemen, that is all I had to tell you.

ON THE MICROSCOPE WITH COMPLEX ADJUSTMENTS, APLANATIC,
SEARCHER, AND OBLIQUE CONDENSER APPARATUS.*

Dr. ROYSTON-PIGOTT, F.R.S. : I shall not occupy much of your time,
but shall confine my attention to a few general remarks, and I think,
perhaps, the most interesting thing for you now would be to hear of
the great improvements which have been made in microscopic glasses
during the last fifty years. Time was, when physiologists were quite
satisfied with the single lens, which would magnify 200 or 300 times.
They then made observations under very great difficulties, and we can
hardly believe now the progress that they made under such circumstances. We should not like to undertake it ourselves, but we can see
that the old English plodding perseverance was as rife then as ever.
It is a simple fact that the microscope at the present moment has
attained such a degree of precision that, whereas fifty years ago, people
were satisfied with a magnifying power of 500 times, and indeed said
that could not be exceeded ; now 5,000 times have been reached. This
has been done principally by a beautiful combination of glasses, and I
believe that Professor Amici was the first to use no less than five
different refracting materials for the formation of one object-glass. Dr.
Goring, the greatest enthusiast in the microscope, perhaps that the
world ever saw, invented tests for the microscope, and the most
favorite of all his tests was one called a " podura "—a little insect
which is found in cellars, and which is getting rather scarce
now. He found this *podura* under the best powers he had, presented
the appearance of little transverse lines, and he found no test so beautiful for objective glasses, and even at the present day amongst
opticians this little scale is much used, and it is their joy and rejoicing
when they can show its wonderful details. Dr. Goring, with the
small glasses then in his possession, was able to get the lines wonderfully perfect. When I was a young medical student, it used to be
thought the highest point of microscopical skill to be able to take a
little diatom called a *hippocampus*, and with a ray of oblique light to
get the transverse lines. If you turn the light the other way those lines
disappeared, and you had lines in a longitudinal direction. For a

---

* Phil. Transactions "On a Searcher for Aplanatic Images." Vol. II., 1870.

great many years all these objects were called *line objects*. It was
little suspected that these lines were no less than beautiful spherules
and beads.   To show the enormous advances that have been made, I
will now hand round for inspection some very beautiful photographs.
In 1845 this object used to be viewed as a line object, and glasses capable of showing these lines were highly approved.   Then there was a
controversy that they were not lines but hexagons, and Colonel Woodward has done me the honor to send me a great number of photographs taken under the microscope, in which he has anticipated some
researches in modern times showing the various lengths of waves, and
he found out that it was the blue light alone which was capable of
photographing the most delicate appearances of microscopic objects.
He found that the red, the yellow, and all the other rays produced a most
damaging effect upon definition.   He therefore adopted a solution of
the ammonia sulphate of copper.   A vessel full of that was placed
against the wall, the sunlight was admitted through it, carried to the
microscope, and he then photographed by means of those blue rays,
this object—perhaps the most splendid piece of photography ever executed at that time.   These lines then resolved themselves into the most
beautiful spherules, and Mr. Slack, the honorary secretary of the
Microscopical Society, sometime ago succeeded in retarding the process of crystallization by means of glycerine, and when that was
retarded, the effect was, the silica in solution began slowly to deposit
itself and formed beautiful glass beads.   These objects are so exceedingly small that some of them are not the one hundred and fiftieth
thousandth of an inch in diameter.   The next photograph obtained
by Colonel Woodward was this very difficult one of the *angulatum*
which he also resolved into beads.   That was for many years a line
object.   I dare say many of you have heard of those wonderful lines of
Nobert which he drew in some unknown manner—112,000 to the
inch on glass ; but these lines are very much more difficult of definition
perhaps than they ought to be.   Here is a specimen of photograph I
am happy to say was produced by an American objective of Mr.
Wales, and I think those who are judges will say it is one of the most
beautiful photographs ever executed.   There are 96,000 lines to
the inch most beautifully executed.   If we had sufficient power
they would resolve themselves also into these famous beads.   I had

the audacity some years ago to send a challenge to America to see if they could resolve the *podura* into beads, and Col. Woodward most handsomely accepted the challenge, and there is the result. That is a photograph of the favourite scale of mine which I operated with. I sent it to America under strong transatlantic promises that it should be returned, but I am very sorry to say that it has not come back.

The next stage of advance in the microscope was this. The *podura* scale was beautifully resolved into little tongues, as it were, and these little marks are still used by opticians to make their glasses by. It is perhaps the most difficult thing we have. This object is most beautifully photographed by Col. Woodward in lines. Dr. Goring's *podura* scales are, I believe, still the *ne plus ultra* of microscopy; indeed there are very few individuals in this country who will take the trouble to resolve it into its component structure. It is too late to go in detail into these matters, but my experience in the use of the microscope has taught me one thing in particular, and that is, the great number of useless rays that pass into the glass which obliterate minute structures and confound the definition. They are hidden in the excess of light or they are confused and striated. The next stage in the direction of microscopical development, if I may venture to use such a term, was this, they managed to invent a series of stops, condensers, and all manner of instruments of various kinds which I think may all be summed up in the simple effect that they cut off useless rays of light and leave the most useful ones, which were found to define the object. As one example of that, I think I may be allowed to mention a fact which I will direct your attention to particularly. I have not the honour to know Dr. Drysdale and the Rev. W. Dallinger personally, but I think I may be excused for mentioning their names, for I believe those gentlemen have done very much towards the development of microscopic science of the most valuable kind. You know there has been for a long time a great controversy upon spontaneous generation. We little thought at that time that there were two gentlemen quietly working at Birkenhead who were making a most wonderful use of the microscope in this way. They used for physiological research the highest power, one-fiftieth of an inch. That is a very delicate objective to use, and the result of their beautiful contrivance was this. They found out that in a given field where ordinary power and an ordinary microscope discovered not the least signs of animal life of even the minutest form, there were myriads of little monads turning and swishing their tails about like long whips, under this remarkable power of one-fiftieth of an inch, so that all conclusions founded upon the non-appearance of animalcules in a given fluid as seen by microscopes of very great cost and power fall to the ground unless these peculiar methods are adopted which have been now invented. Those investigations have brought out in a much better form the instrument which I invented ten years ago. I have here an object glass by which I can direct a light at any given angle; and by a wide spread stage motion, I can get an intense star of light at any point of the field or at any angle I like. Dr. Drysdale and Mr. Dallinger have not only limited the amount of light but they have fixed a lamp upon a bar upon which is placed a mechanical solid rest by which they can move it in any given direction, and by means of a vernier they can bring the lamp to one particular position within 100th of an inch, and out of the millions of positions which you may place the lamp in to get a particular ray of light; they have found out one ray in one particular direction which will give the most superb definition, and in this way they have not only been able to see these monads, but to see them swishing their tails about like a whip in all directions. Those tails are certainly not anything like the 150,000th of an inch in diameter, and I have some little doubt as to the power of the microscope being really limited by "wave lengths." By and bye the little monad grows, and from it myriads of little ones clearly visible come out, swish their tails about, and by and bye the tails join and you get another at the end. These magnificent results can only be shown by this wonderful contrivance, and it shows that there is one particular ray only which brings out the best quality of the lens. An object glass is like a horse, it has its tricks and manners, and if you want it to do its best you must coax it—you must not condemn it. You may spend years over trying a telescope, and so it is with an object glass. That illustrates the truth that by using one particular set of rays, you get out the finest effects.

A microscopist once told me that it would require a couple of hours to resolve the *rhomboides*, though it happened that perhaps by good management I was able to display it directly. If you have elaborate object glasses, if you alter the various degrees of aberration in your condenser, you must also alter the aberrations of the object glass, or they will not fit together, and will not define together.

One other point I may mention, which may be considered intrinsic to the age in which we live, and that is the subject of delicate measurement. I have not had the pleasure of being here at all the meetings, but I have noticed from the reports that a great deal of attention has been given to the subject of delicate measurement; in fact, it may be said that any advance in pure science depends upon accurate data, and those accurate data certainly mean very precise measure on optical principles. Now with regard to this point of measurement, I have taken some little pains on the subject, and I found the first thing necessary to do was to be able in an instant to know what power I was using. I used a *kratometer** which has a lens exactly magnifying ten times; inside is a micrometer lens containing lines 100 to the inch. The moment you wish to know what power you are using, no matter what length of tube you have got, and no matter what objective, all you have to do is to drop this into the tube, look through it, and see how many lines here occupy ten lines there, and instantly you have the power. I can get it comfortably to tens, which is near enough. The next difficulty I experienced was this. Very often one-thousandth of an inch on the stage would occupy very much more than the entire field of view, and how was I to proceed with that? I might look at half the thousandth of an inch, but that was not easy to do. The next thing was to place the microscope horizontally exactly ten inches above the table, then place a camera lucida at the eye-piece, and there you got half a thousandth of an inch. Looking down through the image of it on the table, you begin and draw the half-thousandth of an inch, and then with a pair of compasses you get the exact measurement of how much the thousandth of an inch would look at ten inches distant. It happened to be the nine-tenths of an inch focus I was using at the moment I put this into the tube, and there I got the accurate measurent of everything I was doing. That seemed near the mark, but not quite near enough. I was engaged in measuring how many of these *podura* beads there were to the inch, and I got Mr.

---

* Kratometer. *See* catalogue of South Kensington Scientific Apparatus. Biology. No. 3615. (p. 504. 1st ed. 776. 2nd ed. 919. 3rd ed.)

Browning to make this beautiful micrometer, which reads to the 20,000th of an inch. That would not do; there was a great deal of vibration when it was used, and I could not see what I was doing. I could get a spider line about the twenty-thousandth of an inch, but that is a very coarse rule to measure the 100,000th of an inch by; it is like measuring a minute thing with compasses with blunt points. Then I thought of an idea of forming a miniature of those spider lines, and after a great many experiments I found that a very distinct miniature was one which was exactly diminished seven times. If those spider lines are the 20,000th of an inch thick when you get the image, as I got it, they are 140,000th of an inch, and that I thought was near enough. These lines are never in the way; they always come up at a moment by turning a screw and the same object glass which forms the image also acts as a condenser to illuminate the object, and it therefore gives the power of moving the lines from bead to bead, according to the scale, so that when you are using this instrument, whatever you want to measure, you have only to get it in focus, see it clearly, then move up the lines, move them about and put them at any angle, and you get the most detailed measurements.

In this apparatus (*the aerial spider line micrometer* described in the catalogue) the spider lines are presented on the stage of the microscope as a delicate miniature micrometer completely under the control of the observer; reduced seven times in tenuity.

The question of examining the qualities of microscopes by their powers of defining a brilliant point, has proved one of great interest. In order to accomplish this at first, I constructed, in thin brass, a hole about the one hundredth of an inch in diameter : this was fixed at a distance of 100 inches from the microscope ; and miniatured by an object glass so as to form a minute image in the focal plane of the microscope when strongly illuminated.

I placed this before a very brilliant lamp and produced an artificial star. I could calculate the exact size of this star in miniature, and I think it ought to have been something like the sixty-thousandth of an inch. With a fine glass of $\frac{1}{10}$ focus which miniatured it 1,000 times less, I expected to see this star about the same size it ought to be, and judge my astonishment when I found it to be four times larger than the calculated size, I said, how is it possible for this microscope to define well? I not only saw the brilliant point four times as large as it should be, but a thing like Saturn's Ring round it. Then I found that in every brilliant object there was always a bright point four times bigger than the reality, which was vexing enough, because it was the finest object glass I could get for money. Then a scale of *formosum* was put under the microscope, and I had an instrument made by Messrs. Beck, for the purpose of diminishing the aperture of the object glass to anything I liked. I diminished it one day to the one-hundredth of an inch. I had the sun brilliantly shining, I looked through, and my surprise was greater than I can express, when I saw with the least movement of the mirror an immense blazing sun catching first one of these beads and then the other : they were moving with great splendour, more like a setting sun than anything else. Here, a spherule had formed a magnificent image of the sun, which by the errors of the glass, was enormously magnified. I found every bright object or bright point gave a false image. The next thing now was to investigate it, and in order to do this, I got a chip of the sun as I may call it, and people who came into the room fancied I had an electric light. It was simply an aërial image of the sun, formed by a lens and transferred by total reflection so as to form an image. It was so brilliant that persons blinked at it like an electric light. I formed an image of the sun on the stage of the microscope, and with a high power I obtained the most gorgeous diffraction phenomena that perhaps were ever seen in this world. Sir John Herschel described them as seen in the telescope, by merely using wire gauze on looking at a star ; but the beauty of these phenomena so far surpassed any rainbow in brilliancy, magnificence, and splendour, that I do not think the English language could describe them adequately, and in fact all those who have seen these beautiful phenomena have pronounced them unique. The rings round the object increased so that I had twenty-four of them with an interior black ring the one-fifty thousandth of an inch in diameter, whilst the central disc was nearly the 16,000th of an inch.* The distances from each ring to the next were exactly the same. The idea struck me that I could thus find out the qualities of object glasses by seeing what they could do in

---

* See Proceedings Roy. Soc. "On Circular Solar Spectra," 1873.

this way. You know that no telescope is of any use unless it will show a magnificent black ring around a centre disc.   I aimed at doing the same thing with a microscope, but upon using a great number of glasses in this way I found that with every object glass I tested, this beautiful chip of the sun gave different effects.   I saw engine turned rings, sometimes a series of rings overlapping each other, and sometimes instead of a round centre I got an oblong, differing according to the nature of the object glass.   One thing struck me more than anything else, that after all, however skilful our mathematicians may be in selecting material for forming these beautiful objectives, there is one condition which I fear is past human art except by accident, and that is the perfect centreing of all these lenses.   I can assure you, gentlemen, if you were to shift only one of the component lenses the twenty thousandth part of an inch, you would spoil your image; and how it is these opticians with their delicate fingers can manage to put eight lenses together, some not nearly so big as a pin's head, one behind the other so that they shall all have one common axis is a feat of artistic skill which is beyond art.   It is chance.   It is like making a chronometer. You may make a dozen with the greatest skill and care and three of them may be excellent and the rest are indifferent.   So out of a great many different glasses it is impossible to get them all alike. One alone may be superlative.

In microscopic research the most troublesome thing experienced by biologists is looking at an object through a thick piece of glass.   It struck me that a principle might be adopted which would get over that by introducing a lens to move between the object glass and the image. By that means I could compel the image to go further off.   It succeeded very well, but in doing so I had to encounter a great many difficulties in the way of errors, introducing all manner of aberrations, chromatic and spherical; but by this method I found I could give a great depth to the objective which it had not before, and by means of the "aplanatic searcher" I corrected the definition so as to make it satisfactory.*

Here again is an instrument for finding the refractive index of glass in consequence of the image being raised.   If you look at it, a piece of paper fastened at the bottom appears to be raised, and if you

* *See* Pap. Phil. Tr.   On a Searcher for Aplanatic images.   Vol. II., 1870

measure the exact distance which it is raised above the base and know the length, you can then, by beautiful little formulæ, have the exact refractive index, and in that way this instrument has been able to tell the refractive index of glass the three-thousandth of an inch thick.*

The PRESIDENT : As you know, gentlemen, we have still on our programme another important communication, namely, a demonstration by Dr. McKendrick of some very interesting apparatus relative to physiological acoustics ; but the hour has already passed, and I must bring our sittings to a conclusion. I think you will allow me to say that it is a successful conclusion ; and I have to add that if there are any physiologists present who are specially interested in the interesting questions which this instrument is connected with, Dr. McKendrick will be kind enough to show it—in fact, the state of the light in this room had already placed a difficulty in the way of the demonstration. The light is so strong, that, even if we had had time, it would have been impossible to have exhibited the effects he wished to show you ; but he will show them to any one particularly interested; but it is not possible to show them except to a very small number.

The Conference then adjourned.

---

* *See* Proceedings Roy. Soc. 1876.   Refractometer. Also with plate. Journal. Roy. Microscop. Soc. 1876. Page 294, vol. II.

SECTION—PHYSICAL GEOGRAPHY, GEOLOGY, MINING, AND METEOROLOGY.

President: Mr. JOHN EVANS, F.R.S.

Vice-Presidents:

Herr Professor Dr. BEYRICH.

M. DAUBRÉE, Directeur de l' Ecole des Mines, Paris.

His Excellency Dr. Von DECHEN

M. le Professeur DEWALQUE.

Mr. H. S. EATON, President of the Meteorological Society.

M. le Professeur Dr. FOREL.

Professor N. STORY, MASKELYNE, M.A., F.R.S.

Professor RAMSAY, LL.D., F.R.S.

Major-General Sir H. RAWLINSON, K.C.B., F.R.S.

Mr. W. WARINGTON SMYTH, M.A., F.R.S.

The Baron FERDINAND VON WRANGELL.

Tuesday, May 30th, 1876.

THE PRESIDENT'S OPENING ADDRESS.

In opening the Conferences in connection with this Section of the Loan Exhibition of Scientific Apparatus, it will probably be expected that I should say a few words, if only by way of explanation of the class of subjects that come within our range, which indeed are neither few nor unimportant. Let me first take the general list of subjects which have on the present occasion been grouped together, and which may be said to constitute our domain. These are Meteorology, Geography, Geology and Mining, Mineralogy, Crystallography, &c. Some of these subjects might no doubt with almost equal propriety have been assigned to other sections. Meteorology might for instance have been classed under the head of Physics, and Mineralogy would not have been altogether alien to the Section of Chemistry. There is, however, so close and intimate a relation between all the various branches of physical research, that it is not only difficult to draw exact boundaries between their provinces, but also to determine to which group any given province shall belong when it becomes necessary to map out the whole field of science into some four or five divisions.

Our province may be regarded in the main as comprising the physical history of the earth—the constitution of its mineral parts and the forms and characters they present when crystallised, the geological succession and nature of its component rocks ; the past and present distribution of land and water, and the causes which have led to its modifications ; and lastly, those meteoric influences which not only affect climate, but are active causes in the carving out of the earth's surface, and in the redistribution of the materials of which it is composed. Nor do we only take the purely scientific and theoretical portions of our subjects, but also the application of scientific principles to produce economic results, and to lessen the dangers of those who in the exercise of their calling meet the forces of nature under some of their most destructive aspects.

It is of course only with the apparatus which has been devised for the purpose of carrying on the investigations into the physical history of the earth, and the applications of scientific principles which I have just mentioned, that we are mainly concerned, and not with abstract questions relating to any branches of science. It may, however, be found necessary to enter more or less into such abstract questions if only to show the character of the investigations which have to be pursued, and to elucidate more fully the difficulties with which inquirers have had to contend, or which still have to be conquered. Such questions may also have to be discussed should the history of the gradual development of some of our modern appliances be gone into. Some of the earlier forms of instruments which are now exhibited are indeed of great interest, whether they are regarded in the light of what may be termed milestones on the road of scientific progress or as memorials of the eminent men by whom they were devised or used. The goniometers of Haüy and Wollaston, the nascent safety-lamp of Davy, the blow-pipe of Plattner, the barometer of De Luc and H. B. de Saussure, the thermometer of Gay Lussac, the geological maps of William Smith, the logbooks of Cook, Franklin, and Parry, the instruments and maps of Livingstone, are replete not only with scientific but historical interest.

It is, indeed, as constituting an epoch in the history of scientific discovery, that such a collection as that among which we are now assembled has its highest value and interest. The third quarter of the nineteenth century has just come to its end, and we may venture to compare the advances which have been made during the last twenty-five years not only in our own particular walks of science, but in every branch of it, with the advances which had been made during the previous quarter of a century, the close of which was marked by the first Great Exhibition held in London. Great as had been the progress in scientific knowledge and in the application of scientific principles during that second quarter of the century, and favourably as it contrasted with the by no means despicable attainments of the previous quarter, the advances made during the last twenty-five years both in our knowledge of the principles of the great forces of nature and in the accuracy and delicacy of our instruments for their investigation are such that the present generation has at least no cause to be ashamed of them. Possibly when another quarter of a century has elapsed, those who come after us and those among us who survive as labourers in the field of science, may look back upon some of the processes now in vogue as antiquated, and may even feel surprise at our having been upon the verge of some great discoveries and yet having failed to make them; but I venture to hope that the names of many of those living investigators which we find recorded in the Catalogue of this Exhibition may not only then, but even in after ages, be looked upon with reverence and esteem.

We must, however, turn to the consideration of the branches of science comprised under this Section, and in directing your attention to some of the objects which appear to me of more than common interest, I shall venture an occasional observation on some matters which appear to be well fitted for discussion at an international conference such as the present.

In regard to meteorological instruments we have not only isolated specimens but sets of instruments as supplied to meteorological stations, and to the royal and merchant ships of this country. With the exception of Russia and Norway, however, the means of comparison with other countries are, I believe, wanting. It will be for the representatives of other countries to see whether some useful hints may not be derived from the experience of British meteorologists as embodied in these selections of instruments.

Mr. R. H. Scott in the "Handbook to the Collection" has given so excellent an account of the nature of the meteorological instruments here exhibited that I need add but little to it, especially as he will be good enough to make a communication upon them.

Taking the principal forms, it will be seen that among the barometers there are more than one exhibited which are of historical interest, while numerous examples of modern improvements in mercurial barometers are shown, of which perhaps those intended to facilitate their use and increase their accuracy when employed by travellers by land and by sea, are the most noteworthy. For ordinary use, however, that comparatively recent form of barometer, the Aneroid, seems likely to compete with the older form, and the precision of mechanism which some of these instruments exhibit is marvellous. That extreme delicacy, however, has its disadvantages, and for trustworthy observations the actual weighing of the atmosphere by the column of mercury will long be preferred.

The principal features of the thermometers are their accuracy and sensitiveness. It might be worth while to consider whether any means could be devised for facilitating the adoption of an uniform scale of notation. It will, however, be a difficult matter to supersede the scale of Fahrenheit in this country, where it seems to have taken so deep a hold. The more general introduction of instruments marked with both Fahrenheit's and the centigrade scale might assist the adoption of the latter, but the smaller unit of heat on the former scale gives it practically some advantage.

Of anemometers, both for meteorological and mining purposes, a large number will have been seen, some of them furnished with means of recording both the direction and strength of the currents. Of several of these, details will be given at this Conference.

With respect to rain-gauges but little need be said, unless it be to call attention to the system, which, thanks to Mr. G. J. Symons, is now so universal in this country, viz., for observers who make only one daily entry of the rainfall, to take their observation at 9 A.M. and to enter the amount of rain to the preceding day. The late Meteorological Congress has no doubt discussed this and other points of international interest.

Of hygrometers both ancient and modern forms are exhibited, the hair hygrometer still holding its own among those of the indirect class, notwithstanding the influences of modern civilisation.   One cannot but be touched by the pathetic note of the Geneva Association for constructing scientific instruments.   "The most isolated hamlets have now to be searched in order to obtain hair uncombed," and therefore fit for these instruments.

It is perhaps in the self-recording instruments that the greatest advance made during the last quarter of a century will be observed. The extended use of electricity and photography has aided in this as much as in other departments of science, and the daily weather charts now issued in this country would have been impossibilities but a few years ago.

The automatic light-registering apparatus of Prof. Roscoe will it is hoped be the subject of a communication to the Conferences of this Section ; but this and several other recording instruments are fully described in the Catalogue, as are also various interesting charts illustrative of meteorological influences on mortality and disease. The relation which has been found to subsist between colliery explosions and the state of the weather will form the subject of some observations to the Conference by Mr. Galloway.

There is only one other point in connection with meteorology on which I will say a few words :—that of evaporation.   Two or three forms of atmometers or evaporimeters are exhibited, some of them intended to determine the quantity of water evaporated from different kinds of soil, but no form of instrument is, I believe, in the collection which will serve to ascertain the proportion of the rainfall which percolates to any given depth through a porous soil.   When it is considered how large a proportion of the surface of the globe consists of such soils and how important is the question of the supply of spring-water to our wells and rivers, it will perhaps be a matter of surprise that more attention has not been directed to the subject.   It is not, however, one on which to enter at length in an introductory address, though I hope to recur to it in the course of the afternoon.

The second subject comprised within our section is that of geography, which, thanks to our distinguished African, Asiatic, Arctic, and marine explorers is at the present time attracting so much public attention. Many of the instruments exhibited have much of historical and personal interest, among which may be reckoned the series of instruments belonging to the Ordnance Survey, some of them—like Ramsden's theodolites—exhibiting to what a point the construction of such instruments had advanced even at the end of the last century. What, however, will attract universal attention are the deep-sea sounding appliances, which have so greatly conduced to the success of the "Challenger" expedition, and the great extension of our knowledge of the character of the deep-sea deposits of modern times, which throw so important a light on the history of many earlier geological formations.

This interest is much enhanced by the satisfaction we must all feel in again welcoming among us the distinguished naturalist who has had the scientific charge of that expedition. Let us all hope and trust that the gallant captain of the expedition during the first portion of the voyage, may in like manner return in due course with his present comrades from his still more adventurous exploration of the Arctic regions, crowned with the success which his efforts so well deserve.

Among the deep-sea sounding apparatus that most ingenious invention of Dr. Siemens, the bathometer, which has been exhibited and described in another section, will, no doubt, have attracted your attention, of which many of the levelling and surveying instruments exhibited in this section are also so well worthy.

The collection of maps requires but little comment. The survey of Palestine, the charts of the Arctic regions, the survey maps of India, and the beautifully executed maps sent from foreign countries cannot escape attention. In connection with recent explorations the remarkable section across Southern Africa, executed by Lieutenant Cameron during the perilous journey from which he has just returned, will, I hope, be the subject of comment in these Conferences by its distinguished author. Nor should the ancient maps of the sources of the Nile, exhibited by the Royal Geographical Society, be left unnoticed. It might be a subject for discussion whether some more uniform system of symbols for use on maps might be adopted for general use among all nations.

In the department of geology and mining, it may be observed that the instruments of the pure geologist are but few, and comparatively simple. We have, however, before us a most valuable collection of the geological maps of various countries, showing how vast has been the advance of our knowledge in this field during the last quarter of a century. The principles on which the geological survey of this country has been directed will be illustrated by its present accomplished chief, Professor Ramsay, and we shall, I hope, hear something as to the surveys now going on in other countries. It would be a matter well worthy of consideration in an assembly of this kind, whether for the general geological features of a country, some international system of colouring could not be agreed upon, and in future be adopted. For more detailed maps entering minutely into the subdivisions of formations, such a system might be difficult to devise, much more to carry out; but for the principal formations there ought surely to be no great difficulty. Already, for something like two centuries, the colours in heraldry have been represented all over Europe by a conventional system of vertical, horizontal, oblique, and other lines, and science would not suffer if on this occasion she walked in the wake of vanity.

Among the appliances of the geologist must be reckoned his palæontological and mineralogical collections which, however, are, except in special instances, too bulky for an exhibition of this kind. Some are, however, here; and among them, a magnificent series of rocks, minerals, and fossils from Russia, and the fossil vegetable remains, both from the Continent and England, well deserves notice. We shall, I hope, hear from Baron von Ettingshausen how the genetic descent of much of the flora of the present day may be traced back into tertiary times, and Mr. J. S. Gardner will have something to say on the same subject.

The sub-wealden boring, which has attained a depth of 1,900 feet, without, however, reaching any rocks of Palæozoic age, will also form a subject of comment. The process of the Diamond Rock Boring Company by which it has been carried, has not only the advantage of being more expeditious than the older process, but has the great merit of producing such excellent cores as those which can be seen at the end of this gallery.

The ingenious machines of Mr. Sorby, illustrative of various geological phenomena, and the original drawings of Buckland and Phillips will also attract attention.

The specimens illustrative of Mr. Daubrées experiments on the artificial formation of metamorphic and other rocks, and the minerals formed within the historical period by means of hot springs, will be rendered doubly attractive by the account to be given of them by that eminent geologist.

As objects of historical interest, however, the collections illustrative of the development of Davy's great invention of the safety-lamp, are perhaps unrivalled in this department. Among mining appliances and models, some few will form the subject of communications to the Conferences.

In the remaining department of this Section, that of Mineralogy and Crystallography, there is much of historical as well as scientific value. The improvements in the microscope, the polariscope, and the goniometer, have done much to advance these branches of science during the last quarter of a century, while the application of photography to the reproduction of the images observed in the microscope has most efficiently aided in bringing the results of single observers within the reach of all.

The models and diagrams illustrative of the different systems of crystallography and the various forms of crystals are remarkably excellent and complete, and some questions in connection with the properties of certain forms of crystals, and the method of notation best adapted for international use, will probably be discussed in the Conference.

I have thus briefly touched upon some of the salient points which occur to the mind when taking a cursory view of an Exhibition such as the present. In doing so I have no doubt passed over many instruments and appliances of even greater importance than those which I have thus succinctly mentioned, and have probably left untouched many topics of the highest interest. Among the subjects, however, which will be discussed on each day of our Conferences there will, I hope, be a sufficient variety to give occasion for any one to call attention to any special features of novelty in the collection. What I have ventured to say must be regarded as merely a short introduction to communications of far greater value, from which I will no longer detain you.

The PRESIDENT : Ladies and Gentlemen, the first paper which I have down to-day is a paper on

METEOROLOGICAL INSTRUMENTS IN THE LOAN COLLECTION, BY Mr. R. H. SCOTT, F.R.S.

Mr. R. H. SCOTT, M.A., F.R.S. : Mr. President, ladies, and gentlemen, I need hardly say that the preliminary portion of the communication I was going to make has been already taken out of my mouth by the admirable address which Mr. Evans has given you about the general scope of the Exhibition. However, a word or two is required relative to meteorology, and a special account of some few of the instruments which are exhibited.

In the first place, What is meteorology? It is the science of the air above us, and with all reverence be it said, in the air "we live and move, and have our being." Therefore, it is a most important science for our own immediate life. But there is a special question which should be considered, and that is why it is necessary there should be many meteorological observations taken at different stations, whereas astronomical observations can be taken at a few widely scattered observatories. The reason of this is that the different meteorological elements which we measure, like temperature, pressure, humidity, motion of the wind, and others, are all subject to such very marked influences produced by the local conditions of each observatory, that it is absolutely necessary for us in London to be very careful in choosing the observations which we take as representative of the climate of London. There is not the slightest doubt that at the two first-class observatories that are close to London, Greenwich, and Kew, there is a very material difference between the results obtained, simply because Greenwich is on the top of a hill, whilst Kew is close to the level of the river. This dependence on position is an unavoidable difficulty in the case of all meteorological observations, and you will easily see that it adds very materially and very seriously to the complexity of the subject when we come to discuss the observations, because we have as far as possible to eliminate all local influences.

The subjects which we really discuss in meteorology can very easily be classified under three or four heads. In the first place, we have the measure of the air which presses upon us to a certain extent, the quantity of air about us, and this is Pressure. Secondly, this air may sometimes be warmer or colder, and its Temperature will exert a very considerable influence, as every one knows, on health, on the growth of plants and, in fact, in all biological questions, the temperature exerts a very considerable effect. Thirdly, there is the motion of the air or Wind. Fourthly, you have what we may classify under one group, the accidental constituents of the air. We all know that the air is composed of a mixture of two gases—Oxygen and Nitrogen, and that it contains variable quantities of other matters, the principal of which is moisture, present either in the form of Vapour or in the form of Rain. Then there is Carbonic acid, not a subject to which meteorologists have paid particular attention. Then we have a matter which is going to be spoken about later, but as to which there is a deal of doubt, if Dr. Fox will allow me to say so, and that is Ozone. One of these subjects, the motion of the air, the wind, will be spoken of at considerable length this afternoon, by two or three gentlemen, although I regret to say that Professor von Oettingen is unable to come and describe his instrument ; Mr. Cator and Dr. Mann will also describe various forms of anemometers. As regards moisture, the Rainfall will be taken up by Mr. Symons in a very few minutes, so I shall not speak of it, but will treat now of these other subjects.

How is the pressure of the atmosphere measured? It is measured by the Barometer, and the Barometer as it is usually employed, is in its simplest form, a tube of mercury plunged in a vessel of mercury. You have a tube filled with mercury, and inverted in a vessel of mercury. Then it will be found that the pressure of the atmosphere will keep up in that tube, which should be more than thirty inches long, a column of mercury, the weight of which will be exactly equivalent to the pressure of the air outside. That is the very simplest form of barometer, and that simple form is exhibited in a large standard barometer by Casella. You have got a simple tube, of course in an iron frame work, set up in a vessel of mercury, and then by means of a telescope at a distance you measure the difference in height, that is, the difference between the level of the surface of the column and the level of the surface of mercury below. But it is obvious that it is not the ordinary form of barometer

which you purchase, and which is hung up in your halls or in your observatories.   How do these barometers differ from this crude and simple
form.   In the first place, there is one great matter which has to be
seen to, and that is, how are you to make arrangements for effecting
a constant measure of the height of this column.   Supposing that the
barometer falls, that the level comes down, and falls two inches, the
mercury must come into the lower vessel, and raise the level, so that
no longer can you measure from a definite level below.   There are
several modes to escape this difficulty.   One method is that you move
up the whole of the cistern which holds the mercury, by means of a
screw, which is placed below, until you reach a definite fixed point,
and the scale is measured up from that certain fixed point.   Another
method is, instead of making the cistern below moveable, to make the
scale moveable, so that you can screw the scale up and down, and
thus the bottom of it can always be made to touch the surface of the
mercury below.   The third method which is represented in the collection,
is that, instead of having a cistern, you have a syphon tube.   If the
longer end of this be closed, and the shorter open, and it be filled with
mercury, the level of the mercury in the closed tube will stand
about thirty inches above the level of the mercury in the open
column.   Therefore, if you have a barometer of this kind, which
is called a syphon barometer, it is obvious that it is very easily
managed.   It is very light, and that is the form of instrument which
is usually taken for carrying up mountains, because it is easier to carry.
But it requires a good deal of care in reading, because you must have
the measurement on both legs of the syphon, and for that reason this
instrument is not liked, and is not used much in England.   But it is
used very extensively on the continent, and particularly the Russian
barometer which is exhibited is of this pattern.

   We must next speak of the method in which these instruments are
made self-registering.   In order to avoid the necessity of constantly
having to read the barometer, certain methods have been employed
to make it record its own indications.   The simplest of these is the old
form which all of you know, the wheel barometer, which simply consists
of a weight placed in the open leg of a syphon barometer, furnished
with a counterpoise, and then by means of a string passing over a pulley,
either moving an arm in the wheel barometer or carrying a pencil or

pricker which can be made to mark upon paper. Instruments of this very crude form are down stairs. Luke Howard used an instrument of this nature in the year 1815, which had been made many years before, and was known as the Clock Barometer. This instrument is also exhibited by M. Redier, and the modern form is in use pretty generally in France. But all these mechanical contrivances are very much less perfect than the photographic instruments. Whenever you give the mercurial column any work to do, like lifting a weight, you check it in its motion; but if you simply photograph its position from time to time, of course there is no interference whatever with it. There are two forms of photographic instruments here; one of them is shown by Mr. Charles Brooke, and it is amongst the magnetic instruments, because he attaches to it a magnetograph. He has a syphon barometer, and an exceedingly light float in it which carries an arm extending for some distance, and the position of this arm is photographed upon a piece of paper which is carried round upon a drum. The Kew instrument differs from this simply in the method in which photography is employed, and in the fact that it uses the ordinary barometer, and not a syphon barometer. In the case of the Kew instrument it will be easy to see how it acts. When you have got the glass tube full of the mercurial column, it is evident that no light can pass through it. But at the top of the mercurial column the glass tube is perfectly transparent, and, therefore, light will pass through that. What is done in Ronald's barograph which is down stairs, and in the barographs belonging to the Meteorological Committee is that the space of the vacuum at the top of the mercurial column is photographed from time to time, and so there is no interference with the motion of the barometer. The mercurial column has nothing to do except to move in accordance with the pressure of the atmosphere, it has no counterpoise to lift, and you get a photograph of the true height of the mercurial column.

As to the aneroid, I presume one need not give much description of it. The principle of the instrument consists of a box from which a quantity of air has been extracted in which there is a partial vacuum. When the pressure of the temperature rises, the top of that box is pressed in. When the pressure falls it comes out again, and by this means you get a certain motion. Everyone knows that the aneroid can be made larger or smaller, and even very small; there are some not much larger than a lady's watch down stairs, and they are excessively convenient. But they have this radical defect that the accuracy of their indication depends upon the elasticity of the sides of the box of which they are made, and upon the force of various springs and mechanical appliances used in their construction. For these reasons the aneroid must be compared from time to time with the mercurial barometer in order to see whether it is going right ; just like an ordinary watch which has to be compared with astronomical observations from time to time to see the rate at which it is going. The aneroid has been made self-recording in a great many ways. It is perfectly easy to do so. You have got a certain motion in the box, and that gives sufficient force to move a pencil or a pointer upon paper. So much, then, for pressure.

As regards Temperature, I need hardly say that the temperature is measured by the thermometer. That you have already heard in the physical section. There are various observations, however, which have to be taken, connected with the temperature. We want to measure the highest and lowest temperature of the day, and we want to obtain a photographic register, or a mechanical register of the indications of temperature from time to time. The mode of obtaining the maximum and minimum temperature is comparatively simple in its general form. You have a column of mercury passing up to a certain space, then in Phillip's Maximum there is what is called an index, beyond which is pressed on in front of the mercurial column, and is left at the highest point reached by that column, and this index will not go back until it is set. Just in the same way, in the case of the minimum thermometer, it is a spirit thermometer, and you have an index inside, and that index is drawn back towards the bulb, and never can be left dry.

The points, however, of the greatest interest about the thermometer, which are novel, are the modes of rendering it self-registering. We have not got a photographic instrument here, but there is one promised, I believe, though it has not actually arrived yet, which is very interesting indeed to consider, and that is the thermograph of M. van Rysselberghe, Paymaster of the Belgian Navy at Antwerp. The principle of this instrument is, that supposing you have the thermometric column, the top of the tube is left open, and you have a wire which is brought down every ten minutes or so, which moves up and down mechanically within the tube. As soon as it touches the top of the column there is an electric contact established, a current passes, and the wire passes up again. How is this made self registering? It is made self registering in this way : that whilst the pencil goes down it draws up a point, or a pricker, which engraves a copper-plate, and if that wire in the thermometer tube goes down two inches, it scratches a line two inches long on the copper-plate. Every ten minutes there is a line scratched according to the distance the wire has had to go down till it met the mercury, and you get a plate from which you can immediately print, as soon as the plate is etched. The same process is applied to the barometer, the wind and tide gauge, and various other instruments. In Professor von Rysselbeghe's instrument, when it comes here, will be found the means of recording regularly, almost every observation which can be made in a meteorological observatory.

As regards Wind, I dare say this afternoon the various forms of anemometers will be described, and I shall not go at greater length with that subject, especially as my time is now running short.

We come then to moisture. Everyone knows that the amount of the moisture in the air exerts a most serious influence upon our health. How is this moisture in the air measured, when it is present in an invisible form? Mr. Symons will tell you how it is measured, when it comes down in the form of rain. The way in which it is measured when in the vaporous form is by means of instruments which our president has described as Hygrometers. You all know that if you have got a tent rope or a window cord, and those tent ropes or window cords get wet, or the weather is damp, the rope or cord will shorten and get tight. Therefore, supposing that you had a piece of string stretched by a weight at the end of it, you could imagine that the length of the string might, to a certain extent, measure the amount of moisture in the air. It is not possible to have two pieces of string exactly the same in all their properties, but we try to get two human hairs, or a number of human hairs, and by going into some of the out-of-the-way villages in Switzerland, they say it is possible to find human hair which has not been combed, or troubled too much with oil or grease. The hair is then stretched by means of a light weight, and by means of its length you can determine what amount of moisture there is in the air. The mode of doing this is similar to the method of the wheel barometer. You have a cord passing over a pulley, and according as the cord gets longer or shorter, it is evident that it will move an arm along a graduated arc. This is the form of the old hair hygrometer invented by M. De Saussure, some of whose original instruments we have, and this is an instrument we use in cold climates, (we have supplied it to our own Arctic expedition,) such as Russia and Norway, for the measure of the moisture of the air. There are various other forms of what we call organic hygrometers down stairs, in which, by means of the varying behaviour of substances which attract moisture, and either alter their shape or weight, the moisture is measured; but the ordinary form used in observations depends on a totally different principle. It is what is called the wet and dry bulb hygrometer. It depends on this principle: supposing that you have two thermometers alongside of each other, one of which is coated all round with muslin and kept damp; on a day like this, that instrument will read considerably lower than the bulb which is not moistened, because there is an amount of evaporation of water from the surface of the bulb which is wet. It would be impossible to go into the theory of this at present, but the principle of the instrument which you will see down stairs (and there are a great number of them) is, that one bulb is kept bright, while the other bulb is kept coated with muslin, and this muslin is kept wet by means of a capillary syphon from a vessel of water kept close by the side. The reading of these two taken at the same time, provided the wet bulb be really properly damp, gives you an indication of the amount of moisture in the air. The photographic method of recording this would be similar to the barometer. If you take the mercurial column, and break it, so as to have a bubble of air, you can photograph that, and thereby get the motion of the mercury in the thermometer from time to time.

As regards the other instruments, the ozonometers, they will be described, I presume, by Dr. Fox. As to the evaporation gauges, I have some sort of hope that our President, who is a high authority upon the subject, may give us some communication about them. The difficulty about evaporation gauges is this : supposing I fill this tumbler with water, and leave it, that water will evaporate, but I do not know whether the amount which will evaporate from that glass would be precisely the same as it would be from a pan the size of this table, much less from a pan the size of a large lake, or the sea. Several of the atmometers down stairs may be compared with that glass, particularly von Lamont's. It is a small dish, about three inches in diameter; when that is placed on the dry ground there is a quantity of dry air all about it; whereas if it were on a large surface, where just as much moisture was being evaporated beside it as was coming from the instrument itself, it is obvious there would be no tendency in the vapour to spread out laterally to supply the dry air outside, and therefore probably there will be a greater amount of evaporation from the smaller instruments than from the larger. This affects, as far as we can say, almost all the atmometers yet devised for use in ordinary observations, and the subject of evaporation is most closely connected with hygrometry, with health, and also with rainfall, while its influence on all agricultural pursuits is undoubted; but as yet we wait for a satisfactory mode of measuring and determining what is the amount of evaporation per day at each station.

The PRESIDENT: I am sure the Conference will return its thanks to Mr. Scott for his interesting communication. I do not know whether any one wishes to ask any question; if so, perhaps they will do so in as short a manner as possible, as we have several other papers before us. As no one has any question to ask, I will call upon Mr. Symons to give us his communication upon

THE MEASUREMENT OF THE RAINFALL.

Mr. G. J. SYMONS: Mr. President, Ladies and Gentlemen, I remember very well, some time ago, the Astronomer Royal, when presiding over a section of the British Association at Manchester, to give to a young speaker the advice—You need not begin at the Deluge. I do not wish to begin too early with regard to the measurement of rainfall, but it is scarcely to be presumed that every one here present understands the modes by which it is accomplished, and I think if I occupy a minute and a half, or two minutes, in starting from the Deluge, or from something of almost equal antiquity, it will not be time wasted. We all know the rain falls. The question is, how we are going to measure it, and I think that can be explained in a very simple way. If you have a dish eight inches across, the rain coming in it—assuming the dish to be closed at the bottom—would fill that dish up to a certain depth. Suppose one puts a tray out of doors on a grass plot, it would be filled with water to a certain depth. The depth at which the water would stand in that tray would be the depth of rain falling at that place; but obviously there would be several disadvantages in such an arrangement. The first disadvantage, and the principal one, would be that the water would rapidly evaporate from that tray, and you would never get a measurement of the quantity that there was in it. Consequently, we put in a funnel which leads the water down to a smaller vessel below, in which obviously the water stands at a greater vertical depth in proportion as one area is less than the other. As I have sketched it on the board, the area is perhaps a fifth. If so, the water would stand at five times its natural depth in the lower vessel. I think I may say that all the rain gauges, with the exception of one, are practically different methods of reducing the area over which the water has been collected, and gathering it in a smaller area in order that the natural depth may be artificially elongated, and thereby greater precision of measurement attained. The exceptions are introduced for a special purpose, namely, for that of storing up the large quantities of rain which falls in mountainous districts. It is almost a mistake to run away from the instrument to the results, but you must take it generally that the result is that in mountainous districts a very large quantity of rain, six or seven times as much as in London, falls. Moreover, on mountain tops, we cannot have observers going every day, as the Chairman has said, at 9 o'clock, to measure the rain gauges; the most that we can expect them to do is to go once a month; and consequently the gauge must be large enough to hold the entire rainfall of one month. There are several stations in the mountainous district in the north of England, where upwards of thirty inches fall in one month. It is evident if your gauge be of uniform size throughout, it must be nearly three feet long to store the water collected in a single month. Consequently, for the gauges in mountainous districts.we have them made uniform in size, the precision of measurement not being so necessary as in dryer districts. My friend, Mr. Scott, entirely left the subject of rain gauges in the Exhibition to me, and, therefore, I think I must say a word or two respecting them, because perhaps, otherwise, attention might not be called

to them. There are one or two which are very good. One has come
over from Dublin, and is a modification of a plan which has been
abandoned, because it was found not to work well, and yet we have
that same plan brought forward again, and I who have been one of the
first to attack the old ones, am inclined to support the new one. That
is an inconsistency, but people are inconsistent at times. The water
in these gauges comes down a tube, and goes into a vibrating bucket
set upon an axle underneath it. The rain coming in, one half
of the bucket gradually fills up, and when it becomes of a certain
weight, it tips over to one side, and the other half of the bucket comes
underneath the tube ; the consequence is that as the rain falls you have
a reciprocating motion going on, and the alternate halves of the bucket
are upward and downward. That is connected with a train of wheels,
and in the old Crosley gauge to which I am referring, that train of
wheels indicates like a gas meter. Down stairs there is one of those
instruments. The objection to them was, that being made in somewhat
a rough manner and passing into the hands of persons who were not
skilled mechanists the machinery was allowed to get into a dirty, clogged
condition, and the result was that very frequently the bucket did not
tip at all, the water ran over. and the records were deficient in consequence. I do not know, with the exception of about four, of a single
Crosley gauge in the United Kingdom which has given accurate
results. At Greenwich, where we surely cannot charge them with any
very gross neglect, the Crosley gauge used to be some seven per cent
deficient. There is no inherent difficulty in making its action perfect,
by looking carefully after it and seeing that it works smoothly and well ;
and if it does, then one directly has the means of turning that gauge
to very great practical advantage; and that is what has been done by
Mr. Yates, of Dublin, in a gauge down stairs. He has attached to this
bucket a little metallic contact, and he has so arranged it that an electrical current is sent from the rain gauge every time that the bucket
tips. The recording apparatus can of course be placed at any distance,
and it is very convenient to have in one's study a little *comptoir of
Breguet*, indicating precisely the number of discharges of the bucket,
the bucket discharging once for each one 100th of an inch. There is
another apparatus down stairs, of which I cannot speak so favourably,
because it seems to me to be a frightfully round about way of doing

that which is better done by several other instruments. I allude to one in which there are separate receptacles for the water falling during each hour of the twenty-four. The rain passes into a central tube, from which there is a tube going off at an angle, and the water is delivered into a separate receptacle for each one of the twenty-four hours. The only recommendation of this gauge is its small cost.

I now pass to the more general results which have been attained by the rain gauge experiments which have been conducted in this country. In the first place I call your attention to the fact that many years ago after considerable discussion it was practically settled in Scotland that the smaller the rain gauge was the better, and Scotchmen to a great extent adopted, and to a still greater extent advocated, the use of gauges not more than an inch or two inches in diameter. On the other hand, another equally large party, principally on the Continent, averred that you cannot have a rain gauge too large, and, as a rule, the size of the gauge used on the Continent may be put at something like eighteen inches or two feet it diameter. In order to ascertain the truth of this matter I had a number of gauges made some ten years ago of gradually increasing diameters from one inch to two feet. They were observed in different parts of England, first in Wiltshire, then in Hampshire, and then put on the most exposed part of the Yorkshire coast near Whitby, right on the top of the cliffs where the wind was something frightful; but wherever they were, provided only the observations were made with rigorous accuracy, under those conditions, the gauges give practically the same result ; the difference does not exceed two or three per cent of the entire amount collected. Consequently I think we may say it is quite immaterial what size is used, except this, that if you have a very small diameter, say one or two inches, and put that in a very exposed position, the wind seems to be somewhat more prejudicial to its accurate indications than it is to that of a large one. Consequently for mountain purposes, for placing in such places as the summit of Helvellyn, Skiddaw and Scawfell, where we have placed gauges, we usually employ them eight inches in diameter. There is another point which we also attacked, and that was the question of the amount of rain collected at different elevations above the surface of the ground. It had been known in a vague sort of way for a great number of years that the higher you put a rain gauge above the ground the less would be the amount you would collect. This was first noticed by some experiments made 100 or 150 years ago by Dr. Heberden, who had one rain gauge placed on the top of Westminster Abbey, another on the top of a house in the neighbourhood, and another in an open garden. He found the amounts were something like, approximately, nine inches on the top of the Abbey, fourteen inches on the top of the house, and twenty inches on the ground. The question was raised whether that was due to the buildings or to the elevation, and therefore experiments have since been made in which the gauges have been erected on poles, quite irrespective of any building. One could quite understand that the wind blowing against a building and being deflected over the top would lift the rain away with it, whereas if you stuck up a column it could hardly produce that effect, and we wanted to find whether a gauge on an isolated column would tell the same tale as one placed on the roof of a building. We found that the difference was very nearly identical under the two sets of circumstances. Then came the question what was the cause of it. I do not pretend to have solved the problem, or that my friends who helped me in the matter have solved the problem which has caused many anxious thoughts to many persons for a great number of years. But we have found an almost strict accordance between the decrease per cent. due to elevation and the velocity of wind passing over the place at the time of observation. I think that may be taken pretty nearly as an ascertained fact. The higher you go up above the ground the less rain you get, and the stronger the wind passing over the place the less rain you get. There is one fact I ought to mention, and that is, if instead of an elevated house you have got an elevated piece of ground, a lofty mountain—the conditions are precisely reversed. Height above the ground decreases the amount collected; height above sea level increases the amount. That distinction must be clearly borne in mind, as it is a constant source of confusion. The reason for this increase is, that the ground itself being high is cold, and being colder than the atmosphere passing over it, cast as a condenser and causes the rain to fall. Provided the hill is not more than 1500 feet or thereabouts above the sea level, that is to say, the normal level of the clouds, the increase is found to be greatest on the north-east side of the hill, because the hill acts for a short space beyond the promoting cause, if I may so express it, of the hill; but if

S

the hill be 2500 feet—I am speaking with reference to the United Kingdom—it forms a wall, the clouds cannot get over the top and are precipitated on the south-west side, that being then the side on which the largest amount is collected.   The President reminds me to look at the clock, which is what I have been looking at all along ; and therefore I will conclude by thanking you for your kind attention.

The PRESIDENT : I am sure you will all join with me in returning thanks to Mr. Symons for his interesting communication.   I think it hardly requires any discussion, and I will therefore ask Dr. Mann to give us his communication with regard to lightning conductors.

On Lightning Conductors.

Dr. R. J. MANN : In drawing your attention to the specimens that we have of the apparatus provided for defence from lightning I may perhaps ask permission to start from a very common and simple point of vantage ground.   I will ask you to remember that there was a time in history when in reference to the rain water which Mr. Symons has just been telling us something about, the usual course of procedure was that the mud walls built by our ancestors were washed away pretty well by every fall of rain.   Now with the advance of intelligence this has been changed.   One of the first steps of contrivance in this direction was to prevent the washing away of the mud walls of dwellings by putting up water troughs and pipes, and when these were erected then even the mud walls were saved from the effects of the deluges that occasionally fell on the roofs.   All our houses now have water pipes and rain troughs attached to them, and we should never think of doing without them.   But now with regard to lightning we stand almost identically at the present time in the same position that our forefathers did in regard to rain in reference to their mud walls when they had no water pipes; and in drawing your attention to what I now propose to say regarding the proper employment of lightning rods as a defence against accidents, I will still follow up the line of illustration I have started with on account of the facility it affords me to make my meaning simple and clear.

When we take a rain water pipe and put it up to protect our houses it is obvious we have three conditions to secure. In the first place we must take care that the water pipe is large enough to carry off the heaviest flood that can fall, for if we do not secure that we shall do very little good by putting up the water pipe. In the second place we must take care that the troughs which gather the water to carry it from the roofs into the rain water pipes shall be of sufficient capacity and of good arrangement for the collection of the water. And in the third place we must take care that there is a sufficient outfall from the pipe in order that the water which is collected may get away.

Now, having drawn your attention to these conditions, I propose next to show that almost identically the same conditions are required for the conveyance of lightning from the clouds to the earth in such a way that in its descent it shall not exercise a destructive influence upon material structures or living creatures.

The first thing needed, then, for the safe conveyance of the lightning discharges is ample in the size pipe, or conductor, which in this case however we call the lightning-rod. We have several specimens of rods before us which are now used. I will draw your attention in the first instance with regard to them all to the fact that in the con-veyance of lightning it is not the inside but the outside of the con-ductor that we are concerned with. Lightning has the peculiarity that it runs along the outside instead of down through the interior. Understanding this, you arrive at the position, that it does not matter at all what the form of the rod is which is supplied. There are four different forms in which it is used. Some specimens are constructed in the form of twisted wire rope; others are made in the shape of hollow pipes; others are solid bars; and others have the appearance of flat tapes. All these are equally good as regards form provided they give equal capacity of surface. The rods and ropes and tapes made for this purpose are now universally constructed of one of two metals—either of copper or iron; and the only important point to attend to in making a choice between these metals is, that if you make your conducting rod of iron instead of copper, you must have it five times as large, because copper is five times a better conductor, and then whether you employ copper or iron, the same result will be secured. But another thing which has to be considered in furnishing a lightning conductor is, that a larger area of conveyance must be provided for a greater height. Bear this distinctly in mind, for it is almost always overlooked in the construction of lightning conductors. The efficiency of a lightning rod depends on three things, namely, the specific conducting power of the metal employed, the size of the rod, and that size being in proper proportion to its height. Thus the higher you carry up your main conductor, the larger it must be for the work it has to do. The French electricians, who are very scientific in these matters, employ a copper wire rope or rod, which is from ·4 to ·8 of an inch in diameter—the actual limit for safety is not quite determined, but it lies between those two figures—and they then allow for every eighty-two feet of that rod an equal increase of size. In other words, if you have twice eighty-two feet high to protect, you must employ twice the size of rod, and if you have four times eighty-two feet you must either have four lightning rods, or one rod or rope four times the size. With regard to the material used there is this also to be said : iron has the one advantage over copper—that it is a more tenacious metal. If you use an iron rod it is less likely to be disintegrated by a powerful discharge ; but, on the other hand, it is more readily destroyed by moisture and atmospheric and chemical influences, and therefore, on the whole, copper is the better metal. Copper has also this additional advantage, that it is lighter because only one-fifth the size will perform the same conducting work ; but it has the disadvantage, on the other hand, of being a much more costly metal.

Now regarding the next condition of efficient protection, you know the ample collection of our water pipes is provided for by putting plenty of troughs round the eaves of the house for carrying the water away to the one main pipe which goes direct to the ground. The collecting troughs of the lightning rods are the terminal points. In order that we may be able to collect and convey to our main line of conduction carefully and directly any lightning discharge which may occur, we must have points placed in every direction, and these points will then first receive at their tips the discharge which has to be carried through the rod. This is so well known at the present time, that the best electricians multiply their points and make them dominate every point of the structure that is to be secured. At the Hotel de Ville, at Brussels, where one of the best works of this kind has been performed, the whole building bristles with points all over the top. There are 264 points provided for the reception of the lightning upon this one building. These all lead down to eighteen main rods, which finally terminate in very ample earth contacts.

The third circumstance that has to be alluded to; namely, provision for sufficient outflow below, is perhaps on the whole the most important of all the conditions needed for efficiency and safety. In considering this let us first conceive that we are in the position of having a rain water pipe of say six inches diameter, capable of carrying away the heaviest flood which can fall on the roof of a given house, and that we then leave only a little pin-hole for the water to run out through at the bottom. This is practically what ninty-nine out of one hundred electrical engineers do with lightning conductors. They make a rod which can carry down to the bottom of the house a certain amount of electric force or discharge, and when they have it there they take care there shall be such obstruction at the bottom that it cannot get away. The most important condition of all in providing a conductor for lightning, is that it shall have a very large earth contact. Now it is only recently that this matter has been made the subject of very careful experiment by scientific men, and especially by some of the ablest electricians on the continent, and in the experiments of two of the best of these, M. Pouillet and M. Edouard Becquerel, it was found that pure water conducts very much less freely than copper, which is taken as the standard. Pure water conducts 6,754,000,000 times less readily than copper. That is the theoretical standard of conduction for pure water. It is generally conceived that if a lightning-rod is just thrust into water that is all that is required, but in order to have efficient earth contact through water it must be so large that the water has many thousand million times more size than the rod which comes down into it to convey the fluid away. This is a fact well proved by experiment; but there is the further fact to be added that in dealing with moist earth it is not only pure water which has to be considered, but impure water, and, luckily, every grain of impurity in the water increases its conductibility in an enormous degree, so that when we come to estimate the conditions of water as usually existing in moist earth, we get to the result, that 1,200 square yards of surface contact furnishes an ample earth connection. The result of the electrical experiments of the French electricians indicates that 1000 square metres of contact with moist earth furnish a very good communication for a lightning-rod—that is, 1190 square yards, or, in other words, a square 344 yards on each side, which is very nearly a quarter of an acre. Electrical engineers, however, commonly take a piece of metal, a square foot across, and put it at the bottom of the rod in the ground, and then think they have made a good and sufficient earth contact ; but so far from that being the case there should be nearly a quarter of an acre of superficial communication. Now you may ask me, how is this ever to be secured ? In the simplest way. The experiments of the French engineers have shown coke to be of so porous and so conducting a nature that if three bushels are broken up into small pieces and packed round the bottom of the rod, a quarter of an acre of moist contact with the earth is secured in consequence of the number of pores that are contained in this substance. It is now, therefore, the general practice with the best electrical engineers to make earth contact through the medium of coke. At one time charcoal was in very great favour, but it is not so good as coke, which has simply to be broken up into small pieces. M. Callaud has contrived a very simple and beautiful means of making this contact which, I regret to say, there is no specimen of in the collection ; but a diagram will sufficiently illustrate the construction. It consists of a grapnel, made of galvanized iron, which is turned up into four teeth above, and which has four sharp pointed teeth below, all continuous with the galvanized iron. M. Callaud proposes that this grapnel shall be inserted in a basket of galvanised iron large enough to contain it, with the teeth entangled in the outer meshes of the basket, and that then the whole shall be well packed in with fragments of coke, so as to constitute a kind of bulb to be inserted a certain distance down in the earth. By such an arrangement a quarter of an acre of sufficient contact between the lightning rod and the soil is actually secured to form the outlet into the ground. It is, however, not indispensable to have this construction, because a bore made into the earth, of five inches diameter and twenty feet deep, with the bottom of the rod thrust down into it and well packed round with broken coke, virtually produces as good a practical result ; or, again, a surface trench a few inches deep and twenty feet long, with the end of the rod carried along in it and well packed round with coke will afford the same sufficiency of earth contact.

It is a great pity that this simple mode of procedure should not be thoroughly understood, since without some such provision all other labour is thrown away. This is what really very few people understand ; and even many electrical engineers do not fully appreciate. If I have an electric rod provided for carrying away a discharge, which has at the bottom a quarter of an acre of earth contact, I can put my arms round that conductor during the largest discharge of lightning that can be passed through the rod to the earth, in the full certainty that the discharge will traverse it in a gentle low tension stream quite unable to leave the rod to get to my body ; but if there were only one square foot of earth contact at the bottom of the rod, and I were then to embrace the rod in the same way, the discharge, not then having a free outlet below would be very liable to leap out from the rod, and pass through my body to the earth ; the mere intensity of the force which was passing through the rod would make it dangerous for the work it was expected to do. This principle is one that I insist upon beyond everything ; the lightning conductor must have large earth contact. If it has not, the lightning that passes through the rod, escapes as high tension electricity, which has always a tendency to burst out and make new lateral outlets for itself; but if there is a large earth contact it passes through as a gentle continuous stream, which never bursts out from the main line of channel, and which is, therefore, without danger for objects around. In the two or three minutes that remain to me, I can only draw attention to the model representing the plan contrived by the late Sir William Snow Harris, for the protection of ships. It is interesting, historically, on the ground that it was his own model. In protecting wooden ships his plan was, as shown in the front of this model, a tube of copper, carried up to the top mast, where the points were projected, the tube was brought down to the beams of the vessel, and connected with the copper bolts which pass through the wood, by means of flat copper strips, so that when it got to the hull of the ship it was by means of these copper bolts brought into direct communication with the large copper sheathing in contact with the water of the ocean. This was found with wooden ships to be efficient ; but with iron ships a different plan has to be contrived. The iron masts are generally supposed to answer all purposes in these iron ships, but sometimes in going down through the hull there may be places of imperfect conduction lying between the inner and outer plates. About four years ago, there was a case of this kind in which an iron gun boat called " The Gnat," which had the mizen mast passing through the powder magazine, as is not uncommon in small ships, was actually blown up, by sparks from the mast being discharged through the powder to get to the hull. Sir William Snow Harris, at the time of his death, was still engaged upon experiments, and had devised a plan of jointed copper tubes brought down along the stays and ratlings from the main tops to the outer sheathing, or plates, of the hull. This arrangement, is not, however, necessary in iron ships. In the first place, I do not like the jointed tubes. The joints are necessary on account of the play which this form of construction needs, but in every one of these joints there is a certain amount of resistance like that between the links of a chain. There are interesting specimens in the collection of the influence of lightning, in disintegrating the molecules of the substance it passes through. I now finally just draw your attention to a form of point sent in by Linder, of Basle, which I think is as good as any I have seen. It consists of a rod of gun metal, terminating in copper, which is in its turn pointed with gold. The way in which the gold is screwed in is shown in the section. It is very important to have good points which cannot easily be injured by weather or lightning. That is a very good form, and was primarily designed by Professor Hagenbach-Bischoff, of Basle. The effect in this is similar to the more ordinary plan of tipping the point with platinum, which is also very excellent.

The PRESIDENT : I am sure that we shall all return our thanks to Dr. Mann for his interesting communication. There is one point, however, I should like to ask a question concerning. Is it not the circumference of the rod, rather than its sectional area, that determines its conducting power?

Dr. MANN : Certainly. The mathematical formula for the conducting capacity of a lightning rod is the relative conducting power of the metal of which it is made, multiplied by the diameter or circumference) of the rod in fractional parts of an inch, with the product divided by the length of the rod in feet.

Mr. LIGGINS : I should like to ask one question of the gentleman who has given us this most interesting lecture. I should like to ask him one point of considerable importance; that is, in erecting a lightning conductor on a church tower, would he consider it very essential that the holdfast staples should be insulated? It is very important, because on the principle that the strength of a chain is its weakest link, it might be that a very excellent conductor was erected on a building, and that it would be fatal to the structure of the building if those holdfasts were not insulated; therefore I think it desirable that that should be understood. I have had considerable experience with ships, and I remember perfectly well, when Sir William Snow Harris introduced his admirable system of tape-formed lightning conductors attached to the masts of wooden ships. In those days there were scarcely any iron ships in existence, and certainly there was no iron rigging, but since then the adoption of metal cordage has become quite universal, and the question suggests itself to me whether it is necessary at all to have any conductor to a ship which is fitted with wooden masts, but with iron rigging. The iron rigging comes down from the mast head to the hull of either an iron or wooden ship and is of very considerable size. Is it essentially necessary then that a copper lightning conductor in addition should be attached as in Sir William Snow Harris' plan; or if it is, is not a copper rope which is a common system, particularly in yachts and merchant vessels, where economy is desirable, sufficient? When a thunderstorm arrives, they hoist a copper rope of about the same size that is generally put on a church tower, with a gold point such as has been exhibited this morning to the mast-head, and the other end of the rope is simply thrown overboard into the sea. My life has been dependent on that in some very severe tropical storms; and I should like to know if I was trusting myself to a rotten staff, or if I was under the influence of a good lightning conductor. I may state on several occasions in the Gulf of Florida and other places where storms are very severe, I have never known an accident to happen from that simple form of apparatus. It is very convenient, because when not in use, it is coiled up in a box and put away; and it can be readily run up to the mast-head by the signal halyards in a few moments when a thunderstorm comes upon you.

Mr. R. H. SCOTT : I merely wish to make one or two remarks relative to the very important matter of having proper earth connection for

lightning conductors.   Only the other day I was taken by a gentleman
to see a large house he had recently built.   He has got a very magnificent series of conductors, and the whole thing is so arranged that on
the first thunderstorm that strikes the building it will come to the
ground like a pack of cards.   The rod is carefully taken into the water
pipe which discharges into a barrel below the level of the ground, and
there are a certain number of feet between the bottom of the pipe and
the water, so that there is no earth-connection whatever.   Again there
was a case which Dr. Mann himself explained to me, in which by the
care of his servants he was exposed to the same danger.   He had
arranged that his conductor should be properly managed ; it was in a
town where there are a great many thunderstorms, and he had a
most beautiful arrangement, the conductor being laid along for some
distance in a channel constantly delivering water.   There was a quantity
of mud and water, and a large surface of the rod was constantly
covered with wet mud.   But one day to his horror he found, after it
had been going on for a month or two, that a workman had carefully
taken the rod, or rope, out of the mud, wiped it thoroughly dry, and
packed it away in a place where no water could ever get near it.   I think
I can perhaps save Dr. Mann the trouble of answering the first question relative to the insulation of the holdfasts on a church tower.   The
matter is not very difficult to comprehend.   If you have a conductor
coming down, and if the electricity is passing down along that conductor,
and finds that at the holdfasts there is an easier mode for discharging
itself than by the wire or rod by which it is taken down to the ground, it
would choose that easiest way ;   but provided that you have got a sufficiently capacious conductor to give free passage to the electricity, there
will not be the slightest fear of its passing into the house, and not the
slightest reason for insulating the holdfast by which the conductor is
attached.   I would say in reference to the question of ships that the
danger in my mind would be that the rigging might not be made fast
so as to give perfect metallic connection with the hull of the vessel ;
that there would be a danger that the electricity, even if it were delivered
at the bottom of the rigging, might find a difficulty in passing into the
hull, and thereby getting into the water.   It is only in order to leave a
proper discharge for the electricity into the sea that it is necessary to
have some sort of rope which will afford free passage for the electricity

so that it shall have no inducement to leave it and pass into the vessel where it might meet with obstacles.

Dr. MANN : I can answer the two questions which have been put to me almost in one word.   First, with regard to the insulators.   If there is a large earth contact, insulators are unimportant ; they do no harm, but they do no good.   The insulator is merely a subordinate question, o no consequence whatever if the conductor is properly constructed and terminated below.   With regard to a question of the rope, I should have no hesitation whatever in trusting myself to such a protection, and nothing more efficient could be suggested, provided only that there is enough of the rope submerged in the sea.   Unless there were good sea contact at the bottom, there might be danger to those near the rope above; but having good contact in the sea, no better arrangement than that which was described could be desired in time of need.

The CHAIRMAN : The next communication is by Mr. John Allan Broun, F.R.S.,

ON BAROMETRICAL VARIATIONS AND THEIR CAUSES.

Mr. J. A. BROUN, F.R.S. : As the subject upon which I have now to address you is long, and the time remaining is very short, I can only touch on a few points of it.   Since Pascal first made the experiment, or caused it to be made, of having a barometer carried from the foot of a mountain to the top, showing thus that by rising in the air the column of mercury fell, and that this was due to the diminished mass of air above the barometer—from that time, the diminution or increase of pressure shown at any single station, without moving the barometer, was supposed to be due similarly to a diminution or increase of the mass of air above it ; and this hypothesis has been accepted as the only cause of variation of atmospheric pressure.

It has been supposed that the mass of the atmosphere is increased or diminished at any one station by currents which convey the air from one place to another ; the vapour of water has also been considered by some to increase, by others to diminish, the superincumbent gaseous mass.   But in whatever way the action on the mercurial column was considered, it has always been supposed that a rise was wholly due to an increase, and a fall to a diminution of the mass of air above it.

This hypothesis may serve sufficiently well when we have to deal only with the great irregular movements, which take place in high latitudes ; but when we enter the realm of law we find difficulties not easily satisfied by the hypothesis so long accepted. I will take first the case of the diurnal variations of the barometric height, with which we are but little acquainted here, but which within the tropics are the most marked of all. The barometer there rises with great regularity from near four in the morning till about ten ; it falls from ten to four in the afternoon, and to such an extent that sometimes one two-hundredth of the whole atmosphere would seem to be removed. It rises again from four till ten o'clock at night, and then falls again till near four in the morning. This is a very large and marked variation, and so regular, that as Humboldt said, you may almost tell the time by looking at the barometer. The great difficulty has been to find how this regular semi-diurnal movement can be explained by any hypothesis of the translation of the air from one spot to another.

A distinguished German meteorologist, about forty years ago, proposed the hypothesis, that these movements were due in the first place to the ascension of masses of air within the tropics, which pass from thence to other countries ; and in the next place, to the great increase of the pressure of aqueous vapour, with increase of temperature during the day. By means of this double cause of variation he believed that he explained the semi-diurnal oscillation in the pressure of the atmosphere. Unfortunately it has been found that this hypothesis is not supported by the facts. In no place within the tropics is there the smallest evidence of the assumed currents : I have sought for them in vain during days and weeks, from the summit of the South Indian Ghats, where the semi-diurnal movement of the barometer is most marked. With regard also to the varying tension of vapour, that is frequently so small that it can make little difference in the semi-diurnal movements : and it has been shown by me that in places where it is little or nothing, the barometer rises and falls with the same regularity, and according to the same semi-diurnal law, as in places where the diurnal variation of vapour tension is greatest. It has also been shewn by me that at places within a few miles of each other in South India, where the change of vapour tension during the day has differed by nearly 0·1 inch of mercury, the barometers at the two stations have gone exactly together. Dr. Lamont of Munich, has shown the same fact by a laboratory experiment, so that this hypothesis fails completely.

We find then, when we have the question of the semi-diurnal law of barometric pressure, that there are no known causes existing which can produce it by a double variation daily in the mass of the atmosphere. When we examine the large, apparently irregular, movements which take place in our latitudes, there are several curious facts to be noted. One of these, studied by Quetelet, is the movement of atmospheric waves across Europe. Quetelet has shewn that the crest of one of these waves may extend from England to Pekin, and that this wave with its preceding and following hollow of minimum pressure has advanced as a whole southwards. This immense stretch of wave he supposed might be caused by the passage of upper currents of air from the tropics flowing northwards and descending at the poles, whence on moving southwards the great increase of pressure constituting the wave crest is produced. Here again, unfortunately, the supposed currents do not appear to exist. Having observed the upper cirrus currents for years in our latitudes, I have never been able to see a trace of them, and Quetelet himself could find no evidence of the supposed lower currents in the observations at the time of the great wave.

It has appeared to me that if we examine the movements of the barometer from day to day within the tropics, where the causes of irregularity existing in high latitudes are less intense and less conflicting, we shall have some better chance of finding out what these causes really are. I have found that all the great variations in the daily mean intensity of the earth's magnetic force are due to the sun's rotation on his axis, and to the revolution of the moon : the question then occurred whether these movements had any influence on the pressure of the atmosphere? To answer this question, I took the barometric observations made simultaneously at Madras and Singapore, and these seemed to show a marked variation of the daily mean pressure in the periods of the sun's rotation, as shown by the diagram put up here. It is not to this result however that I desire to draw the attention of meteorologists, but rather to the irregular movements from day to day.

For the study of this subject I have employed, in addition to the observations at the two stations mentioned, those made by General Boileau at Simla (7,000 feet above the sea), on the Himalayas ; so as to include a large area and a great difference of climate.   There are storms with snow, and hail, or rain at Simla, when at Madras and Singapore there is calm and sunshine.   The diagram shows the simultaneous variations of the daily mean barometric pressure at the three stations.   You will observe that there is scarcely a movement which occurs at Simla, for which there is not a similar movement at Madras and Singapore.

In order, however, to determine exactly the epochs at which the maxima or minima of daily mean pressure occurred at each station, I obtained the daily means corresponding for their middle points to each hour in each day (that is to say the means for twelve hours before and twelve hours after each hour).   I give an example for a single week on this diagram, on which I have included the corresponding means for St. Helena.   It will be seen that the minimum barometric pressure occurred near six a.m. at Simla, Madras, and Singapore, and about eight hours later at St. Helena.   Although St. Helena is nearly on the opposite side of the globe, yet most of the movements to be seen at the other stations are to be found there, but with differences due to local position if not also to what may be termed polar causes. I have determined the epochs for each maximum and minimum to be seen in the movements at the three places in the same way as for the minimum in the first week of April on this diagram ; and I find that they occur on the average within thirteen hours of each other.   In several cases the minima occurred at all the stations within six hours of each other ; sometimes first at one station, sometimes at another. In fact we may say that the movements from day to day are simultaneous all over India, and even over a large part of the earth's surface.

It may seem rather extraordinary that all the movements at one place should appear also at the others, and indeed there are a few exceptional cases.   One of the most remarkable facts in connection with this question, is that we find at Simla one of the greatest falls of the barometer (0·3 inch) in twenty-four hours during the whole period examined, which however is not to be seen at all at the other stations. On examining the register of the weather at Simla, I found there had been a great thunderstorm during the whole of that day, but there was no storm at the other stations. We have here a fact of great importance, that when the greatest disturbance of the atmosphere occurred at one station, this was not propagated to either of the other stations. This great movement which was not seen at the other stations was due to a local cause. I have examined several of the more marked cases of difference of movement at the three stations, and whenever these have occurred, I have found that a storm had happened at one place and not at the others. All the other movements, however small they may have been (down to 0·01 inch, from maximum to minimum), have been shown at all the three stations.

I desire then to draw the attention of meteorologists to these facts. We have seen here that variations of atmospheric pressure occur simultaneously (or nearly so) over millions of square miles, under different circumstances of climate, where, according to any existing theory, it is wholly impossible that the mass of the atmosphere could increase or diminish simultaneously. I say we have no facts to shew that this is possible. There is no mode of propagation of atmospheric waves that we know of that could produce such effects, and we know of no waves which have crests or hollows covering such areas. The greatest velocity of propagation of atmospheric waves in Europe is about thirty miles an hour, and it would take days for one of these movements to proceed from one station to another, the distance of Singapore to Simla being nearly 3,000 miles, while it is nearly 10,000 to St. Helena.

We cannot then easily suppose that the mass of air is increased over all this area, unless we can assume that the mass of the whole atmosphere may be increased or diminished by some unknown cause. This is an assumption that we cannot easily make ; the question however may be put, whether meteorologists have not been too sure that no other cause but variation of mass is connected with variations of barometric height. Thus I have found a variation connected with the sun's rotation, which produces also variations in the sun's magnetic or electric action on the earth, and we may inquire whether an electrical attraction of the sun on our earth and atmosphere may not produce simultaneously irregular variations in the atmospheric pressure over large areas, if not over the whole globe. We are acquainted with the existence of atmospheric electricity ; we have also the Aurora Polaris which many physicists now acknowledge to be an electrical phenomenon ; and M. Becquerel has even indicated a source of atmospheric electricity in the sun, conveyed from it by solar emanations. Altogether, I would suggest that I have shewn grounds for seeking for another cause of varying atmospheric pressure besides that of variations in the mass of the atmosphere.

The PRESIDENT: I am sure you will be all glad to return your thanks to Mr. Broun for his communication.

The Conference then adjourned for luncheon.

The PRESIDENT : I will first call on Dr. Cornelius B. Fox for his paper

ON THE EMPLOYMENT OF ASPIRATORS IN ATMOSPHERIC OZONOMETRY.

CORNELIUS B. FOX, M.D. : The words "Atmospheric Ozonometry," which strictly mean the measurement of the amount of *Ozone* in the air, have been employed to include the estimation of the quantity of the other air-purifiers occasionally contained in the atmosphere, namely, peroxide of hydrogen and nitrous acid which have been proved to influence test papers in a manner which cannot be distinguished from the effects of ozone. The old fashioned mode of detecting ozone was to suspend a test paper in a box or cage and there submit it to the action of the air. It was found that tests thus exposed rather acted as anemometers than ozonometers, on account of the continual changes in the force of the wind and the ever varying quantities of air to which they were in this way exposed. Of course the higher the wind the deeper the coloration, for the simple reason that more air passed over the test paper. To obviate this important fallacy aspirators have been employed for the purpose of passing a known quantity of air over a test. Large water butts were constructed for this purpose, and I believe Mr. Smyth, of Banbridge, was the first to use aspirators of this cumbrous form. I hope he will give us the result of his experiments in this direction. I think that Dr. Mitchell, of Edinburgh, in 1855, in Algiers, was one of the first to work an aspirator in ozonometry. It was a dry aspirator with a capacity of four and a half cubic feet, and was employed in experiments undertaken in Edinburgh in 1871. This dry aspirator consists of a cylinder of mackintosh cloth closed at each extremity by a disc of wood. In the upper wooden disc is a stopcock through which the air enters, and a valve through which it makes its exit. When it is wished to make an observation, the lower disc is pressed upwards against the upper so as to empty the aspirator of air through the valve. On the under surface of the lower disc is a hook to which is sometimes attached a weight. Equally cumbrous is the water aspirator devised by Professor Andrews of Belfast. This apparatus consists of two copper cylinders, the smaller being inverted within the other which contains water. The inner and smaller one is supported by weights which are more than sufficiently heavy to counterbalance it. The base of the larger cylinder is perforated by two tubes that rise above the surface of the water, one being the entrance and the other the exit pipe. When one stopcock is opened air enters and the inner cylinder rises. When the inner cylinder is full of air the entrance pipe is closed, the exit pipe is opened, and the inner cylinder is pushed downwards so as to empty it of air either by the hand or by a large annular weight allowed to descend on it. A more portable instrument is Dancer's Reversing or Swivel Aspirator, which is generally made in glass. Some anonymous individual has had one of these instruments constructed of tin, which is exhibited in this Loan Collection. This reversing aspirator consists of two very large metal bottles or jars, fixed mouth to mouth, on an axis, one of which is filled with water. When the upper vessel is full of water, the water flows into the lower, causing in its descent the entrance of air through one of the taps. The air of the lower vessel, as the water enters, passes out by the other stopcock. When the upper vessel is empty it is simply turned around until the lower one stands uppermost. The ozone box, of course, is attached to the stopcock through which the air enters. Now these three varieties of aspirators are all constructed on the ever-changing—velocity principle, on the same principle, indeed, as the common jar aspirators used formerly in chemical laboratories. The laboratory aspirator consists, it will be remembered, of a large jar of known capacity, with a hole at the summit, and another hole at the base. The jar is filled with water, which is allowed to run out at the

T

base, whilst air streams in at the top to take its place The opening at the summit is, of course, connected by a tube with anything over which it is desired to pass a known quantity of air. The aspirators to which I have directed your attention in this Exhibition are constructed, I say, on the same principle as the jar aspirator, namely, on the ever-changing-velocity principle. When one commences to set either of these aspirators in action the velocity of the entering air is much greater than when the operation is nearly at an end. From beginning to end the velocity of the air is continually changing. To remove this fault Dr. Andrews has had a clockwork arrangement attached to his water aspirator, in order to regulate the pace at which the instrument shall work. This addition adds to the expense and bulk which is already sufficiently great. It may be asked, what is the objection to aspirators which pass air over test papers at a great velocity at first and at a slowly diminishing velocity afterwards. I have shown by experiments, an account of which has been published, that the velocity of the air passing over a test paper has much to do with its coloration. When the velocity of the air is great there is less colour than when it is little, because chemical action takes time, and because when the air is moist or hot the free iodine or the iodide of starch is volatilized when air of great velocity is passed over a test.

What actually occurs when an aspirator constructed on the ever-changing-velocity principle is used is that the air purifiers do not act on the test during the commencement of an experiment. As it proceeds the test paper begins to be influenced, and towards its termination is decidedly acted upon. If, perchance, the test paper during the com-mencement of an observation, when the velocity of the air is at its greatest, should be to a slight extent influenced, as for example when the air purifiers are *very* abundant, the free iodine or iodide of starch will most certainly be volatilized if the air be damp or hot, and will not be volatilized when the air is dry or cold. Here, then, is a fallacy which it is desirable to avoid arising from variation in the velocity of the air and from differences in the temperature and hygrometric condition of the air, when that velocity is great. What we want is an aspirator which will pass a known quantity of air over a test at an unvarying and low velocity. I employed the simple tube aspirator for years in connection with a small cistern, the water in which was always kept at the same level by a constant supply from the main. These tube aspirators of mine transmitted about 3000 cubic inches of air per hour. Mr. Dewar's water aspirator is in reality the same instrument, the water passing in at the end through which I allow the air to enter. Christiansen's water aspirator is essentially the same in construction, consisting of a short glass tube inserted into a small hole in the side of an india-rubber pipe. These tube aspirators all require for their working, water. In towns this is to be obtained, but in the country, where many of us live, anything like a constant or intermittent water supply is out of the question. Many of my correspondents inform me that they have not sufficient water for domestic purposes, and there-fore the expenditure of water for working aspirators is impossible. I am myself in a similar predicament where I live in Essex, having to rely directly on the clouds for my supply. To meet my own wants and those of travellers abroad, and the requirements of those who have expressed a wish to carry on observations in ozonometry in our colonies and at other foreign stations, I have devised a portable clockwork aspirator which transmits a known quantity of air at an unvarying velocity, that velocity being low.

The ordinary way of measuring the amount of air which passes through these instruments is by an ordinary gas meter. I can strongly recommend this aspirator as being one of the most portable for carrying on ozone observations, especially when they have to be made at foreign stations where the use of such cumbrous instruments as are before you, is out of the question. I hope, Mr. President, you will kindly allow the Conference to have the benefit of Mr. Smyth's experience on the subject of my address, as he has done so much work in this depart-ment for many years.

The PRESIDENT: I am sure you will all agree with me in thanking Dr. Fox for exhibiting this very simple and ingenious instrument; and, having given our thanks to him, I will ask Mr. Smyth to make any emarks which he may feel disposed to on the subject.

Mr. JOHN SMYTH, JUN., M.A., C.E., F.C.S.: With regard to the first point just mentioned, I was led to experiment on atmospine ozone by means of an aspirator from having observed that the curves of ozone increased with the intensity of the wind, and that the ozone intensity and force of wind corresponded for a number of years, as can be seen from the curves on the diagram in my paper on the Ozono-meter, published in the proceedings of the Meteorological Society for June 16th, 1869. I was led, therefore, to expect that ozone is constant in quantity in normal air, and sought the proof of that by means of an aspirator, which, for my first experiments, was an ordinary water cask. It was very troublesome to use however, as a fresh supply of water had to be pumped into it for every experiment. I mentioned the matter to Dr. Andrews, who had experimented on ozone for many years, and he showed me the aspirator described by him in the "Philosophical Magazine" for November, 1852, and used in his experiments to estab-lish the identity of the body in the atmosphere which decomposes iodide of potassium with ozone. I gladly adopted it at once, had one made of a capacity of five hectolitres, and made by its means a great many ozonometric experiments. In these experiments the velocity of the air entering the aspirator had little effect in the result, but when the same quantity of air was passed over the test-paper, it had always the same effect except in cases of fog, which Dr. Fox has alluded to. Then the moisture seemed rarely to clog the paper and to absorb the ozone. I also found in using long tubes for admitting the air to the ozone box no effect was produced as the moisture seemed to condense in the tubes and absorb the ozone.

Mr. G. J. SYMONS: With regard to the remark made by Mr. Smyth as to the correspondence between the wind force and ozone, I may mention that the officers of the Scottish Meteorological Society many years ago drew out a table something like an ordinary distance-table, giving numbers for correcting the amount of ozone indicated by the papers according to the velocity of the wind passing the particular place during the time the experiment was conducted, and by that means they got rid, to a great extent, of the difficulty arising from the velocity of the wind.

A GENTLEMAN: I should like to know if Dr. Fox has anything to add on the application of ozone papers in sick rooms.

Dr. FOX: As the subject of ozone is a very large one, I pur-posely restricted myself to that particular branch of it relating solely to its measurement. I think, perhaps, it would be more convenient for the Conference if I communicated with this gentleman in private.

The PRESIDENT: The next paper is by Mr. Daubrée, who will give us his communication with regard to experimenting in geology.

ON SYNTHETICAL EXPERIMENT IN GEOLOGY.

M. DAUBRÉE: During the last century the knowledge of the history of our globe has made most important progress. Accurate and profound investigations into the structure and composition of the soil, carried on in various localities, have led up to fundamental and positive facts. In this rapid advance made by geology, England has had a rich share, thanks to fortunate natural circumstances, but more especially owing to the sagacity of eminent men, in whose wake have followed numerous and useful investigators. Let me be permitted to render due honour to the memory of the great geologists that England has produced, from Hutton and William Smith to Sedgwick, Sir Roderick Murchison and Sir Charles Lyell ; the names of these men are engraved in your memories, as well as the names of their successors, whom you still possess among you.

It is chiefly by the observation of natural facts, coupled with reasoning, that that knowledge of geology, which we now possess, has been acquired.

This method, however, which leads to general conclusions, does not always suffice clearly to prove their certainty ; nor can it accurately explain the circumstances of the phenomena.

As far back as the beginning of this century, Sir James Hall supported Hutton's new ideas by means of experiments which have become classical.

The use of experiments is equally manifest from works, twenty years old, relative to many of the principal questions of geology.

As I am forced to keep within bounds, I will but mention three examples, and I will first of all take one in the artificial formation of minerals, which has been called the synthesis of minerals.

Careful observation of the rocks, known under the name of metamorphic rocks, has led to the recognition of the fact that heat alone could not have produced many of the minerals which have developed themselves to an immense extent in the rocks of the Alps and elsewhere ; it was to be supposed that heat had found a fellow-worker in some chemical agent, and very likely in the water which pervades the pores of all rocks. It is natural that it should be so ; for on account of the increase of temperature which is proved in each following vertical, the lower regions are hotter than the neighbouring parts of the surface.

The minerals, of which it was desired to explain the formation, particularly quartz and a series of anhydrous and crystallized silicate, had not been reproduced artificially under these conditions. The question therefore was to prove them by experiment ; and that is what I have endeavoured to do.

The chief difficulty is in finding partitions and fastenings which will resist long enough the enormous tension which aqueous vapour acquires when the temperature approaches a dull red. The water and the substances which are to react are placed in a glass tube which is then sealed. This glass tube is inserted into an iron tube, divided by extremely thick partitions and one of the ends is closed by fastening at the forge. The other end used often to be closed by means of a long screw cap, arranged with particular care ; but I have adopted, in preference, a strong iron stopper, so forged as to form one with the tube. But in order to succeed, it is necessary to have a most skilful workman ; for it is essential that the greater part of the tube should remain cold, so that, by vapourizing, the water inside should not frustrate the operation. To counterbalance the tension inside the glass tube, which might make it burst, water is poured on the outside of the tube, between the partitions and the iron tube which forms its envelope. By this means the greatest strain is thrown upon the latter tube which is capable of much greater resistence.

But even if all these precautions be taken the experiment often fails. Generally speaking the tube loses its water, which appears, by the tension, to have freed itself through the screw. When the fastening is strong enough to prevent this escape, the water produces a violent explosion ; the partition of the iron tube, although it is more than a centimètre thick, is torn as a sheet of paper would have been, and you may see this by the tubes which lie before you. Gunpowder is far from having the power shown by water under circumstances in which the elements are perhaps dissolved.

When in exceptional cases, it has been possible to keep the water at this high temperature, it can be seen that even alone, without any chemical help, it is capable of most energetic action with regard to silicates.

At the end of a few days the ordinary glass is completely transformed : in its place is found a white mass perfectly opaque, porous, and which would have the appearance of kaolin, had it not a very marked fibrous formation. It is a new silicate which has been formed by fixing water, and which belongs, by its composition, to the family of the ziolithes. Moreover a portion of the alkali of the glass has dissolved, carrying with it alumina. What is more worthy of notice is that there appear innumerable colourless crystals, of perfect limpidness, which streak the glass, and have the ordinary bipyramid shape of quartz ; they are merely anhydrous and crystallized silica.

Sometimes these crystals, which at the end of a month have attained to millimètres, are isolated ; at other times they are implanted on the partition of the original tube, where they form true eagle-stones which it would be perfectly impossible to distinguish, to a nicety, from those which are so often to be seen in crystalline rocks.

And what renders this transformation of the glass still more remarkable, from a geological as well as from the chemical point of view, is that it is brought about by a very small quantity of water, the weight of it not being even equal to a third of that of the glass transformed.

In two experiments I have obtained, besides, upon the surface and in the interior of the whitish mass, which results from the transformation of the tube, crystals of a different kind. They are very small, but perfectly distinct in shape, and are very brilliant and transparent; their colours are various shades of green, among them olive green. Their shape is that of a dissymetric prism with modifications on the edges. These crystals scratch the glass ; they are unaffected by boiling and concentrated chlorhydric acid. They belong to the species pyroxene and to the variety diopside, of which they recall the best-known appearance, such as may be seen in the crystals of Ala in Piedmont, which are to be found in all collections.

Pieces of wood have been transformed into a black mass, of great brilliancy, perfect compactness and having the look of a pure anthracite. This stone-coal, although infusible, is entirely in the shape of globules, hence the substance has been melted in course of transformation.

Without bringing forward other examples, it can be seen that over-heated water has a very powerful influence on silicates. It causes some to form, which are either hydrated or anhydrous. And moreover it makes these new silicates crystalize much below their melting point. If silicious acid be thrown in while the process is going on, it will detach itself in the form of crystallized quartz.

Now, although over-heated water is to be met with everywhere, and necessarily in the interior of deep rocks, it is by no means easy to obtain it in experiments. Even putting aside the danger of explosions, which are often of surprising violence, I have not been able to repeat these experiments as often as I could have wished. The facts, however, which have already been observed, are conclusive and show with what ease common minerals can be produced in over-heated water ; and they throw light upon the formation of these minerals, as they are presented to our view in many rocks, and particularly in metamorphic rocks. Such are the well-known development of pyroxene and amphibole in certain lime-stones, the production of various minerals in blocks of "Somma" lime-stone, the growth of felospath in strata of different kinds, and the appearance of chiastolite or macle in phyllades. Such, likewise, is the manner in which quartz has detached itself in silicated rocks, in slates and in others, in varied and many other forms. Such is, finally, the combination so frequently met with of anhydrous and hydrated silicated [chlorite with turmalin, and feldspath "*adulaire*," etc.]

It is known that pyroxene had been obtained in crystals by means of melting and slow re-cooling. But this experiment of Berthier did not explain the formation of pyroxene in very many bearings. This is not the case with regard to the process which forms it at a degree of temperature far below that of its melting point. It is the first instance of the production an anhydrous and crystallized silicate in the midst of water.

In a treatise on geology, which enjoyed a wide spread circulation and much authority, some twenty years ago, it is said that if crystallized silica cannot be seen in laboratories in the state of quartz, it is because the help of centuries is undoubtedly necessary to bring about this crystallization.

Every day the complete resemblance between ancient and actual facts is being more fully established. Time, no doubt, plays a great part in the crystallization of minerals; but too much must not be made of this auxiliary in difficult questions.

The contemporary formation of minerals in nature can likewise be included in the domain of experiment, when it takes place under circumstances which can be determined with as much certainty as if the laboratory were the place of investigation. In this way many kinds of minerals have been formed in the basin of certain thermal springs, at such a depth that they could be got at.

The Romans have regulated in the most skilful manner the management of the chief thermal springs of Europe: they have, in many cases, laid down layers of concrete masonry round the mouths of the springs in order to secure them and keep off all superficial water. In several places, notably at Plombières (Vosges), at Luxeuil (Haute Saône), and at Bourbonne les Bains (Haute Marne) it has been necessary to cut through the ancient masonry, in order to carry out new works for obtaining the pure water ; it was then perceived that a very remarkable change had taken place ; there is to be seen in the substance of which the bricks are made geodes covered with crystallized silicates of the family of zeolites, chabasie, and others. These zeolites are so developed in swellings as to call to mind by their shape, and by their nature, even in the smallest details, those which cling to many rocks of an eruptive kind, such as basalt, trap, or melaphyres. It is like a faithful representation of amygdaloid rocks ; and consequently the history of these rocks thus reproduced is most instructive. These minerals disseminated from human constructions of a known date, belong, of course, to the contemporary period ; they have evidently been produced by the thermal water, which, during the course of centuries, has slowly filtered through the pores of this mass of brickwork. This water has a temperature equal to, or less than seventy degrees ; it contains but very little soluble salts (at Plombières o gr. three per litre)—that is to say less than in many drinkable waters.

It has, nevertheless, produced very marked effects. Its very weak action seems to have been multiplied by the number of centuries, during which it has continued. It is like an experiment instituted by anti_quity for our benefit. I will but mention the numerous metallic minerals which have been discovered very clearly crystallized in the same kind of springs—chalkosine, chalkopyrite, philippsite, tetraedrite, phosgenite, and many others.

Among the many other geological questions of which the solutions have been attempted by experiments, I must draw your attention to one relating to the theory of volcanos. It is well known that water is the most constant as well as the most abundant product of all the volcanos of the globe. This water sometimes acquires, in the reservoirs of these huge machines, a pressure of a thousand atmospheres or even more. This fact cannot be doubted when at Etna we see a column of fluid lava thrown to a height of more than 3000 mètres above the level of the sea. Some geologists have supposed that this water, set free by the volcanos, had been originally—that is, since the consolidation of the earth's crust—shut up in deep regions ; but it is far more natural to enquire whether these increasing losses are not made good, at least partially, by some aliment derived from the surface, and if such is the case, in what manner this infiltration takes place.

It would be difficult to understand how it could be by open fissures, for, considering the enormous pressure from within, the water ought to be repelled towards the surface and return by the very same fissures that had brought it in a liquid state, without there being any necessity for it to make new and special outlets for itself. I have been led to consider whether the water could not penetrate into these hot and deep volcanic regions by availing itself of the porosity and capillarity of the rocks. By the help of this tolerably simple apparatus, which I have the honour of placing before you, experiments have been made which prove that, by means of capillary force acting together with weight, water can be forced from the basins of seas and continents into the deep and hot regions, and that in spite of strong pressure in the opposite direction.

Although the time at my disposal is most limited, I cannot but allude to the way in which the experimental method has thrown light upon purely mechanical questions, such as the cleavage and the foliation of rocks ; the formation of clay, sand, and shingle.

Synthetic experiments have also been able to be applied to bodies, known by the name of meteorolites, which come to us from celestial spaces; they have been imitated by certain terrestrial rocks which have simply been placed under the influence of a reducing action.

The preceding examples are sufficient to show how useful experiments can be for giving accurate information regarding several geological phenomena. But what has already been accomplished is but a very insignificant fraction compared with what yet remains to be done in this branch of science, and what, some day, assuredly will be effected.

Geology is, in these days, in such a state that observations, collected in immense numbers, have been applied in every possible manner, with the view of throwing light upon certain questions, but without, if I may say so, gaining much result.

I need only touch upon one as being in this condition, and the importance of which is very great, namely, the formation of granite, that is, the rock which in all regions of the globe forms the general base of stratified soil. Granite has been the subject of much study, both with regard to its bearings and its mineral constitution. The researches which have been made from the latter point of view, upon thin slices, have been most useful. Nevertheless dissertations and hypotheses upon the origin of granite have for a long time been, and are still, matters of every-day occurrence. No satisfactory result will in all probability be arrived at regarding this fundamental problem until granite, complete in all parts, shall have been made by a synthetical experiment; and not only the quartz, but also the feldspath and the mica which are likewise crystallized in it: let us hope that this moment is not far distant.

In dealing with the question of the method of formation of all kinds of rocks, whether stratified or eruptive, it is most useful to consult contemporary phenomena—which are but the continuation of ancient phenomena. This is what Hutton has so clearly shown, and Sir Charles Lyell applied with such remarkable shrewdness. The actions which are at present taking place on the surface of the globe are, in point of fact, to be compared to experiments which are being prolonged before our very eyes.

They are, however, far from being able to take the place of the experiments originated by man. It is true that these latter can act but within very limited bounds; but they have the advantage of being able to give rise to the facts, keep them under our observation, and vary the conditions which produce them: they are accordingly most useful, even as addition to the study of contemporary phenomena themselves. It is, as Sir John Herschel expresses it, *active* observation following in the wake of *passive* observation.

Experiments ought not to be made until careful observations have been taken—nay, until these observations have been reasoned upon, so as to deduce, as far as possible, the general principle.   Then comes the question of giving an opinion upon the inference which, we feel, must be drawn from the facts.   It is the method which has been defined by Mons. Chevreul as "experimental *a posteriori;*" moreover, uncertain or vague explanations are thus revised and expressed more precisely.

It devolves therefore upon the geologist himself, who has observed the facts upon the spot, to carry out the experiments.   For example, in the synthesis of minerals, what concerns geologists and mineralogists, is not only to reproduce a particular mineral species, but to obtaining this result according to methods which appear similar to those employed by nature.   When we read Hall's classical works, we cannot help noticing to what an extent this learned man personally possessed all the data of the observation, which he afterwards submitted to a judicious experiment.

The magnificent exhibition, of which we are now witnesses, shows us in vivid colours, how frui ful for the human mind, the experimental method has been, since the time of Galileo ; and to what an extent it has led up to happy and unexpected results.   It is this experimental method also, which must day by day occupy a larger share in geological investigations.

Setting aside the history of organic fossil remains of ancient periods, the phenomena, which are embraced by this science, are mechanical, physical, or chemical.   However remote may be the age of the history of our planet to which they refer, they are within the province of the so-called experimental sciences.   The foundations of geology will always be laid in observation and in reasoning ; but the science must also become experimental ; it will then enlighten the world, according to Bacon's expression: "Under the iron and the fire of experience."

The PRESIDENT: I am sure all those present in this room will join me in expressing our thanks to M. Daubrée for the most interesting communication he has made to us.   I will now ask any gentleman who wishes to make any remark upon the subject to do so.

Mr. R. H. SCOTT, M.A., F.R.S.: As one who has for many years wondered at M. Daubrée's experiments, and his successful prosecution of a branch of science which, I regret to say, is not half enough prosecuted in England, and to which the Geological Society owes its foundation,—for Wollaston, and some of our most eminent geologists, some sixty years ago, were all mineralogists, while you can now count on your fingers the mineralogists in the Geological Society of London. —I think we are extremely indebted to M. Daubrée for coming here to explain these investigations he has been so long prosecuting. It was at the very end of his speech that one remark came up, which would perhaps appear strange to most of the audience, that is, that no man living is able to say how the granites that we walk upon were formed. Geologists are divided into two great camps upon this question; chemists maintain that the granite must be an aqueous rock, while geologists maintain that it must be an igneous rock. Now a remark which fell from M. Daubrée to-day seems to me to throw some light on the difference between these two. One of the chief constituents of granite is quartz, and there is also mica. Now there are questions connected with the constitution of the constituents of granite which are not explicable simply. In the first place, quartz, as is well known, like many other substances, appears in different states of density. Quartz, which is obtained by crystallisation from a solution, is very different from that obtained by fusion, and it appears to me that, as M. Daubrée puts it, quartz may have crystallized, by a sort of semi-fusion in the presence of aqueous vapour, and that fact throws a light on the way in which granite might have formed itself without the actual necessity of its ever having been brought into a perfectly liquid state. There is again the question of mica. The difficulty about mica is that it is a mineral which does not contain as much silica as it might possibly contain—the base is not saturated; and also that it contains water. Chemists ask very fairly how is it possible to take a piece of mica and melt it without causing the composition to assume a more definite form, and driving off the water. In all these questions it is only by experimenting, and by actually making the minerals again, that we can ever hope to throw light on the question. I regret most sincerely that our great English synthetical mineralogist, Mr. David Forbes, is not present to-day. He has not made the history of the composition of granite his study, but has dealt with other metamorphic rocks, for I call granite a metamorphic rock, and not an aqueous one. He has proceeded by taking a mass of soapstone and leaving it for two or three months in a furnace, never allowing it to reach a temperature of fusion, and finds that by the gradual molecular arrangement of the particles this massive soapstone will become converted into a crystallised talcose mineral. In that way we have gradually found a number of minerals which are found in the earth and which we are able to reproduce; and as chemists maintain that they hope to be able to reproduce in their laboratories every crystallizable organic compound which is formed in the human organism, such as urea and various other substances, so I hope, under the teaching of M. Daubrée, Mr. Forbes, and other gentlemen who have studied in this way, that we shall gradually be able to fill our museums, with a set of minerals on one side, as they occur in nature, and on the other side, as they have been made in our laboratories.

The PRESIDENT: I am sure it must be very gratifying to all the ladies present to hear so eminent a mineralogist as Mr. Scott holding the hope that possibly we may have diamonds made by artificial means. We must now go on with our programme, and reverting to the old system of four elements, as we have been treating of fire, earth, and water we will now turn to the wind, and take up a series of papers on Anemometers. As Mr. Gordon is not present I will call on Mr. Cator.

ON ANEMOMETERS.

Mr. CATOR : The principal subject of this paper is a description of the " Lever Anemometer," exhibited here, No. 2846 in the catalogue ; but I propose first to offer a few preliminary remarks on Anemometers in general.

Anemometers may be divided into two classes :—

      1st. Those for measuring the *velocity* of the wind.

      2nd. Those for measuring its *force*.

If the relation between velocity and force were accurately known, one class of these instruments might in some cases be sufficient, as, if one element were known, the other could be deduced from it ; this, however, is not the case. The relation subsisting between the two elements is expressed by the formula, $P = p\ V_2$, where P represents pressure or force ; V, the velocity ; and p is a constant, the value of which is not exactly known, but it is of such a value that a pressure of one pound on a square foot is=fourteen or fifteen miles per hour, or thereabouts. It must be assumed, therefore, that for all purposes, the two classes of instruments are required ; and, in particular, the first mentioned class is better adapted for registering the aggregate amount, and the second class for registering the details of the wind, sudden gusts, &c.

I. Of the first above mentioned class, several kinds have been made, varying principally in the method of registration ; they may, however, be resolved into two, viz., Whewell's and Robinson's, the difference depending on the form of the receiving surface.

(*a*) Dr. Whewell's anemometer, No. 2880 in the catalogue. The receiving surface is a windmill-fan set vertically, and connected by means of an endless screw and cog-wheels with a pencil, which registers in a box below the horizontal motion of the air, and the direction.

This instrument has not been at all generally adopted, and is now superseded by

(*b*) Robinson's anemometer, which consists of four hemispherical cups revolving in a horizontal plane, and fastened to the ends of four arms of equal length rigidly connected together, and set at right angles to each other. This instrument is now in most frequent use. The result of Dr. Robinson's experiments was, that the rate of the cups is always one-third of that of the wind, whatever may be the length of the arm, or size of the cups. He has lately, in a paper read before the Royal Irish Academy, treated thoroughly of the constants for determining the relation between the velocity of the wind and that of the cups, and he is apparently now not quite clear in his opinion that the deduction before referred holds good for all relations between the cups and the arms, and he indicates a line of research which he thinks will lead to useful results, both in theory and practice, and to obtaining values which will be a close approximation to the truth. I understand that he has lately obtained a grant from the Royal Society, and is now prosecuting these researches.

The mode of registration in this class of instruments may be either by indicating the total number of revolutions of the cups, and thence the number of miles, by a series of wheels like a gas-meter, or it may be a self recording mode, and this again either by a pencil taking a curve on paper moved by clock work, or by Beckley's method of a spiral thread round a drum pressing on metallic paper, or again by means of a long strip of paper, moved at a uniform rate by clockwork, and on which a prick is made for every mile of wind. Perhaps I should here mention another of this class, viz., one by Professor Von Oettingen, an excellent instrument, entitled "The Self Recording Wind Components Integrator." He was to have been here himself to-day to describe it; it will, therefore, suffice if I say that it is for the purpose, not only of measuring the velocity and recording the direction of the wind, but it measures by electricity the component parts of the wind from the several points of the compass.

II. We now come to the second class of instruments, viz., those for measuring the *force* of the wind. The amount of force is generally calculated at so many pounds pressure per square foot.

There are various anemometers of this class, their difference depending mainly on the *media of resistance* to the moving air which have been employed. In some cases, the medium is a set of springs, in another case water is used, in another mercury, in another weights, and in the last hereafter described a system of leverage.

(*a*) Lind's Water Anemometer. This is of very simple construction. It consists of an inverted syphon, one end of which is turned at right angles, its mouth being always kept opposite to the wind by means of a vane. The syphon is partially filled with water, and the strength of the wind is measured by the difference in length of the columns of water in the two arms. This instrument has been but little used. Sir W. Snow Harris, some years ago, suggested some modifications of it, the principal feature being that the hinder branch of the tube should be of smaller bore, so as to give a more open scale.

(*b*) Professor Wild's Anemometer, see No. 2784 in catalogue. This is also of simple construction, consisting of a plate hanging in the air like a sign-board, and the force of the wind is indicated simply by the angle through which the plate is driven from the vertical. It is found that a comparatively moderate wind will cause the plate to move through an angle of ninety degrees, and a strong wind can do no more, and, therefore, this would be of no use for strong winds; and evidently it could not be accepted as a standard instrument.

(c) Howlett's Anemometer, No. 2881 in catalogue, has some similarity to the last-mentioned. It consists of a rod, suspended somewhere about the middle with a sphere at the top and a weight at the bottom, and it swings in any direction, consequent on the wind blowing against the sphere. It records the pressure and direction of the wind on a paper under the weight, so that from this instrument only the maximum pressure from each direction can be shewn, every record in the same direction being covered or overlapped by the succeeding one. The apparatus is thus not applicable where the pressures vary very much, and because an additional weight must be applied for recording a gale, and because there is no time scale.

(d) Osler's Anemometer.—This was, until lately, the most perfect instrument for recording the force of the wind. The surface exposed to the wind is a square plate set vertically equal to one square foot (and sometimes equal two square feet). The resistance to the wind is a set of springs behind the pressure-plate; weak springs come into action first for light winds, and as the plate is acted upon by stronger winds, stiffer springs are brought into play, and according to the space through which the plate is moved, so a corresponding motion is communicated to a pencil below, which records every movement. The direction of the wind is also registered on the same paper as the pressure.

Different kinds of springs have been adopted, some lengthening out by pressure, and others undergoing compression, but virtually the principle is the same. But wherever springs are employed, they must be subject to variation in their elasticity, both from changes of temperature, deterioration from exposure to the weather, and other causes, and so the results cannot always be depended upon.

(e) Another kind of instrument for recording the pressure is by Ballingall, No. 2,832 in catalogue, where a plunger is pressed into a vessel of mercury by the wind, and the space through which it moves measures the force.

U

(*f*) A Balance Anemometer, by Francis Ronald, No. 2,839.—The surface exposed to the wind consists of a plate of one foot square, kept at right angles to the direction of the wind, and which, according to the pressure exerted, lifts up a lesser or greater number of a series of weights. This will, of course, only register different intervals of pressure referable to the intervals between the weights.

(*g*) The most recently devised instrument of the second class, above mentioned, is Cator's Lever Anemometer, No. 2846 in the catalogue. In its construction, the pressure-plate exposed to the moving air is of a circular, instead of a rectangular form, and has an area equal to one square foot. Attached to it at the back is a horizontal rod, which runs on friction-rollers. The plate is kept at right angles to the air-current by the usual method of a vane, or by windmill fans: there is something to be said in respect to each of these two methods. The vane, I am inclined to think, is the best, but sometimes this is apt to move through too large an angle; and windmill fans hardly cause the pressure-plate to move quickly enough in azimuth; so that, with either method, the full force of the wind can scarcely be reckoned upon in all cases. To the horizontal bar, above mentioned, a small articulated chain is attached, which passes over a wheel and is then joined to a wire, which wire passes through a perpendicular tube down into a room below, where the leverage and recording apparatus are fixed. Springs form no part of the medium of resistance; but instead thereof an arrangement of what may be aptly termed "curve-leverage" is adopted. This arrangement consists of two "snails," or two curved plates, of differing and different radii, firmly joined together; each has a grooved edge, and they move together in vertical planes, and are supported on a common horizontal axle. Thus an adaptation of the principle of the lever is secured, when on the one hand a fixed weight is attached to a cord passing over the larger curve, and on the other hand the pressure-plate is brought into reciprocating action with the smaller curve by another articulated chain passing over the periphery of the latter. This chain is fixed to the lower end of the wire connected with the pressure-plate above mentioned. When at rest, it is so arranged that the lower chain is a tangent to the smaller curve at a point furthest from the axle or fulcrum, and the cord to which the weight is attached is a tangent to the larger curve *at* the centre or fulcrum from which it hangs. As the pressure-plate is acted upon by the wind the system of leverage rotates, the chain is unlapped from the smaller curve, and thus becomes a tangent to it nearer to the centre, and at the same time the cord becomes wrapped round the larger curve, which it touches further and further from the centre; so it will be seen in effect (supposing the fixed weight is 7lbs) that for all pressures less than 7lbs, power is gained; and for all above 7lbs, power is lost.

When in action, the pressure-plate and the recording pencil move through equal spaces, and these are severally equal to the length of the chain unlapped, so that the corrections due to relative spaces do not require to be made.

By means of this system of leverage, friction is reduced to a minimum. An extremely small force of say one or two ounces can be registered, and, in consequence of the shape of the curves, larger spaces on the scale are devoted to the smaller pressures, where fractions of a pound are required and smaller spaces for the greater pressures;—the whole scale from zero to 50lbs being comprised within a space of about five and a half inches.

The pressure-plate is conified at the back to minimize the possible error due to a partial vacuum. The recording pencil is moved to and fro on a paper fixed on a drum, which revolves by clock-work, the pencil being connected by means of a string with the wire above mentioned.

Beckley's method is adopted in order to give a continuous diagram of the direction of the air-current on the same paper as that on which the pressure is recorded.

The only parts of this instrument exposed to the weather are the pressure-plate and the vane, all the other parts being protected and under cover in a room below, and therefore not subject to its disturbing effects; whereas when springs are adopted, they being fixed behind the pressure-plate, are exposed to and affected by the weather, and necessarily give uncertain results.

It has been mentioned above that the pressure-plate is kept at right angles to the direction of the inpinging wind, but it must not be forgotten that this cannot always be the case in practice, for the inpinging force itself is not always exerted horizontally. Sometimes the wind strikes the plate obliquely either in an upward or downward direction, and in such cases the force registered is not the full force, but only the resolved horizontal component part of it.

As to the best form and size of pressure-plate, it is still a moot question, and one deserving of much consideration, *i.e.*, whether it should be square or circular, and whether its area should be equal to one, two, or more square feet; a suggestion has even been made that there should be a hole in the middle of it; and another that a projecting rim should be fixed to the edge of the plate to counteract the error arising from cases in which the wind strikes the plate obliquely.

Compared with other meteorological instruments, anemometers are not in very general use, chiefly on account of the difficulty in obtaining a suitable position, and the trouble and attention involved in reading them, and reducing the observations; this applies more particularly to costly self-recording anemometers, of which I do not suppose there are above eight or nine in use in the whole country. On account of their rarity it is almost impossibible properly to compare results, but so far as comparisons have been made they have nearly always shown divergent readings.

One of the practical difficulties in fixing anemometers, is, to find a really good position, where the instruments will be free from the disturbing influences of buildings, of trees, and of the undulating surface of the country.

The PRESIDENT: As there are several more papers on the same subject, we will take them before entering upon any discussion, and I will now call on Dr. Mann.

LOWNE'S SERIES OF ANEMOMETERS.

DR. MANN: I have selected the small series of instruments out of the Exhibition which I have before me to bring under your notice, chiefly on the ground of the interest I have had in using them myself. I think they are exceedingly good instruments, and very handy. The principle of the whole of them is involved in the one I first take up. You observe this instrument has a fly wheel placed on an axis which operates on a set of wheel work, recording at right angles to itself.

The inclined fans of the fly wheel are made of light aluminium metal. It is very delicately centred on jewels, and great pains have been taken to make it sensitive to the movements of air. The excessive sensitiveness of the instrument may be seen in this way. If I hold it over the flame of a candle, you will see the current of heated air rising from the flame will cause the registering wheel to go round. The amount of motion necessary to put this in action is really thirty feet per minute. Any wind moving at a less rate than that, does not produce motion, the reason being that the friction of the fly wheel resists up to that extent. When it amounts to thirty-one feet, the fan wheel immediately begins to turn, and the extent of the motion is registered. The effect of the friction is a constant loss of thirty feet per minute which always has to be applied to the indications of the instrument. It is possible to have these fan-wheels so carefully adjusted that a current moving at the rate of 150 feet in a minute, shall always cause 120 revolutions in the same time, giving thirty feet difference between the two. Whenever they are so adjusted a constant is obtained which applies to all velocities. When this adjustment is not made perfect. the error of each instrument is ascertained by experiments and recorded, as a constant of correction. The instruments therefore all practically give very exact and accurate results. This is the form which was first produced, and manufactured by the inventor Mr. Lowne himself. Here is, however, another form in which the registering dial is placed in the same plane with the fan-wheel. In all the instruments there is a contrivance by which the connection between the indicator wheels and the fan-wheel can be thrown out of gear by shifting the position of a small lever. In using the instrument it is placed in position and put into gear, and the number of revolutions in a minute is noted by a watch; this then gives the velocity of the air-current. The object of having the facility for shutting off the connection, is to prevent the impulse which is given by the fly wheel from causing a false indication. When the indicator is thrown out of gear, the register is that which had been reached at the instant by the fan. Here is another form of the instrument which I think is very ingenious. In this form there is actually no material connection whatever between the recording wheels and the fans which take up the impulse, and yet the wheels notwithstanding record the motion of the fans. The object of this arrangement is to

allow the recording wheels to be hermetically sealed up in an air-
tight case, so that the instrument may be safely employed where there
are corrosive vapours which would injure the delicate wheel work.
The object is obtained by a small magnet, carried on the axis of the
fan, and a soft iron bar on the axis of the sealed-up movement. This
specimen is intended for use at sea, and is mounted on gimbals. You
will observe that there is a small cap which can be taken off or put
on, to increase or diminish the effective diameter and the air-currents.
The reason for this is that the inductively magnetized soft iron bar
cannot follow the magnet at very high velocities. But if the inlet to the
fan-wheel is then diminished one-half, the magnet arrangement continues to record the number of revolutions correctly, and the velocity of
the air-current in feet per minute then only requires to be multiplied by
two. This is another form still newer, intended for ventilation purposes,
and to be used in mines. The improvement in this consists in having
the fans made of vulcanite, which is even lighter than the aluminium
metal. The sensitiveness depends on the lightness of the moving
parts of the apparatus. In this instance the fan-wheel is immediately connected with the first wheel in the registering part of the
dial, and is so contrived that all dust and dirt are entirely excluded,
so that the instrument is nearly as good as if it were hermetically
sealed, and it can hardly get out of order. The fan-wheels are manufactured by machinery which produces identically the same angle
and weight in this part for any number of instruments.

Here is another piece of apparatus which I will only just mention,
in which the same arrangement is employed, in order to get the
cubical value of the expirations of the chest. It is called a spiro-
meter, and by blowing into a tube, anyone who uses the instrument,
ascertains how many cubical inches of air he is able to expire, or, in
other words, what is the effective respiratory capacity of his lungs.

The various forms of the simple anemometer are exceedingly convenient instruments for ascertaining the amount of ventilation that is
secured in a building through any given aperture of inlet, or outlet.
The superficial area of the section of the aperture, either in inches or
in fractional parts of a foot has first to be ascertained, then after the
anemometer has been applied, this is multiplied by the number of feet
which the air current passes through in a minute. The square of the

diameter of a circular aperture in decimal parts of a foot, multiplied by the velocity of the air current in feet per minute, if again multiplied by 7854, gives the number of cubic feet of air passed through the aperture per minute. This method of testing the ventilating capacity of an aperture is practically shown by the application of the anemometer to the air inlet of George's ventilating gas calorigen, exhibited in operation in one of the small rooms of the exhibition.

The PRESIDENT : I am sorry that Mr. Gordon and Professor von Oettingen are neither of them present, but we must first of all thank Dr. Mann for this further communication on the subject of anemometers, which I am sure you will all do, and then I will ask any gentleman who wishes to make any observation on the subject to be good enough to do so.

If there is no one who has anything to say on the subject of anemometers, I will call on Mr. Wenley, who has a communication to make on

A NEW METHOD OF FORECASTING STORMS AND FLOODS.

Mr. WENLEY : It is some thirty years since I was a witness of very great destruction of life and property while on board a vessel in the North Sea, and my attention was turned very forcibly to the question, whether there could not be a means devised by which gales and storms could be foreseen. In imagination I placed myself above the atmosphere, and in doing so I supposed I should see it, if it were visible to the eye, much in the same way as we should see fog, and that it would not remain perfectly level, but that it would take the form of waves or undulations, that the earth would represent an inner circle with the atmosphere lying around it, and if it were allowed to remain perfectly at rest, lying at an equal distance around it, and pressing with a weight equal to about thirty-two feet of water, or twenty-nine-and-half inches of mercury. But the atmosphere is not allowed to remain at rest, but is generally undulating with a quick or slow movement. To-day the air, according to the reading of the barometer, is about six-tenths above its normal pressure, taking each one-tenth of an inch on the scale to represent seven pounds pressure on the square foot, that would give something like forty-two pounds per square foot more than there should be. That has continued for a long time, and at any rate it extends over the whole of the British Islands; in many cases, the movements of the barometer lately have been reported to be simultaneous all over Western Europe. The idea in my mind was this, that if we found the air line very high, it would naturally try to reach its level, but it must take time to do so. If you have forty-two pounds too much it cannot get down to the level for some time. The quickest time I have been able to ascertain was in a hurricane that was recorded in the *Times* sixteen years since. It was reported by the captain of the "Porcupine" man-of-war. At the commencement of his observations the barometer stood at 30˙32, which is about fifty-eight pounds per square foot too much air. The captain reported that at 8 p.m. it had fallen to nineteen pounds too much, that is, it had lost about three-and-a-half pounds per hour. When it reached that distance he reported heavy squalls—a more rapid fall, at the rate of about nine pounds per foot per hour took place for seven hours. Then for three-quarters of an hour the barometer stood level, and it was nearly calm. During the next three hours there was a very rapid up-cast, and the inrush of air come at the rate of twenty-two-and-half pounds per foot per hour, at that point the hurricane began. When it reached the normal line it slightly eased off, and the wind fell from a hurricane to a gale. I took the record of the barometer from the *Times*; at 8 a.m. on the 2nd of October it was 30˙32; at 8 p.m. it had fallen to 29˙75. During the next quarter of an hour it fell 0˙4. The next record was at 8˙45, when it stood at 29˙62. At 10˙20 it was at 29.34, and so on, until 2˙40 when it stood at 28˙87. At 3˙20 it still stood at 28˙87, and six minutes afterwards the hurricane began. At 5˙30 the barometer had risen to 29˙52 ; the next movement was at a rather slighter angle, and the hurricane fell to a gale.

There is another thing to which I should like to call your attention, and that is, that for seventeen months since 1874 the air has been remarkably in excess. I have been noticing the weather on this plan for twenty-seven years, and I never before saw one month like any one of the last seventeen, as regards the extent to which the air has been in excess from December, 1874. By means of the local press in Essex, I warned the public that there was a probability that we should have very disastrous gales in conseqeunce. That was also obtained from another consideration, which was to this effect. Near the equator there is a north-east trade wind blowing pretty regularly for ten months in the year; and on the other side there is a south-east trade wind blowing still more strongly for eleven months in the year; these two currents are always taking the cool air from the Polar regions to the equator. These two currents converge to the equator, and the question is: Where does the air go? Lieut. Maury has explained, as the result of observations taken by thousands of sea captains, that the probability is that the south-east trade comes from the south as a vapour-laden wind, which has been raised over the South Atlantic and South Pacific, that it crosses at the calm belt near the equator, and then having acquired the velocity of the earth at the equator it finds itself travelling quicker than the earth, and becomes the south-west counter-trade—the wind which should be the prevailing wind over England. That wind has been kept back for many months, we have had very little of that south-western wind, and when we have had it, it has not been able to lower the barometer. From my knowing that this wind was kept back, I knew there was a great amount of danger, and as the north-easterly wind last June was in great force, I was able to predict the floods which took place early in July. This prediction was made when the weather was very fine, with a high barometer averaging 30·5. The danger I pointed out last summer I consider intensified, for we have never yet had a true settlement. If we take 29·5 to be the fair average, the barometer ought to work fairly right and left of it, but it has not done so. All the barometers in England have for many months continually risen, and but rarely fallen. Previously to the beginning of last year I never saw one month in which it did not cross over this line six times at least, and sometimes as many as twenty, but during the whole of last year the barometer never crossed more than four times, and for seven or eight months it never went below at all. In November it went sharply below, and for about a week gave us the worst gales of the half century; but whatever we had then I fear we are likely to get very much worse.

There is one other point to which I should like to allude, which was mentioned by the gentleman who addressed us this morning (Mr.

Broun). He pointed out there was a difference in the height of the barometer over hundreds of miles of country with very little wind. I have observed that myself. I have sometimes seen the barometer come down in a single night, so that 600 or 700 tons of air per acre disappeared. It was there at night, and it was gone in the morning. That has taken place in some cases all over Western Europe, and yet there has been no wind. Where has it gone? As Mr. Broun said this morning, the barometer shows that it has gone, and he asked whether meteorologists could supply a reason for it. My belief for a long time has been that it goes in this way. Mathematicians tell us that water does not move forward with the wave lines; that if you see waves running along the sea at the rate of thirty miles an hour, the water is not going, but simply the wave line, the same as you see it passing over a corn field. That happens in the case of water by the particles taking a circle and coming back to the point whence they started, so that a cork upon the surface simply rises and falls, and does not travel with the wave. If that be the case at the top of the atmosphere, and we have this wave line, it will explain why it is that the barometer shows these differences. The day before Good Friday there was in the Midland District of England rather a strong hurricane, reported to have blown down nearly a mile and-a-half of new telegraph posts. I watched that hurricane on my barometer, and at 1 in the morning it had fallen really more rapidly than in the hurricane of the "Porcupine"—4 lbs per foot per hour; and if that had continued, it would no doubt have resulted at Chelmsford (where I was observing) in a gale of wind; but instead of doing so, when it got to the average line it eased off until about 4 in the afternoon, and then it began to rise. My contention is, that by using this plan, a seaman can go just as near as he likes to the gale of wind, and there is no occasion to be frightened. If he finds the barometer goes down a part of the distance, he has no occasion to be frightened, because the force will not come until it gets low down. It is the depth to which you go which regulates the force of the wind when rising again. The time you have been at the upper level is also a point of importance. If you have had fifty pounds or sixty pounds pressure more than you ought to have had for fourteen or fifteen days, that would be likely to result in a bad gale of wind.

The PRESIDENT: I am sure it will be your pleasure to return your thanks to Mr. Wenley for his communication, the moral of which appears to be that a lull comes before a storm. I do not know whether any one wishes to make any remarks upon it.

Mr. R. H. SCOTT: I have listened with a great deal of interest to Mr. Wenley's paper, which is entitled "A new method of Forecasting Storms and Floods;" but I cannot find any method at all in it. Mr. Wenley announces that there are waves at the upper surface of the atmosphere, but has he ever been there to see them, or to measure them? What on earth do we know about it? We simply do know on earth the indications given by the barometer, and we do not usually count those as pounds on a square foot, but as inches of mercury. As regards the storm, he might have taken from an observation taken at Chelmsford on the 12th March last an illustration quite as good as that taken from the "Porcupine." The barometer then fell nearly as rapidly as it did in the case he cites. I forget the precise figures, but I remember that from 2 o'clock in the afternoon till 4 it rose ·43 of an inch, which is precisely at the same rate as he gave for a rapid rise in the case of the Porcupine, which was ·2 per hour. As regards the statement that you cannot have a gale unless the barometer falls below a certain level, I cannot characterise it as being any thing else than entirely false. In the worst gale they had in Liverpool last year the gale blew the hardest when the barometer was standing at 30·1, and it never fell below 30 inches at all. The force of the wind does not in the slightest degree depend on the height of the barometer, but on the gradient,—on the barometrical tension—there is the difference between the barometrical readings at adjacent stations. Whether the barometer be high or low, if there is a high gradient there will be wind. I could give abundant instances of what Mr. Wenley and Mr. Allan Broun have stated of tremendous falls in the barometer over extensive areas—over 200,000 square miles in the space of twenty-four hours without any wind; and because there was such a general fall there was no wind. I may tell him that on one of these occasions there were four vessels—the "Scotia," the "Foam," the "Inverness," and the "City of Brooklyn," on the 20th November, 1869, when there was this fall of over an inch over 200,000 square miles, and those vessels were on the meridian of 16 W., and they got gale enough. That gale did not reach Corunna, but on account of the gale the "Foam" was driven out of her course and delayed a fortnight. The gale blew where the barometrical gradient was severe, but at the centre, where there was no gradient, there was no gale at all. There is no doubt whatever that we do want a very great deal of light to be thrown on the method of forecasting storms and weather, but I fail to find any in the communication that has been laid before us to-day.

Mr. LIGGINS : I should like to make a remark or two consequent on the remarkable statement of Mr. Broun this morning, because as far as my experience goes in the tropics I must differ entirely with that gentleman's conclusions. I have been, not in the East Indies, Simla and those places to which he alluded, but in the West India Islands, in sixteen of which I have been at different times ; and I have closely noticed the barometers, several of which I have of my own, made by Dolland, for many months together, and there has not been any depression of the barometer nor any rise. You may look until you are perfectly tired of watching it day after day and not see a twentieth rise or fall in the barometer. It is perfectly wearisome day after day to see a strong wind blowing nearly always as it does in the West Indies, without any effect being produced on the barometer. As a rule it is only in the hurricane season in the West Indies that the barometer indicates the progress of the wind, and then it does it with a vengeance. It comes down from thirty where it usually stands perhaps two inches or a little less, and of course you expect to be blown into the air, and you will be if you go into it, for no power of man can withstand the terrific force of the gale ; which, as I heard the governor of one of the West India Islands say, is equal to the velocity of a cannon ball. It is always indicated in very good time to enable you to take every possible precaution you can. It gives seamen warning to do the best they can under the circumstances, and it is, I believe, only in the hurricane season that it is of that use to them, for generally in the tropical regions as far as my experience goes, it does not indicate the strength of the wind that may be blowing. It may be blowing a four knot breeze to-day and a fourteen knot to-morrow, and the barometer does not indicate that you are going to have such a change. With regard to the forecasts of Mr. Wenley I thought it was one of the most glorious prospects I had ever listened to; but. I must say, according to his theory, either he has not explained himself or he has quite failed to satisfy my mind that he can forecast anything at all. Many of us can recollect Mr. Murphy foretelling that the 19th January in a certain year a good many years ago should be the coldest day ever known in the world, and he made a fortune because he was lucky enough to make a fortunate guess, but he did not do it on any scientific principle, and the means by which he did it have never been utilised by any gentlemen who have succeeded him. I disagree with this gentleman, and I should be very much alarmed if I were in the "Porcupine" or any other ship to find the barometer going at the rate he indicated, for I should be preparing for a severe hurricane. I should prefer to be guided by the law avoiding for hurricanes which was so beautifully introduced by Colonel Read, of the Royal Engineers, when he was governor of Bermuda, and to take the course he recommended of working my way out of the centre of it. I should not like to keep down in the calm region of air, because I should be very much afraid that the hurricane might come with double force. I should certainly get out of it as fast as I could; and that is the only advantage I see in being guided by the barometer. I think this gentleman's wave line theory is not one I should like to see supersede that most beautiful and perfect standard, the warning of the barometer.

Mr. WENLEY: Mr. Scott said that I had not been at the top of the atmosphere to see whether it took a wave form, but you may see if you refer to the plan adopted by the *Daily Telegraph* that the air does take something very much approaching to a wave form, and if it is opened out it really does become a wave or curvilinear form. I certainly never intended to represent that if the barometer did not fall below the medium line you would not have a gale of wind. I say you don't have the same force as you do if it comes lower down. The gentleman who spoke last admitted that in the worst hurricane he had experienced in the West Indies it went down two inches very quickly. My belief is that in the neighbourhood of the West Indies the air is abnormally above its true level, and is always struggling to get down, and whenever it does get a chance, down it goes, and then up again uic kly. I will give an instance from our own latitude. There was a vessel lost four years since—in December, 1871—when the barometer had stood too high for something like twenty days, it then took twenty-four hours to round the corner and begin to come down. It came down to the average line, rested a day and a half, then it took twelve hours to go still lower down, and at the upcast the gale of wind was so strong that the vessel was lost with a quarter of a million of property and forty-eight lives. The Board of Trade cannot issue warnings against such occurrences. That was a large vessel, and yet one of the daily newspapers the next day had an article saying that it the ship had been ten miles further away from the land it would have been saved. To my mind there was at least a fortnight's warning, and all I should have done if I had been on board would have been to take care not to be caught just at the bottom of the descent of the barometer. Not being accustomed to public speaking, perhaps I did not weigh my words as I should have done, but I never presumed to say that no gale of wind takes place when the barometer is high. I know they do, more particularly when a current is over due, and you may get a strong wind at any height of the barometer. It does not matter at all where you are. It is the gradient that does it; but you are more likely to get it when it has fallen below the line. Neither of these gentlemen addressed themselves to the remark I made respecting the movement of the air—that the particles of the air take a circular form as mathematicians tell us the particles of water do in a wave. My contention is that the air does the same, and that with a low barometer these circles come much nearer the earth, and that this is the reason the worst gales come with a low barometer.

The PRESIDENT : The only other communication on the list this afternoon is a short paper by myself, and I will, therefore, ask Mr. Eaton to be kind enough to take the chair.

ON DALTON'S PERCOLATION GAUGE.

JOHN EVANS, F.R.S.: It was at the end of the last century that the philosophical Dalton, in conjunction with Mr. Hoyle of Manchester, carried on some experiments with the view of determining the amount of the natural evaporation of water from the surface of the earth—their object being to ascertain how much of the rainfall was carried off by the processes of evaporation and vegetation, and not merely how much water might be evaporated from a given surface of ground by the heat of the sun and the circulation of the air. As a means of determining this, they conceived the idea of catching so much of the rainfall as percolated to a given depth through some absorbent soil covered with vegetation, and of comparing its amount with the total rainfall.

The instrument they devised for the purpose has since been termed Dalton's gauge, and in principle appears excellent, though as originally planned it is susceptible of some improvements as to which suggestion will be offered in this brief communication.

The original gauge is described as being a cylindrical vessel of tinned iron, ten inches in diameter and three feet deep. There were two pipes soldered into it, the one at the bottom, the other at the top, which conducted any water that came into them into bottles. The vessel itself was filled with gravel, sand, and soil, and subsequently the soil was covered with grass and other living vegetables. It was nearly buried in the ground in an open situation, and provision was made for placing the bottles in connection with the two pipes. In this manner it was exposed to receive the rain and to suffer evaporation from the surface, the same as the surrounding green ground. A register was kept of the water which percolated through the soil and gravel into the bottle, as was also a rain gauge of the same receiving area, for the sake of comparison. The overflow water does not appear to have been particularly noticed, and the observations do not appear to have extended over a period of more than three years.

Experiments on the same principle, but with some variation in the size and nature of the gauge, have been carried on by the late Mr. John Dickinson and myself during a period now extending to forty years. Others have been conducted by Mr. Charles Greaves, C.E., for upwards of twenty years, at Lee Bridge; and others, again, by Mr. Lawes and Dr. Gilbert, at Rothamstead. Some account of all these experiments will be found in the Proceedings of the Institution of Civil Engineers.

In Germany some experiments of a similar character have been conducted, and their results published by Ebermayer.

It seems highly desirable that such observations should be verified, and their scope extended, and the following suggestions as to the construction of Dalton gauges and their application may possibly be of service to those who are willing to turn their attention in this direction. As to the size and form of such gauges, ten inches appears to be too small for the diameter, as it limits too much the area for vegetation; eighteen inches, which is the diameter of my own gauges, is better; but the area of a square yard, as adopted by Mr. Greaves, is better still. If, however, the depth of the vessel filled with soil is only three feet, it is a question whether, from the contraction of the soil in the larger area, fissures may not be formed during dry weather which will unduly increase the amount of percolation.

A depth of three feet only appears moreover to be insufficient for carrying the rain below the influence of evaporation, as Dr. Gilbert's experiments seem to show that with deeper gauges the amount of recorded percolation is less than with shallower. In fact, with a gauge only three feet deep, the actual depth of soil through which the water percolates is less by some inches, as allowance must be made for the necessary under-drainage of the soil. There can be no doubt that a depth of five or six feet, or even more, would be preferable, especially in cases in which there is great capillary attraction in the rock or soil on which the experiment is to be made, as for instance Upper Chalk.

As to the construction and materials of the gauges, cast-iron seems preferable, as though somewhat liable to rust, it is less brittle than slate, which has also been used by Mr. Greaves. In any case, the upper edge of the sides should be brought to a bevelled edge, so as not unduly to increase the receiving area. The overflow pipe of Dalton may be neglected in the case of porous soils, which are free from floods caused by rain. The bottom of the receiving vessel should be somewhat concave, with a pipe of lead or of galvanised iron leading away from it. In filling the vessel, a layer of coarse washed gravel will suffice for the bottom of those of small dimensions. In those of large size, some perforated drain pipes leading towards the centre may be used in addition to the gravel. Above the gravel thin layers of coarse and fine grit should be spread, and above these the soil to be experimented upon. At the upper surface a little mould should be added, for the sake of the grass with which the surface is to be covered.

In filling the gauge, it will be well to wet the contents from time to time, and even to ram them in, but of course anything like puddling must be avoided. After the whole has been filled, it should for a time be kept watered, until some percolation takes place into the lower tube—but until after the lapse of a year it is doubtful whether the soil has sufficiently resumed its natural conditions to place much reliance on the results of the gauge. Dr. Gilbert has attempted to fill gauges with undisturbed soil, but it seems doubtful whether some channels are not in such a case left between the sides of the vessel and the included soil.

As to position, the receiving vessel must of course be in an open situation, and, if practicable, buried in a level piece of ground with a pit a few yards away for the bottle or gauge which receives the percolating water. Should a mound have to be erected for the reception of the vessel, it should be of sufficient diameter thoroughly to protect the soil around the vessel from the desiccating action of the sun and air, or the evaporation from the gauge will be unduly increased. The surface of the soil in the vessel should be covered by the same vegetation as that around it.

It seems needless to dilate on the various conditions under which such gauges might be used. Not only should experiments be made with vessels of different depths and filled with different materials, but the cropping upon the surface might also be varied with advantage. The amount of the rainfall which finds its way through a few feet of earth, varies so much in different years and in proportion to the total fall, that the one quantity affords hardly any guide to the other. By an extension of observations by means of the Dalton gauge, some approximate rule for determining the probable amount of percolation from a given rainfall may be discovered. In the meantime it seems highly desirable that such observations should be multiplied, and the favourable opportunity which this Conference affords of bringing the subject before meteorologists of all countries will, I hope, justify me in having brought it forward, although in so brief and imperfect a manner.

Mr. H. S. EATON : I beg to propose a vote of thanks to our worthy President for the very interesting address he has given us. The rain as it falls is by no means a pure substance or uncontaminated. It acquires a number of impurities in passing through the air, which

on reaching the ground are partly taken up by vegetation, while a
portion are absorbed and got rid of in their way through the soil ;
but the rain after it has been thus filtered acquires other impurities,
so to speak, from the ground, and these render it agreeable to the
taste.   Pure rain water, as I dare say many of you know, is insipid
and by no means a pleasant thing to drink, but in filtering through
a chalky soil for instance, it acquires a quantity of carbonate of lime,
which is dissolved and taken up in solution, and it is that, amongst
other things, which gives the sparkling appearance and pleasant taste
to chalk water.   These are very interesting questions for geologists, as
well as meteorologists, and others interested in physical enquiries.
Then, as the rain falls, much of it is at once evaporated.   Another
portion enters into the tissues of plants and helps to build them up.
Only a comparatively small remainder is left ; and during the summer
months nearly all the rain that falls is restored to the air again by
evaporation, so that very little escapes below to replenish the reservoirs
in the ground.   It is in winter that the large stores of water which are
available for the use of man are collected.   We shall be glad to hear
any remarks upon the paper.

Mr. G. J. Symons : There is one fact I might be allowed to mention
with respect to the very able paper which has been read by our President,
that is with respect to the percolation through similar areas under different crops.   That is already in operation at the Montsouris observatory in Paris, under the direction of M. Marie Davy, who goes
largely into the question of meteorology in relation to agriculture.
He has a large number of cubes in brickwork containing soil, and on
that soil he has various crops of vegetables, grasses and so forth,
so that he is actually determining the very point which Mr. Evans
suggested should be taken up.   I fully agree with him that it should
be taken up, not only in France, but in this country, but, as far as I
know, there is only one set of observations which have been made for
that purpose, namely, those of Mr. Lawes, at Rothamstead.   There is
no doubt it is a most important point, which ought to be carefully
worked out, and Mr. Evans and his predecessor at Nash Mills have
done more to show us the value of the Dalton gauge in its simple form
than any one else.   It is, therefore, a very great advantage that this
Conference has had the opportunity of hearing from himself the results

of his experience, based on some twenty or thirty years continuous observations.

The PRESIDENT : I can only express my pleasure at hearing that these experiments are now being undertaken in France, and I hope something may be done in this country, and in others, to carry them out.

(The Conference then adjourned until Thursday).

SECTION—PHYSICAL GEOGRAPHY, GEOLOGY, MINING AND METEOROLOGY.

Thursday, June 1st.

Mr. JOHN EVANS, F.R.S., IN THE CHAIR.

The PRESIDENT : Ladies and Gentlemen, the Conference of to-day was intended more especially for geographical subjects, and though I am afraid there is some doubt whether we shall get Lieutenant Cameron here this morniug, still I have some hopes that we shall see him. We have some interesting communications which will be brought before the morning's sittings, and the first on the list is that by Captain Baron Ferdinand von Wrangell, on Self-registering Tide Gauges, which I will now ask him to communicate to the Conference.

Captain BARON FERDINAND VON WRANGELL : Mr. President, Ladies and Gentlemen, the object of all self-registering tide-gauges is two fold : 1st, to give us the vertical distance between the sea-level at any given moment and a certain horizontal plane, which corresponds to the zero of the gauge ; 2nd, to enable us to deduce the average sea-level for a given interval of time.

Some of the self-registering tide-gauges exhibited in the lower gallery of this collection, fulfil both above-mentioned conditions in a very efficacious manner, particularly so the beautiful apparatus devised by Mr. Reitsch, of Hamburg, and constructed for the Prussian Geodetical Survey.

I think, however, that all of them have one disadvantage, namely : in all the tide-gauges exhibited, the fluctuations of the float are *mechanically* transmitted to the registering apparatus, and this mode of transmission necessarily requires that the registering apparatus should be put up in the immediate neighbourhood of the float—a condition which in many instances it is difficult to fulfil.

Indeed, the well or tube which contains the float of a tide-gauge must by necessity be close to the sea or river, in a position where in most cases it is decidedly inconvenient to put up the registering parts of the apparatus, with its clock-work and other delicate parts, requiring a solid foundation, good shelter, and constant superintendence.

For these practical reasons it seemed to me very desirable to construct a tide-gauge in which the registering apparatus could be removed to any distance from the float, so that we might be free to choose for each part a convenient position. This can be attained by electrical transmission of the motion of the float to the registering apparatus.

An instrument of this kind has been devised by me for Nicolaef, in Southern Russia, and if it works well, as I have no doubt it will, similar apparatus will be used in other ports of the Black Sea. The registering part of the tide-gauge consists of two electro-magnets and a drum covered with a strap of paper and turned round by clock-work. The anchor of each electro-magnet is attached to a lever, which holds on its end a small glass tube filled with liquid colour. When the electro-magnets are not excited, the two glass tubes are held side by side by their respective levers at a small distance from the upper part of the cylindrical drum. The moment a galvanic current passes through one of the electro-magnets, it attracts its anchor and the corresponding glass tube touches with its lower end the paper that covers the cylindrical drum, and makes a dot on the paper.

By means of a brass wheel and commutator which will be presently described, every time the float has risen for one inch, an electric current is led through the coil of one of the electro-magnets, let us say electro-magnet No. 1. When the float has fallen for one inch, the current passes through electro-magnet No. 2. In this manner we get on the paper two parallel lines of dots, the one corresponding to the number of inches the sea has risen, the other corresponding to the number of inches the sea has fallen. The time at which the rise or fall has taken place is directly seen on the divided paper, which is turned round once in twenty-four hours.

The construction and working of the wheel and commutator mentioned before, will be best understood with the help of a diagram.

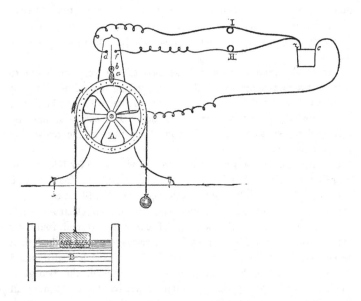

The motion of the float B is transmitted directly to the brass wheel or pulley A, to which thirty-six metallic pins ($c, c, \ldots$) are fixed perpendicular to the plane of the pulley. The distance between the pins is one inch, so that the whole circumference measures thirty-six inches or three feet. Above the wheel the metallic commutator a swings on an axis C attached to an isolating board. A thin ivory plate is screwed on each side of the lower half of the commutator, the upper half of it, above C, consisting of a metallic spring.

When the float B rises, the wheel A turns in the direction indicated by the arrow, and the first pin on the left side of the commutator will press its upper part to the metallic button d, to which is attached one end of the wire from electro-magnet No. 1. As long as the pin touches the ivory cheek of the commutator, there will be no current, but the moment the lower metallic end of commutator a passes over the pin, the galvanic circuit will be complete, the current passing from one pole of battery cz, through the metallic pulley, the uppermost pin, the commutator a, the button d, the coil I, to the other pole of the battery. The current lasts only one instant, because the upper, elastic part of commutator a forces it to pass over the pin and to resume its vertical position on the other side of the pin it has just passed. With the further rise of the water for one inch, the following pin establishes the current for one instant, and each time electromagnet I. attracts its anchor, the glass tube No. I. makes a dot on the paper.

When the water falls, the first pin to the right will press the commutator to the button f and the dimensions of all parts are so calculated that the moment the fall amounts to exactly one inch, the commutator passes over the pin and consequently the current excites electro-magnet No. II., which attracts the second glass tube to the paper. In order to secure good contact, the edges of the pins and the lower part of the commutator are made of platinum. It is evident that with the data given by this instrument we can easily draw a diagram of the tides on divided paper to any scale, or make a table containing the height of water for every hour of the day to the fraction of an inch. The mean level of the sea for any interval of time may be calculated by measuring with a planimeter the area contained between the tide curve, the line corresponding to the zero of the gauge, and two ordinates corresponding to the beginning and to the end of the interval. This area, divided by the length of the base (or time-line), will give the average sea-level during the chosen interval of time and in the unities of the scale.

In this rough sketch of my tide-gauge, I have only mentioned its essential parts, not entering into details. Of course the apparatus could be easily improved, particularly the registering part of it, which is rather primitive in its construction ; however, it answers its purpose, and has the great advantage of being very simple and inexpensive.

If this communication should give the impulse for the invention of a more perfect electrical self-recording tide-gauge, my object in drawing your attention to this method will be fully attained.

The PRESIDENT : Ladies and Gentlemen, I am sure you will all return your thanks to Baron Ferdinand von Wrangell for this very interesting communication with regard to the self-registering tide gauge, which appears to be of great simplicity, and at the same time of great value. I hardly know which we shall admire most, the simplicity, and ingenuity of the apparatus, or the admirable manner in which a foreigner has been able to place it before an English audience. It is of considerable interest, and I shall be glad to hear any remarks upon the subject. I am afraid that owing to some misunderstanding and partly to a loss which has been sustained by Mr. Clements Markham, the members of the Geographical Society who have to read papers have not been apprised of the time at which their papers would be brought before the Conference, but I hope we shall see some of them here very shortly. As Mr. Galton has some ingenious devices to bring before the Conference I will ask him to be good enough to do so.

ON MEANS OF COMBINING VARIOUS DATA IN MAPS AND DIAGRAMS.

Mr. F. GALTON, F.R.S. : Geographers want above all things an improvement in their methods of combining various data upon the same maps. The whole object of geography is to show the physical features of the ground in combination with the facts of which those features are the stages, but this cannot as yet be effected without a great confusion of lines and tints. The limits of applicability of shading and colouring, are very soon reached. I show you here a most beautiful specimen of shading ; it is as perfect as can be, being taken by a simple photograph from a relief model. But beautiful as it is, it shows hardly anything more than the mountains; the shades are so dark that what is written and engraved in the shaded parts is barely legible, and if we added such colours as geologists use, the shading would be still further interfered with, and there would be great confusion. I don't propose to enter into questions as to how far data may be superimposed upon the surface of the same map by means either of shading or of colouring, but I wish to speak of two other methods of combining data which appeal to quite different faculties.

One is by the use of stereoscopic maps. In using the stereoscope, the notion of relief depends upon the varying convergence of the optical axes of the eyes to the different parts of the picture. In this beautiful model of Mont Cenis the eyes converge more nearly to the peaks near the eye than to the depressions farther removed from them. As you are perfectly aware, a stereoscopic picture consists of a pair of photographs taken from two slightly different points of view, corresponding to the distance between the two eyes, and upon which the eyes are made to converge through the interposition of the ordinary stereoscope. I proposed this method of stereoscopic maps eleven years ago before the Geographical Society, and a small paper which I then read is published in their proceedings. The plan has not been adopted, but I think I may venture to suggest it again to geographers, because there are yearly increasing means of obtaining stereoscopic pictures. We want, first of all, good models, and every year increases the number of them. Those of you who were at the French Geographical Exposition will have noticed how large was the number of topographical models, and from any of those photographs might be taken. The Royal Engineers attached to this Loan Exhibition have taken for me some photographs of various models which perfectly illustrate what I desire to show. Models require a table to stand upon, they are of great weight and are very costly; but a stereoscopic picture taken from a model gives nearly all that the model can shew, and costs only a few pence.

The other method which I am going to bring before your notice is one I have not hitherto described. It is a plan of utilising the element of time by presenting different data that we desire to superimpose in rapid succession before the eye. Most of you are acquainted with the old fashioned instrument used for accomplishing this object; there used to be, and perhaps is now, one at the Polytechnic. By turning round a disc, separate pictures of the object, but in different positions, were brought in quick succession into the field of view, and the object of the picture seemed to move. For example, a wheel was drawn in several consecutive positions; these were rapidly brought in succession before the eye and the effect was that the axis of the wheel appeared stationary while the wheel itself appeared to revolve round that axis. But the instrument did not act well, it produced an unpleasant jerking effect, and it has never come into general use.   The late Sir Charles Wheatstone gave great attention to the subject, and Professor Clerk Maxwell has also devised an instrument, but none of these " Wheels of Life " as they have sometimes been called, have come into use, either from being too cumbrous or from some other cause.   The plan I propose is a very simple one—it is like a dissolving view ; each picture fades into the next one, and there is no reason why a perfectly continuous change of appearance might not be produced by it.   If you look through an ordinary telescope (not an opera glass), and cover half of the object glass, you still see the same objects as before, the only difference being that a certain quantity of light is cut off and the objects are less bright.   If instead of covering a part of the object glass by the hand or a card, you use an inclined mirror, then on look- ing through the telescope you will see two images superimposed on one another, namely the image of the object in front of the telescope seen through the open portion of the object glass, and that of whatever may be seen reflected in the mirror.   In short, by means of a telescope we are able to superimpose two or even more separate pictures on the same field of view.   This is the first of the principles used in the arrangement I am about to explain.   The second is this : suppose that we have a small carriage running on a tramway to and fro, and that we fix a lens into the roof of the carriage and a picture of the same size as the lens, on the floor below it at its exact focal distance.   Then if we fix a telescope vertically above the tramway, looking down upon it, and bring the lens exactly below the telescope, we shall see the picture through it just as though it were an extremely distant object.   Now let us push the carriage a little, so that the lens and picture cease to be immediately below the telescope, then the only alteration in the image as seen through the telescope will be a diminution in its bright- ness.   As we push the carriage further, the image will wane more and more in intensity until it wholly disappears, but it will never alter its position, which is absolutely stationary.   In the apparatus about to be described, the two principles just mentioned are used in combination. I have here a carriage on a tramway, moving to and fro beneath a fixed telescope.   In the roof of the carriage is a row of six similar lenses side by side, and below them on the floor of the carriage at the focal distance of those lenses is a row of six pictures, one picture below each lens. When No. 1 lens is brought below the telescope the picture No. 1 comes into view just as if it was an extremely distant object. Now let us push the carriage until only a portion of No. 1 remains below the telescope while a portion of No. 2 has come under it. We shall see, on looking through the telescope, that the image of No. 1 has faded and that of No. 2 has been superimposed on it. Push the carriage still further and No. 1 will gradually fade into extinction while No. 2 grows to its fullest brightness. Continue pushing, and just the same will take place in respect to Nos. 2 and 3 that has been described in respect to Nos. 1 and 2, and so on for all the lenses in succession. By this means a series of geographical data printed in maps may be successively superimposed. It affords a peculiarly suitable method for picturing changes, whether in physical or political geography. I will not describe the mechanism by which complex and powerful instruments of this kind might be constructed; where the images should be thrown by a lime light on a screen, and a string of perhaps only three large achromatic collimators should serve for an indefinite number of pictures. I think this is all I need now to say to you upon the subject. My present object is merely to explain a general principle and to show it in action on a small scale. The method of putting the idea into a good practical shape has still to be worked out, and I hope some gentleman will try to do so, for it appears to me to be one that has many important consequences.

The PRESIDENT: Ladies and Gentlemen, I am sure you will all return your thanks to Mr. Galton for his interesting account of a method by which you may see three or four things through one glass. I am sure you are all glad to have heard his remarks. I will now call upon Professor von Oettingen to give us a description of his Anemometer which we were obliged to postpone on Tuesday.

Professor Von Oettingen entered into a detailed explanation of his ingenious and elaborate apparatus for recording the action and force of the wind. It would, however, be impossible to give a report which would be intelligible except in presence of the machine itself, or with the aid of numerous diagrams.

The PRESIDENT: I am sure we all feel grateful to Professor von Oettingen for the description of his exceedingly interesting and complicated apparatus for marking the action and the force of the wind.

He has entered into great detail in explaining the devices which he has adopted in this curious machine, and seeing it for the first time there is some little difficulty in understanding them all; but it is at work below, and has been at work at Dorpat, as he says, for some time. I believe that Baron von Wrangell has studied the instrument to some extent; and as he has such a wonderful command of the English language perhaps he will say a few words as to the principal points of interest in this ingenious invention.

Capt. BARON FERDINAND VON WRANGLE: I will mention the principle on which it is founded in a few words. Generally we get the direction and the force of the wind in a certain lapse of time. Professor von Oettingen gives the components of this wind upon two vertical drums, north and south, east and west, during any interval of time. We can adjust it according to our wants, and can integrate the direction and force of the wind for ten minutes, for half an hour, or for an hour, just as we wish, and can get the precise amount of the components. If the wind is blowing in one direction, say north east, with a certain force you get the components to east and to north with mathematical precision. This he attains by means of the four rollers which he has explained. They roll upon a disc which is put in motion by the wind vane. The motion of the disc is transmitted to the four rollers which are in contact with it, and, according to the relative position of the disc and the roller, the motion of the rollers will be faster or slower. Supposing this is the disc, the four wheels are put in this condition, the two wheels which are parallel to the tangent of the disc will roll with a full motion, while the wheels perpendicular to it will not roll at all. Supposing the direction of the wind is north, the roller which represents the north will move with its full motion, while east and west will not roll at all; so that you will have the component of the north wind in its full extent by this roller. By an ingenious arrangement of some small discs which are above, where the wind is north the south roller is lifted, and this and other details will easily be seen upon an inspection of the apparatus. If the wind is in any intermediate direction, the position of the rollers will be at an angle with the disc, and, upon the well known principle upon which the plani meter of Amler is founded, the rolling of the discs will depend upon the angle at which they stand; one roller will give you one of the components of the wind's velocity, and the other will give the other component of the wind's velocity and direction. That is the principle upon which it is founded, and being a strictly mathematical one, I think it has a great advantage over all other anemometers. There are many details which are of great importance which it would take too long to explain now, but the instrument may be seen at work in the basement, where Professor von Oettingen will be happy to supply any information that may be required.

The PRESIDENT : We must all feel that our thanks are due to Baron von Wrangle for his description of the instrument invented by Professor von Oettingen. One of the papers on the list for this morning consists of notes upon the maps of Palestine, by Major Anderson. Although Major Anderson is not here, Lieut. Conder is present, and will give us a few remarks upon the maps illustrating the survey which has been so admirably carried out.

Lieut. CONDER : Ladies and Gentlemen, I have been called, upon very short notice, to give an account of our work, and I hope if what I say is not very connected that I shall be forgiven, because I hardly expected to have the pleasure of addressing you.

The work which I have had the honour of taking the charge of for the last four years is a continuation of the work commenced ten years ago by the Palestine Exploration Fund. Their main object was to obtain all the information which could possibly be obtained of the topography and ethnology, and of every existing detail of archæology in the country, principally with regard to the illustration of the Bible. The work, which was commenced by excavations at Jerusalem, conducted by Captain Warren, produced many results of importance, especially with regard to the discovery of the Temple site. Captain Warren also made discoveries in the Jordan Valley and in Philistia. He was almost the first European who had ever traversed the whole length of the Jordan valley. His health failed after three years, and he was obliged to return to Europe, and it was then determined that the whole of Palestine should be surveyed, in order to give the world a complete and trustworthy map, and discover the sites which would prove of most interest for further explorations. A great number of the ancient sites had been recovered, but the travellers' routes as a rule passed over and over again the same parts of the country,

leaving districts which had never been visited, and it was determined
that a regular triangular survey on the one inch scale, the same
as our own Ordnance survey, should be run over the whole country,
so as to make it certain that hardly an inch of ground had been left
unvisited in Palestine by the party. In the end of 1871 the party
started, being probably the smallest party which ever undertook
such a large work. It consisted of Captain Stewart of the Royal
Engineers, of two non-commissioned officers, and of a Civilian,
Mr. Tyrwhitt Drake, who was to act as linguist and archæolo-
gist. They went out in the autumn, and the consequence was that
within a fortnight of Captain Stewart's arrival he was attacked with
fever; he was obliged to give up the command, and I having already
volunteered to go out under his command, was appointed by the committee to succeed him, and to go out immediately to take up the survey.
We have been at work four years, and the whole survey is supposed to
extend from Dan to Beersheba, and to consist of about 6,000 square
miles; the sea being the western limit, and the banks of the Jordan
the eastern. Of that district we have now surveyed 4,600 square miles,
leaving only 1,400 miles unsurveyed. It includes the whole country
from Beersheba in the south up to the latitude of St. Jean D'Acre,
Nazareth, and the sea of Galilee on the north. Over the whole of that
portion of the country proper triangulation has been run, and the
triangles vary from five to ten miles in the side. The whole of the
detail has been filled in by the use of the mathematical compass and
by means of points fixed by theodolites over the various trigonometrical
stations, and all the heights, both by the use of the instruments and by
the use of aneroids which we could collect have been taken throughout
the country, numbering 3,000 or 4,000. Every native name that could
be collected has been collected, and we have now one native name to
every square mile of the country. The maps have been brought home
as far as they have been completed, and are in process of being prepared
for publication. I believe it is the intention of the committee to have
them published on the original scale of one inch to the mile. We have
sent all our calculations down to Southampton, where they have been
worked out by the Royal Engineers, and, I believe, with satisfactory
results, so that the new maps are now in course of construction at the
Albert Hall, which enables me to come here at a moment's notice. The

object of the survey is perhaps not quite clear to those who have not followed the subject from the commencement. It was thought by many persons that as a great number of maps had been made of Palestine, it was not necessary to begin a new one. But the reason for so doing was, that the identification of a great number of famous places mentioned in the Bible would throw a deal of light on the Scriptural narratives, it being impossible to crowd these places on the ordinary maps. In the parts of the country which have been gone over by Dr. Robinson and others, the names of Scripture sites was enormous, but it was known that there were other districts, such as the low lands of Palestine, and even the Jordan valley, where an equal number of names and ancient sites might be collected if they were looked for in the same manner. It has always been a difficulty also how it should occur that the ancient nomenclature in the country should remain so unchanged to the present day. It has been supposed that whereas the old nomenclature was either Jewish or Canaanite, the modern must necessarily be Arabic. The survey has thrown a great deal of light on this point. First of all I think we may now state that the nomenclature of Palestine, such as it appears in the Bible, is not Jewish but Canaanite ; for this reason, that Marriette Bey has lately brought to light two new pylones of the temple of Carnac, and there are three pillars upon which are written a list of the towns conquered by Thothmes III. ; there are 120 towns, and the list is thrice repeated, so that any chance of error can be immediately detected by a comparison of the three copies of the list. These 120 towns were all in Palestine, and out of the 120 we are now able to fix on the survey sheets something like 100 as now existing. So that, as Thothmes III. lived after the time of Abraham and before the time of Joshua, it follows that the nomenclature in the list was Canaanite and not Jewish, as appears in the book of Joshua. We find that more than half of these Egyptian towns are indentical, not only in sound, but also in position with those mentioned in the book of Joshua. Therefore we have in this existing monument, which cannot have undergone any change by the fault of copyists or the imperfection of manuscripts, a check as it were on the correctness of the nomenclature as now found in the book of Joshua, and we find that not only did the words agree, but that the places could be so identified at the present day, as to give a consecutive order to the lists which appear in the book of Joshua. This is a discovery which could not have been made unless a very detailed survey had been run over the country, and a great number of ruins found within a mile or two of one another, which consequently could not be shewn on the smaller scale, and the names identified and their position and antiquity fixed. Then besides that, it comes out clearly from the survey that the language of the country is not what is ordinarily called Arabic. We found some forty or fifty topographical words used in the bible, Hebrew words relating to all the natural features of the country, to its cultivation, &c., we found that all those words exist almost unchanged in the modern language of the country, and I think, that finding that the topographical language of the native peasantry is unchanged, it is not a wild assumption to suppose that the whole of their language approaches much more closely to the Hebrew than it does to the modern Arabic, such as we find in Morocco or Arabia to the south of Palestine. The number of these discoveries that have now been made all over the country is so large that I can hardly give you any idea of them. I think before the survey is finished we may fairly claim to have set at rest the position of over three-fourths of the biblical towns. Besides that, there are a large number of towns mentioned in the early christian times, and the famous places in the chronicles of the crusaders, and these we have been able to fix and have visited their ruins. One or two of our discoveries have become pretty well known, and I may just mention them. One of these was the site of the cave of Adullam. No satisfactory site had been found for this place, and the monkish site of the middle ages which had been accepted, has become to be discredited amongst scholars. We knew to a certain extent the district in which we should look for the cave of Adullam, because we knew it was next to the royal city of that name, and if we accepted what has now been laid down by us as a canon, that there is a consecutive order in these topographical lists, we knew exactly the district and the towns between which we were to look for the site. The name also was recovered by M. Clermont Gauneau, a French explorer, about three years ago, and last spring we passed over the whole of the low country between Joppa and Philistia, and we visited this site. We found it stands just at the turn of the great valley of Elah, where

David and Goliath fought their famous duel. It shews every signs of being an ancient city of great importance ; there is a very strong natural position, and the heights above have been scarped. There is sufficient extent for a city as large as any to be found in Palestine. We found that along the sides of the city and on the opposite side of the valley, there exist a great number of small caves in rows. The caves are inhabited at the present time by the fellaheen or native peasantry ; in some of the caves they keep their flocks and herds, and in some they live themselves. We calculated that, supposing David's band was 300 strong, there is no doubt the whole might have been accomodated in these caves, and that the real cave of Adullam, where he himself lived, might be identified with the great cave which stands outside the city on the top of the hill. This was rather a surprise to us, because it had been generally supposed that we should find one large cave, but when I came to consider the matter, it seemed that the discovery of a number of small caves rendered the site far more probable ; because there are two kinds of caves in Palestine ; there are very extensive excavations, partly formed by water and partly artificial. These are extremely unhealthy, you can hardly go into them without coming out with fever. They are generally very dark, full of vermin and bats, and not fit for habitation, whereas the small caves are inhabited at the present time. There are signs that they were inhabited in early Christian times, and there are also indications that they were inhabited in still earlier times. In South Palestine, where the Horites and Troglodites lived, we found a great number of these caves, and also in districts where, according to the early Christian fathers, the Horites lived till the fourth century. We considered this the site, not only by its position and name, but also because the character of the caves around it answered exactly to what was required for the scripture Adullam. The second site was the site of our Lord's baptism, and this has been sought by a great many travellers, but unfortunately not in the right direction. We know that there must have been a ford over Jordan near which Bethabara stood, and therefore we hunted up every ford throughout Jordan. One day I rode a distance of seven miles through the thistles, and in that time we fixed twenty-one fords, being three to a mile ; we discovered some forty or fifty fords, where only two were shown on the previous maps

Y

of the country, but there did not seem to be any chance of recovering the name " Bethabara." It was only when we got at the end of the survey into the neighbourhood of the sea of Galilee, that we found the name "Abara" without the "Beth," which means a house, applied to one of the principal fords leading into the country of Gilead. This was quite an unexpected position, because it had always been sought near Jericho, but if you compare this description with the biblical narrative, you will find it is the only part of the Jordan valley in which the ford of Bethabara could properly be looked for, because our Lord, having been at Bethabara for two days, on the the third day was in Cana of Galilee, and it was utterly impossible to get in a day, or a day and half, from Jericho to Cana of Galilee; but from the ford as now fixed by us it would be an easy day's journey either to Nazareth or to either of the two cities considered to represent the ancient Cana of Galilee. Then there are a great number of similar discoveries, but I hardly know which would be most interesting to treat upon. Another subject, which perhaps has excited more interest than any other, is the question of the ancient condition of Palestine; and I think, with regard to that, we have also thrown a great deal of light on the subject. It has been supposed that the climate is extremely changed; that the soil has become barren, and that a great amount of forest has disappeared from the country. We have now collected all the data which I believe exist on this subject, and we arrive at several conclusions: first of all, that the cultivation of the country has certainly immensely decreased. Wherever we go we find on the wildest hillsides, and amongst the channels at the top of Carmel, the remains of ancient terraces, ancient vineyards and watch towers, showing originally the cultivation of the vine to have extended over a much larger portion of country. In the same way, when we get down nearer the plain, we find the hillsides cut into terraces, so as to economise every inch of ground for the cultivation of corn, and many of those terraces are now in ruins covered with wild growth—not planted at all. On the other hand, with regard to the existence of forests, there seems an indication that in the north the forest has in many places extended further than it did in the original times; whereas in the south, in the country of Ephraim and Judea, forests have been swept away. As late as the seventh century, we find the forest of Hareth existed as a fir forest between Hebron and Jerusalem. That has now disappeared. In the same way the wood of Ephraim, which contained oak trees of some size apparently from the Bible account, has also disappeared; and in that plain the great forest mentioned by Pliny as *ingens sylva,* which appears to have covered the whole of the north part of the plain of Sharon, now only shows by stumps of trees in the neighbourhood of Jaffa, and by very open barren woodland forests to the north. But in other parts of the country, as on Carmel, the forest has increased. The words that are rendered in our English version "forest" and "wood," do not always refer to timber trees, certainly from the derivation of the words, but to a kind of thicket with dwarf trees in it, which still exist. I do not believe there is a single tree mentioned in the Bible that is not to be recognised in modern Palestine. The change in Palestine, it appears to me to be very clear from the result of our work, is a change of degree and not of kind ; the want of good government, and the want of security for life and person, have certainly caused the Fellaheen to neglect the cultivation of the country, and so a great deal of the wealth which could be obtained from Palestine at the present day is lost ; but that there is any great change in climate, in water supply, or in the natural products of the country, from those that are mentioned in the Bible, I do not believe. The reason why we are now at home in England is perhaps known to some of you. We had worked for four years with comparatively little trouble. We had been over some of the most fanatical country—the Hebron hills and the south country, where no Christians exist; but we had never met with any interruption; but, unfortunately, when we got into higher Galilee, we came amongst some very influential Mohamedans who were not natives of the country. These men picked a quarrel with our servants, and a very serious affray ensued, in which every member of my party was more or less severly wounded, and we were for about a quarter of an hour in imminent danger of losing our lives. Fortunately, we had a faithful Mohamedan with us, whom I sent off immediately to bring up some of the Bashi Bazouks some little distance from us, and when our efforts at calming the people and staying the row were just being defeated, and Lieut. Kitchener and myself had been surrounded, they arrived just in time to assist us; but the consequence of the wounds, excitement, and fatigue, was that the whole party was laid up with Syrian fever.   At the same time the Bedouin rose in the south, and the cholera appeared in the north, and therefore the committee recalled us to England.   During the winter we were again attacked by returns of Syrian fever, so that in the spring, when we were hoping to go out, we found that none of us were in a fit condition to tackle field work, and therefore we remained preparing what we have done, the 4,600 miles, which we hope will be ready by the end of the year; and that next spring we may be able to go out and pitch our tents again on the very place where we were attacked, and see if they will do so a second time.

The PRESIDENT: I must ask you to return double thanks to Mr. Conder for coming forward at such very short notice, and for giving us an interesting account of the Palestine survey.   It is a subject which has occupied much public attention, and in which all must take great interest.   I think it is hardly one which will lead to any discussion, and therefore I will not ask for any remarks upon it.   I am sorry that we shall not, at all events this morning, have Lieut. Cameron here to describe his passage across Southern Africa, but there is another portion of the globe to which as much public attention has been directed, and which is at least as deserving of interest as Africa—I mean the Arctic regions, where we have a number of our fellow countrymen, we know not in what condition they may be or how they may have passed the winter.   I am glad to say we have amongst us an Arctic traveller who is willing to say a few words to us upon the subject of these Arctic maps—Dr. Rae; and I will now ask him to do so.

Dr. RAE: I appear here most unexpectedly to myself and still more so to you.   I have been a good deal along this Arctic coast—in fact I may say I have seen more of the coast of Arctic America proper than any man living.   When I bought this handy little handbook which has been compiled by a number of eminent men I naturally expected that I should have found the details as far as they went perfectly correct, and that the facts cited had been stated perfectly.   But, speaking of the portion of the book to which I am going to allude, which is the only part connected with the Exhibition of which I know anything, I was excessively surprised to find on page 250 this statement: " Franklin, Richardson, and Back in their wonderful land journeys were tracing the coast line of Arctic America while Parry made his attempts by sea. In 1819-22 they explored from the mouth of the Coppermine eastward to Cape Turnagain, and in 1825-26 their discoveries extended westward from the mouth of the Mackenzie to Return Reef." But the principal part of that wonderful land journey is quite omitted. When Franklin and his party went down this river, the McKenzie, they separated—Sir John Franklin and Captain Back—now Sir George Back—went west nearly to Point Barrow. They intended to have got through but the ice stopped them. Captain Beechey, R.N., was waiting at that point expecting to see them, having been sent out for the purpose. However they fell some 160 miles short of that and had to go back again ; but my friend Sir John Richardson (who nearly sacrificed his life in a previous expedition to save the party), and Mr. Kendall made this survey which very largely exceeds the other. They went from the mouth of the McKenzie to the Coppermine river ; there is another curious fact connected with this, that at the time this exploration was made a large reward was offered by the Admiralty to any one who made twenty degrees west longitude along the coast of Arctic America. This was over twenty degrees, but the claim was ignored because the exploration was made in boats instead of ships, although the boats found plenty of water for any ships to pass along, and it has been done since by Admiral Collinson. That is omission number one. The next I shall direct attention to is this : in land journeys Sir John and Sir James Ross explored both the east and western shores of Boothia, and they made the wonderful discovery of the magnetic pole. They got to a point where they found the compass was no good—that the needle pointed straight up and down, which was an indication that the Pole was immediately under that spot, deep down in the earth, I believe. The hand-book further says that a survey of the whole of this coast of Boothia was made, which was called after the person who sent out the expedition ; but this is not exactly true. This portion (more than 100 miles) was not surveyed by Sir James Ross. Here again we have another error. It says :—"Between 1837 and '39 Messrs. Dease and Simpson almost completed the delineation of the coast of Arctic America." They went from Cape Herschel to the Coppermine, and they also completed the portion from the Mackenzie to Point Barrow, the part left unsurveyed by Franklin in a previous year. They did an immense deal of work, but the most wonderful voyage is that alluded to here. They came down the Coppermine and from this point went all the way to Back's great river, making the longest voyage ever performed on the Arctic coast and returned safe to the Coppermine and went inland ; but the statement here is that they surveyed from the Coppermine to Cape Herschell. That was not exactly the fact, because this part here (pointing on the map) was not surveyed by them. But it does not give them sufficient credit in another way, because they actually surveyed beyond Back's River. I will mention now that one of my own expeditions should come next in date, though I feel some diffidence in speaking of myself, and would not do so were it not that credit is due to those who sent out the expedition, and to my gallant fellows who supported me so well. It is further mentioned that Dease and Simpson had nearly completed the survey of North America. People have different ideas of what is "nearly completion." I may mention that when Dease and Simpson left off the survey there were fully 900 miles left unsurveyed. And those were naturally the worst parts, because they were points where nobody had been able to get to ; but this large piece here (describing) upwards of 600 miles was not completed. Three Government expeditions attempted it. Captain Parry, with two ships, tried two years to get round, but not understanding sledge travelling, he failed to get through. Sir John and Sir James Ross came down part of the way in the Victory but they failed, and this was left unsurveyed. The Hudson's Bay Company thought they could do it and asked me if I would undertake it. I took two small boats which would only carry three months' provisions, and we obtained food by hunting and fishing for twelve months. We lived in a stone house without a particle of fire for twelve months, and by boat and sledge journeys we first surveyed this side and came back along the coast, making over 1200 miles. I had only ten men with me, and we suffered a great deal because the country was so rough and the ice so bad, that we had to carry everything on our own backs for upwards of 500 miles. We were reduced, from having miscalculated the time we should take, to eating bones and skins and pieces of anything we could get. When we killed a Ptarmigan we ate it, bones and everything, except claws and beak. We were perfectly healthy though we got very thin. The whole account of that part of the survey is omitted in this handbook. Had the surveys of others been omitted I should not have objected, but similar surveys made much more easily are described in this account. I am only asking for fair play in the matter, and because anyone who reads this book, which will go among many thousands of people, would be led to believe that nothing had been done but what is there described. Again, at page 254, it says : " Sir James Ross in 1848 was the first to make an extended sledge journey with Lieutenant M'Clintock, and thus to establish the true and only efficient system of Arctic exploration." The date is wrong in the first instance, because it was in 1849 that Sir James Ross made that so-called " extended journey." He went out in 1848. But what was the extended journey? He had at his command two ships' crews amounting, I suppose, to 100 men or more, and he went along here where I am pointing on the map. He put his ship in winter quarters at Port Leopold and surveyed this part of the coast, a journey measuring, on the most elastic principle, not more than 500 miles. Two years before that, with a few men, I had done journeys of 650 and 540 miles. And this journey of Ross's is what is called the first extended system of sledge travelling. The system is now being used in the present Arctic expedition, and the sledges are such as you see on this diagram. They run beautifully and smoothly over the hard snow, but when they get amongst rough ice they go with difficulty, and in soft snow they sink down, so that eight men can scarcely drag them through, and they carry such an amount of dead weight that each man of the expedition has to haul eighty pounds dead weight, giving for a sledge with eight men a weight of 640 pounds to drag after they have consumed all their food and fuel, giving unnecessary work—enough to knock up any person. By the system I used we could travel with less than forty pounds of dead weight per man. Our sledges were well constructed on low runners, and would only sink about three quarters of an inch, even in soft snow, whilst they ran equally well on hard snow and ice. The consequence was in our several journeys, amounting to nearly a thousand miles each, we made on an average from eighteen to twenty miles a day, and on one occasion equal to twenty-four miles a day ; the Government expeditions on the other hand were capable of doing only from ten to twelve miles a day. They are on the horns of a dilemma. A finer set of men than those who have gone out this year could not be found anywhere, either physically, morally, or in any way; therefore there must have been either some superiority in my system or in my men, and I leave people to decide which it was. All my men slept comfortably and had shelter when they were cooking, and in every way were better off, because we built snow huts instead of taking tents with us.

Going back to another place where my name is honoured by being mentioned, at page 256, I was asked by the Admiralty to go to look for Franklin in any direction I pleased. I was in a difficulty. I was then in charge of this district here on the Mackenzie river, and the only place I could select was to go round towards the Coppermine river. When I came to this channel, as I was the first to discover it, I named it Victoria Channel, knowing it was a channel because the flood tide came from the north and the ebb tide went back again. At this point Sir John Franklin abandoned his vessels, so that I was very close indeed to where he had been, but I did not know at the time he was there. That channel is now very properly called "Franklin Channel" instead of "Victoria Channel." I would not mention this specially, except that it is distinctly stated in the handbook that Captain Collinson in May, 1853, made a sledge journey still further east until he reached within a few miles of the position where the Erebus and Terror were abandoned. Now, 1853 was nearly two years after I had been there. Admiral Collinson came round here very near from where he left his ship, but owing to the rough ice I suppose he never saw this large bay of 130 miles, which I went in with boats and sounded. It is mentioned that two years afterwards he got close to the place where Admiral Franklin had been lost, but it is not noticed that I had been there two years before and had seen it. Then there is another little error where it is said that during 1854 Dr. Rae was employed to ascertain the connection of Boothia with the American Continent, and thus to join the work of Sir James Ross to that of Parry. Now that is exactly what I did seven years before. That was not my object in going in 1853, but I proposed the scheme to go and explore, and to join this survey so as to make the whole survey complete. I did so. I went and I lived at my old quarters and connected these points, and showed that King William's land was an island. But I did something more. There is not a word said that on that occasion I brought home information of Franklin, which was so authentic that we got the reward of £10,000 from the Government. The way in which that is noticed is this. It says that" The news that the Eskimos had tidings of Sir John Franklin's expedition in King William's Island, and at the mouth of the Great Fish River led to the dispatch of the *Fox*." That is all that is said about my having discovered news of Sir John Franklin's expedition. My information was completely verified by a great quantity of things, gold and silver marked with the crests and initials of fifteen or twenty of the officers. These are the principal points I wish to allude to. I do not say that there are not errors in other parts, but these are things which I think it right to bring before those who have not had the opportunity of knowing the facts, and who may read this book and take it as entirely correct. The eminent names attached to it must make people believe it, and I can scarcely imagine that these things can be done intentionally, but they show great inadvertence. Had other people's work been omitted as well as mine, it might have been different. Though we have seen so much of this coast, yet to read this article one would think that I and my party had hardly ever seen a bit of the Arctic coast at all. All that I have said can be verified by any one who will look at that admirable little circum-polar map of Stanford's.

The PRESIDENT: I think we must feel thankful to Dr. Rae for having pointed out some omissions which appear in this otherwise admirably compiled handbook. There is no doubt whatever that any one who writes an account of what others have done is hardly in the same position to give a full detailed narrative as the person who has made the exploration himself. I cannot imagine that there was any intentional neglect of Dr. Rae's most valuable services to this country, nor of these discoveries of which he is so justly entitled to be proud. I am only sorry that the author of this article, in consequence of a loss in his family, is unable to be present to take his own part. If there is an omission, or if there is any possibility of a misunderstanding, I think those who are the subject of the omission or misunderstanding are at perfect liberty to bring the matter before a Conference of this kind. I do not at all complain, therefore, of Dr. Rae, or of the manner in which he brought it forward. It seems to me he has done so in a temperate manner, and has offered the fairest means of having the facts verified by a reference to public documents. It is a subject on which I have no opinion of my own whatever. I was in hopes when Dr. Rae addressed us that he would have gone to some extent into the prospects of the Arctic expedition about which we are all so anxious, and I must confess that in that respect I feel somewhat disappointed. With regard to the question of the sledges that there are, as is usually the case, two sides. It may be that you may have a lighter sledge; that will give less weight per man, but possibly for lengthened journeys and where you are bound to take the utmost possible precaution for the preservation of those concerned a greater amount of weight may be desirable, although under ordinary circumstances it may not be proved to be absolutely necessary. I am sure from what I have seen with regard to the Arctic expedition now gone out, that no human pains have been spared in order to ensure its success, and we must all hope in the course of the next year, or perhaps this, that we hear of its having to some extent at all events accomplished its object. With these few remarks I will ask you to return your thanks to Dr. Rae for his communication.

The Conference then adjourned for luncheon.

The PRESIDENT: The first paper for this afternoon is by Captain Evans, R.N., Hydrographer to the Navy, on the present aspects of hydrography, and I think at a time when the observation of the sea is occupying so much attention, any remarks which can fall from so distinguished a hydrographer will be well worthy our attention.

HYDROGRAPHY :—ITS PRESENT ASPECTS.

Captain EVANS, C.B., F.R.S.: In connection with hydrography there are many interesting objects in this Exhibition which mark notably the advances in that branch of applied science during the past hundred years; advances not confined alone to the marine maps or charts in ordinary use by seamen and to the many and varied instruments employed in their construction; but in the further rendering, by graphical illustration, clearer and broader conceptions of those physical elements which bear so materially on the art of navigation.

Hydrography is commonly defined as the art of measuring, mapping and describing the sea-board of a country, together with the depths of water near the shore and in the immediate offing : this is ordinarily known as marine surveying. Hydrography, however, at the present day, takes a much wider range : it demands in the interests of natural science on the one hand and in that of the rapid interchange of thought by the aid of telegraph cables on the other, the contouring and describing the great ocean bed in the interests of commercial enterprise, by mapping the boundaries of the trade winds, monsoons and currents of the several seas so as to speed the trader on his voyage ; and specially in aid of the security of iron-built ships by the mapping the earth's magnetic elements as distributed over its surface.

The journals of the voyages of the great navigator, Cook, fairly illustrates hydrography as it was rendered one hundred years ago. That period was one of discovery rather than for the detailed examination of sea coasts ; although Cook himself had been engaged in making a marine survey of Newfoundland just previous to his first voyage (1768-71), and other hydrographers were at the same time similarly engaged on our own shores, and on those of our North American possessions.

The charting of coasts and placing soundings in position at this time were chiefly executed with the compass, a process obviously slow and imperfect. Hadley's quadrant, the original type of the reflecting instruments now mainly employed in surveying afloat, being then alone used for astronomical observations. It is not till 1771 that we find mention of the superior value of Hadley's quadrant on board ship in the preparation of sea charts,—both for facility of execution and accuracy of results,—as compared with the compass.

Alexander Dalrymple, the author of an "Essay on Marine Surveying," wherein this advice and caution is to be found, says: "Experience has fully convinced me that bearings taken by compass cannot be safely trusted to in making an exact draught. I have found not only a difference of three degrees or more in different compasses, but in the same compass at different times: I do not say the effect had no cause, but there was no visible one which I could discover ; and I have heard other people say their observations gave room to believe that there is a casual deviation consequent to the state of the atmosphere, or some other occult influence. Hadley's quadrant is as much preferable to the compass for taking angles in facility as in exactness."

We thus find in this essay an early suspicion of the now well known deviation of the compass on board ship, as also an early development of exact methods in marine surveying. Dalrymple was no ordinary man; although not trained as a seaman, he became one in instincts, and executed in the interests of the East India Company several marine surveys in the China Sea and Indian Archipelago.

On the formation of the hydrographic department of the British Admiralty in 1795 Dalrymple was rewarded for his nautical labours by being appointed its chief.

At the beginning of the present century, voyages purely for discovery —excepting those to the Arctic Seas—may be said to have ceased; and those for the exploration and charting of little known regions, but within defined limits, to have commenced. Flinders, a name in the annals of scientific navigation, second only to Cook, now broke ground on the comparatively unknown shores of Australia, colonization just then having taken root on its south-eastern coast.

In 1808 Beautemps Beaupré, an accomplished French hydrographer, who had accompanied D'Entrecasteaux in his voyage in search of La Perouse (1791-3), as his chief marine surveyer, published as an appendix to the "Narrative of Rear-Admiral D'Entrecasteaux's Voyage," an analysis of the hydrographical operations made therein; and it is to this able account of the many points of geodetic detail connected with what is technically called a "running survey," combined with the essays of Dalrymple and Murdoch Mackenzie (1771-74) that we may even now look for instruction, and certainly with admiration, on the solid foundation on which modern hydrography is based.

From the beginning of this century to the present date Great Britain has without stint employed her naval officers in marine surveys over all parts of the globe; expedition after expedition has during this interval left our shores with the twofold purpose of determining the accurate geographical position of the principal ports, or points, of distant lands and among the islands of the several oceans; and then gradually to fill up the details of the seaboards as the demands of commerce or colonization, or, unhappily, war, prevailed. Our own East India Company pressed forward in the same direction, and their hydrographic labours extended from the Red Sea to China. Other European nations with America have largely contributed in the opening up of distant regions; and perhaps the sum of the united efforts of all nations in the field of hydrography, as represented by the charted seaboard of our globe may be best illustrated by my giving the number of charts now published by the Admiralty hydrographic department:

Their sum total is 2650*

of which 880 relate to European coasts and seas.

270	...	African „ „
385	...	Indian, China and Japan.
280	...	Austria, New Zealand and Pacific ocean.
800	...	America.
Some 35	...	Arctic, Antarctic, and miscellaneous.

It is well now to look at hydrography in its present aspect. Important questions to ask are: what may be the relative values of these manifold charts? Are they all sufficiently full in detail for the daily wants of seamen?; do they in short meet the requirements of the time, and if not, in what directions should our efforts be put forth? The answers can here alone be given as general as are the terms of the queries.

Primarily it may be said, that the value of a seaboard region considered in its ordinary relations commercially and politically to mankind, is represented by the fullness or otherwise of its charted details: this I apprehend will be found a fair general rule of application, though I admit open to exceptions.

It may be further said, that a great degree of accuracy prevails over the whole series of these charts in their broad features; for example, it would be a refinement of no practical value to amend the geographical positions of probably four-fifths of the navigable regions of the globe; similarly the descriptive character of prominent coast features would

---

* This large accumulation of charts, of which at the present moment not one can be considered superfluous, cannot be looked on with unmixed satisfaction, the numbers are so large that considerable difficulty is experienced in keeping them individually connected to meet the growing improvements and wants of the day. They further become cumbersome and expensive to the shipmaster, although sold at little more than the cost of paper and printing, and yet their reduction requires great discrimination and judgment. This reduction has, however, been in progress for some time. One of the chief points for consideration in the publication of new forms is that of scale. Charts, and especially plans of ports if given on too large scales are apt to embarrass the seamen as they convey erroneous ideas of distance and space, and not unfrequently give a fictitious and undeserved value to a port or anchorage. The distinction has not in many cases been drawn between what is necessary for nautical and what for engineering purposes.

come under the same rule. This brings me to one of the growing wants of the day in a matter of chart improvement, and at the same time in the interest of hydrography to protest against this cause.

Navigation by steam has brought with it a certain amount of boldness, amounting too frequently to recklessness in the hugging of the sea shore, and this without adequate reason even as a question of saving in time and therefore of cost. Numerous wrecks on uncharted hidden rocks have resulted, but these uncharted dangers more frequently than otherwise are found to be far within those offing limits that prudent navigators would have adopted. The chart is then denounced as incorrect.

Now it by no means follows—as is too generally understood—that because a chart may not have some unsuspected, or newly found danger marked, that it is inaccurate. On the contrary, the chart may be a model of accuracy so far as the details given are concerned, and an example of labour and care in so far as the execution of those details have been carried out; and yet some rocky pinnacle, that would prove fatal to a ship, may have escaped the most diligent surveyor.

Such a chart, we must maintain is not an incorrect one; it certainly is incomplete, and it is this completion to the hazardous navigation of the day that has become a pressing want.

If the general views on the present collection of charts I have placed before you are correct, and the abundant information thus gathered together, weighed, as to the demands made in the interests of trade and government; we are assured that the direction for marine surveys now to take is that of the streams of commerce. Those shores which were considered sufficiently charted for trade when carried on by sailing craft require working up in detail—notably this is the case on the coasts of China consequent on the opening up the several treaty ports. Those channels, or highways through which the passing trade directs itself, require to be cleared up, notably now special parts of the Red Sea, and the many passages among the islands of the eastern archipelago. Here, however, it must be borne in mind that certain regions cannot be rendered safe to navigation by the most detailed and accurate charts; such are parts of Torres Strait, among the Barrier reefs of North East Australia, and indeed among coral reefs generally; as a rule the general sea boundaries of these masses would suffice; their interior navigation must be left to the vigilance of the navigator.

Again we must keep in mind that many regions fair to the view, and possibly at no distant future destined to be peopled by civilized races, are now, from geographical position, associated with conditions of seasons, winds, and currents, widely separated from lands where civilization and its many appliances are to be found : such are the extensive groups of New Britain, New Ireland, and Eastern New Guinea. Detailed marine surveys of these lands might well be left for their higher development. Similarly among the comparatively imperfectly known coral groups of the Pacific Ocean, minute details of the areas of the many islets and lagoons are not so much needed, as an accurate knowledge of their several boundaries and the security of the intervening ship channels.

Bearing on the hydrography of the immediate sea-board of the maritime states of Europe and the United States of America, great activity has prevailed for years towards its perfection, and in the interchange of that special information relating to the changed boundaries of shoals and their buoyage: also in the institution of new lights and the re-adjustment or improvement of old established ones—in short in all that tends to the safety of navigation. Chili and Brazil, in emulation of this worthy object, have recently established special hydrographic departments, presided over and managed by seamen. It is well to record this latter arrangement, ensuring as it does, that in the manipulation of charts for mariners the draughtsman's art will be subordinate to that of the hydrographers.

Passing from Cartography to those auxiliary subjects in the art of navigation which now demand the attention of the hydrographer, Ocean meteorology takes a prominent place ; and, as is well known, great efforts have been made in this and other countries within the last twenty years to extend our knowledge of the distribution of the winds and currents, and their associated phenomena over the several seas and oceans.

The first idea of collecting and combining information respecting ocean statistics, originated, I believe, in the early part of the present century, with Mr. Marsden, who then filled the important post of Secretary to the Admiralty. He proposed dividing the several oceans into squares of ten or five degrees of latitude and longitude,—a proposal subsequently adopted—as a convenient method for arranging or grouping the observations made within their several areas. A systematic monthly collection for the Indian Ocean was in fact commenced, but the limited means then applicable, and the pressure of other duties, prevented the hydrographic department from following the measure up.

It is to the genius and untiring energy of the late American seaman, Maury, that we are mainly indebted for the impulse given to existing organizations for the collection of sea observations. The well known Brussels Conference, held in 1853, resulted from his efforts, and may be considered the starting point of Ocean meteorology in its scientific aspects, as then the mode and extent of the necessary observations were regulated, and the exactitude and intercomparability of the several instrumental appliances provided for.

Thanks to the indefatigable labours of the accomplished navigator Fitz Roy, followed by those of the existing meteorological department connected with the Board of Trade, a large amount of ocean statistics have been gathered—indeed is still accumulating—and many valuable contributions, resulting from the analysis and discussion of special areas have been given to the nautical world.

This large collection of ocean statistics, to which may be added the collections made by other maritime states, appears to me to deserve consideration from two distinct points of view, the practical and the scientific. Practically put, the wants of the seaman have been tersely described, as the desire "to know when to find a fair wind, and where to fall in with a favourable current." Science desires to grasp the laws under which the winds and currents of the ocean are distributed. Hydrography more immediately concerns itself with the wants of the seaman and it is to these that I would now briefly address myself.

The question as it appears to me, is : How can the hydrographer present in a succinct form and clear character the statistical information now acquired? I use the term succinct form, from the belief that any undue expansion in rendering this information, whether by graphical or textual description, will defeat the desired object. Maury, in his zealous efforts for seamen, contemplated the preparation of no less than 120 sheets of large size to represent " Track," " Trade Wind," " Pilot or Wind," " Thermal," and " Storm and Rain " charts. Seventy-six of these were published accompanied by two volumes of some

1300 pages of explanatory or descriptive matter. Assuredly these were grand storehouses of information; but I venture to agree with those scientific contemporaries of Maury, who advised their government to discontinue the issue of this collection as constituting a practical aid to navigators.

With this experience, the Admiralty Hydrographic Department prepared, in a tentative form, the two folios now before you, one under the name of "Pilot Charts for the Atlantic Ocean;" the second, "Wind and Current Charts for the Pacific, Atlantic, and Indian Oceans." The many contributions from all nations to which I have referred, were examined in the compilation of these charts, and the broad features of ocean meteorology fully and fairly represented to meet the daily wants of seamen. In symbolical illustration as also in arrangement of details, they may be open to improvement: (in the expansion of their seasonal limits they certainly are) but this is a question for the future; when the time arrives that this expansion can be effected, I trust the convenience of the seaman will be recognized in the direction of simplicity, clearness, economy of space and cost.

But little time is now left me to touch on other interesting questions affecting modern science and as bearing on commercial enterprise; I refer especially to contouring the great ocean bed by ascertaining the depths of water, and bringing up for examination portions of the material forming its surface. Here, again, some of the maritime states have done good service. The United States are now working up the Pacific Ocean in a comprehensive manner; and our own good ship the "Challenger," has just returned home from the circumnavigation of the globe, after spending three-and-a-half years in this laborious, but highly enlightening service. We must leave the unfolding of what these deep sea soundings have given us to the physicist and naturalist. The practical appliances used by the seaman in making them are not without interest, and I shall be followed by an able expositor in this field.

I now reach my closing subject, the existing relations of terrestrial magnetism to hydrography; this, I fear, must be dismissed more briefly than its interest deserves.

Those thoroughly scientific hydrographers to whom I have referred, Cook, Dalrymple, Beautemps-Beaupré, thoroughly recognised the, to

z

them, treacherous action of the mariner's compass, and the cause was certainly known to them, though the laws of action were hidden. Flinders [1801-3] mastered the problem, or nearly so, so far as the wood-built ships of the day were concerned, during his Australian voyage of exploration ; and the essay on the subject, forming an appendix to his published work, may be studied now with interest. The changing of the earth's magnetism, as represented by the dip of the needle, when the navigator passed from the Northern to the Southern hemisphere—or the opposite—and the consequent changing of the polarity of the iron used in the construction and equipment of the ship was clearly developed by Flinders, as the great function of disturbance in compass action ; and thus the charting of the values of the dip of the needle became recognized as a useful adjunct to navigation.

The charting of the lines of equal variation of the compass had been recognized as of inestimable value to the seaman, so far back even as the seventeenth century.

The change in the material, iron for wood, in the building of ships, involved new conditions far beyond those elucidated by Flinders ; and here scientific research of an advanced order has unfolded other and more complicated laws. Suffice it here to say that in addition to charts of the navigation of the compass and dip of the needle, we now require the earth's magnetic force—as expressed in certain arbitrary values—to be mapped; and it is not too much to say, that by the aid of such charts, the navigator reasonably versed in the scientific treatment of the magnetism of iron ships, can within close limits, predict the changes of his compass needle, owing to the varying action of the ships iron, in any part of the navigable globe.

My well-trained colleague, Staff Commander Creak, in this field of science, will describe some of the instrumental appliances in the several lines of magnetical research.

The PRESIDENT : I am sure you will all return your thanks to Captain Evans for bringing this subject before us in so able a manner. I think a paper of this kind coming forward in the geographical section shows how intimately connected are all the branches of science, because some of the matters of which he has treated might certainly with equal propriety have been treated in the section of physics, or possibly under some other sections. The points he has dealt with mainly are the provisions adopted by our own navy in regard to hydrography. I do not know whether any representatives of foreign countries would wish to make any remarks upon that subject, for I believe we have some officers connected with other navies here, and if they wish to say anything, we shall be only too happy to hear them.

If no one wishes to make any remark, I will ask you to return your thanks to Captain Evans, and will call upon Captain J. E. Davis, R.N., to read his paper on sounding apparatus. It will be particularly interesting at the present moment, in consequence of the great attention now devoted to the discoveries made by H.M.S. "Challenger," which has returned from its circumnavigation of the globe, bringing with it such satisfactory results of the labours of those who were in command of the expedition, and charged with carrying out the details of the observations.

ON THE VARIOUS FORMS OF SOUNDING APPARATUS USED BY HER MAJESTY'S SHIPS IN ASCERTAINING THE DEPTHS OF THE OCEAN AND NATURE OF ITS BOTTOM.—EXHIBITED BY THE ADMIRALTY.

Captain J. E. DAVIS. R.N.: The sounding instruments exhibited may be classified as those that can be used only in sounding in moderate depths—say not exceeding 200 fathoms—and those that are applicable for deeper water.

The necessity of obtaining a vertical sounding, on which, not unfrequently (in the absence of astronomical observations), the safety of a vessel depends, and the difficulty attending such sounding without stopping the vessel for the purpose, and also the uncertainty even then at times attending the operation—for instance, on a dark and stormy night—led to the invention of several instruments to overcome the difficulty, and give confidence to the seaman in navigating in what is, or was, considered to be "*within soundings.*"

Of these instruments adapted for moderate depths three are exhibited, viz.:

3014. Massey's self-registering sounding machine, invented in the year 1800, it is still in use, although different forms on the same principle have been introduced.

Vanes are fitted on a tube at an angle, which causes it to rotate in passing down through the water: the length or distance required to cause one revolution being known, the number of rotations are conveyed by multiplied cogged discs to an indicator. On the instrument reaching the sea bottom, a stop, which had been kept up on its passage through the water, falls and prevents the tube rotating, on the sounding line being hauled in.

These instruments *have* been used for deeper sounding, but the results are not so conclusive as for the lesser depths.

3013. Burt's bag and nipper was invented by a citizen of the United States about 1812, and has been much used in the Royal Navy of this country.

The bag is first soaked, and then inflated by blowing through the wooden tube, and pegged to keep the air in: the sounding line is then placed in the nipper attached to the bag, at some distance from the lead. When the lead has been hove from the bow of the vessel, the bag and nipper are thrown overboard, the bag keeps immediately over the descending lead, and the line passes through the nipper. On the lead reaching the bottom, the bag—which had been dragged downwards by the friction of the line passing through the nipper—springs up, and no more line will pass through. On being hauled in, the external pressure, in being drawn through the water, causes the bag to collapse (the canvas being porous), and the vertical depth is denoted by the nipper on the marked line.

3015. Ericsson's self-registering sounding machine.

The vertical depth with this instrument is ascertained by the compression of air within a glass tube, the amount and value of the compression, in accordance with depth, being denoted by the quantity of water passing into the tube as the air is compressed. A scale is adapted coinciding with the compression and the water introduced.

A guard moves round to protect the glass tube, and the stop-cock allows the water to run out when the observation is completed.

DEEP-SEA SOUNDING APPARATUS.

Previous to the introduction of instruments for detaching the sinkers at the bottom of the sea, the mode of obtaining a deep sounding was extremely crude and unsystematic, and generally consisted in attaching a thirty-two or sixty-eight pound shot to marked spun-yarn and letting it run out. When it reached the bottom—a point rendered doubtful from the friction of such rough material through the water— the line was hauled on until it was apparently vertical, and the depth thus obtained recorded : it was then hauled on until it broke ; but the requirements for laying telegraph cables caused deep-sea sounding to be considered as a necessity, not only with regard to depth, but also as to the nature of the bottom on which the cables were to rest, and this necessity soon gave rise to various modes of obtaining the desired information.

I believe the first machine invented to combine the act of detaching the sinker and bringing up a specimen of the bottom, was by Mr. Brooke, then a midshipman in the United States navy, in 1856, No. 3005 in the catalogue.

The principle of construction is a perforated sinker supported on a tube or rod by a ring and wire, to a tumbler hook, the hook keeping the weight suspended by the wire and ring as long as there is a strain on the sounding line, but the moment that strain ceases, by the rod touching the sea bottom, the weight of the sinker turns the hook, the wire falls off, and the sinkers slide off as the rod is drawn upwards. Those first constructed had no valve for securing the sea bottom specimen, but quills were inserted in the tube which brought up small portions of the mud or ooze.

No. 3006 is Brooke's machine, with another mode of detaching. It was used in sounding the North Atlantic in 1857.

The principle of Brooke's is again seen in the Hydra machine, No. 3011, designed in 1868, the difference consisting in the mode of detaching the sinkers. In this instrument a sliding rod works within the tube, and instead of the supporting wire being placed over a tumbler hook, it is placed over a button which protrudes through a steel spring on the sliding rod, the weight of the sinker when suspended keeping the spring back. On reaching the sea bottom and the sounding line slackening, the rod slides down and the sinkers then resting, the steel spring throws the wire off, and the weights are left behind as the tube is drawn through them. A portion of the lower end of the tube unscrews, and a butterfly valve secures a portion of the bottom. This instrument was used in sounding in the Atlantic and Indian Oceans.

No. 3012. Baillie's sounding machine, designed by Navigating-Lieutenant Baillie in 1872, is another form of Brooke's principle. A short sliding rod with two shoulders works within the upper end of the tube, and on these shoulders the suspending wire is hooked. On reaching the bottom, the weight of the sinkers draws the sliding rod downwards, and the shoulders passing within the tube, the wire is thrown off and the sinkers released.

A tube for containing the bottom specimen is proposed to be fitted with a glass window to observe the strata.

The form of sinker originally used with Brooke's apparatus was round,—a bad form for sinking quickly through the water,—this gave place to an elongated cylindrical form, but in both the weight of the sinkers could not be increased beyond that in which it was cast. The combination of three forms of equal weight have been found most convenient, as it enables the sinking weight to be increased in accordance with the expected depth, and other circumstances connected with the sounding. Four to five hundred weight can be used on the Baillie tube.

3010. Fitzgerald's sounding machine is entirely a new form of instrument, the sinker being hooked to the side of the bar and detached by the bar being reversed. The sounding line is attached to a short lever, which is connected by a rope or chain with the scoop at the lower end of the bar, a hole and vane being at the upper end of the bar. When prepared for sounding, one end of the lever is inserted in the hole in the bar, and is kept in that position as long as the strain, caused by the suspended weight, is on the sounding line; when the strain ceases, owing to the instrument reaching the bottom, the hook slips out, the bar falls, and the act of hauling in the line reverses the bar, the sinker falls off, and the scoop brings up the bottom specimen.

3007. Skead's weight detaching apparatus is an ingenious appliance for disengaging the sinker on reaching the sea bottom. It was successfully used in sounding in the Mediterranean in 1857. The small leaden weight at the end of the wire is kept up by the weight of the sinker on the hook at the other end of the wire, the sounding line being attached near the centre of the wire which forms a fulcrum. On reaching the bottom, the hooked end being free, the small lead weight falls by its gravity, and in so doing disengages the sinker. The lead weight is roughed in order to bring up evidence of the nature of the bottom.

3008. Bonnici's claw is a double tumbler hook that locks when the lever arms are extended, in which position they are kept by the weight of the sinker. On reaching the bottom, the claw, being relieved of the weight, opens by the arms falling, and the sinker is thus released.

No. 3004 is a claw for bringing up a portion of the sea bottom; it was designed by Sir John Ross, and used by him in Baffin Bay in 1818. Two ringed arms within the claw keep it open and the outer case up. A spike descends from the hinged arms, and projects below the claw. On the spike striking the bottom, the hinged arms are forced up, the claw closes, and thus releases the outer case, which slides down and secures the contents.

No. 3009. Bulldog claw brings up about seven pounds weight of mud or ooze; it was used in H.M.S. "Bulldog" in 1860. The claw is kept open by a sinker resting on the four horns. When the weight is detached on reaching the bottom, the india-rubber bands contract and close the claw, thus securing the portion of the soil contained within.

Nos. 3016 and 3017 are cup and tube leads, which have proved most useful when the depth expected does not exceed about 1000 fathoms. The cover of the conical cup (of the cup lead) is lifted and kept up by the action of passing through the water, and on the cup being filled with the soil at the bottom as the lead ascends the cover slides down and prevents its being washed out. The tube of the tube lead is similar to those on the Hydra and Baillie machines, having a butterfly valve to secure the bottom specimen.

A few words may be said respecting the lines used in sounding No. 3018 in the catalogue. The two exhibited were adopted after many experiments had been made to ascertain which were the best. The ordinary deep-sea line, cable laid, one inch in circumference, weighing nearly twenty-four pounds to the hundred fathoms, formerly used in the navy, broke at a strain—after soaking—of only 630 pounds; the one exhibited of the same size, which weighs eighteen and-a-half pounds per hundred fathoms, will bear a strain, under the same conditions, of upwards of 1500 pounds, being about 140 per cent.

stronger with one-fourth less material. The smaller line, eighth-tenths of an inch in circumference, will bear a strain of half a ton. These lines are made with the best Italian hemp, and are well rubbed down with equal parts of olive oil and bees wax.

Connected with the lines, it would be remiss not to mention one great adjunct to success in their use. When employed sounding in the " Porcupine," we found that a constant source of our lines breaking was the sudden strain and jerk brought on them by the ship rising to the swell or sea; this has since been remedied by the application of a number of India-rubber accumulators capable of stretching five times their normal length ; these are arranged to take up that sudden strain on the line, which they effectually do by their elongation as the ship rises to the sea, returning to their normal length as the ship falls. They may be considered to serve also as a dynamometer.

There have been many other sounding instruments invented in addition to those described—some are within this building; but I have confined my observations to those that have done work in Her Majesty's ships, and of which types were preserved at the Admiralty. I may, however, add that the " Challenger " has most successfully worked during her cruise with the Baillie machine and these lines, with scarcely a failure, and I much doubt if that would have been the case with any other instrument that I am cognizant of.

Whether we have attained the most perfect mode of obtaining a deep sea sounding and securing a sample of the bottom, it is scarcely for me to say, and I may readily admit that under some conditions other modes than any of those I have described may be preferable—I can only point to results by the return of the most important expedition of scientific investigation and deep-sea exploration that ever left our, or any other, shore. In the course of the three and-a-half years the " Challenger " has been absent, she has obtained about 350 deep soundings; and, rejecting all under 1000 fathoms, 116 were between 1000 and 2000 fathoms, 171 between 2000 and 3000, 11 over 3000, and one over 4000; the deepest being 4475 fathoms, or a little over five statute miles.

The PRESIDENT : We must all feel grateful to Captain Davis for giving us this very interesting account of those instruments which are in use in Her Majesty's navy, and which have been so fruitful in results upon various branches of natural history, and also upon geology, which comes within our section. He has alluded to some of the other instruments which are exhibited below, among which I may just refer to the ingenious contrivance of Sir William Thompson, which, I believe, has not yet been brought into use, and also that most ingenious apparatus of Mr. Siemen's—the bathometer, for ascertaining the depth of the sea without any actual sounding. The principle upon which that rests is one which is well known, but the application of the principle to the present purpose is, I think, one of the most marvellous instances of ingenuity which this exhibition affords. I do not know whether it is desirable to enter into that, but if the Conference will bear with me, I should like to say a word or two upon it. The principle is this, that the attraction of the earth at any point depends upon the mass below. The specific gravity of the outer surface is about two and-a-half times, or rather more than that of sea water. Now it must be evident that if that denser matter comes near the surface or at the surface, the attraction upon a column of mercury, or of any other mass will be greater than if in the neighbourhood of the mass to be attracted there is a body of less specific gravity. It, therefore, must be evident that the attraction of the earth would be less, if for instance there is a mile of ocean between the denser portion of the globe and the body to be attracted, than there would be where there are only half a mile or a quarter of a mile, and by an ingenious contrivance, by which some allowance is made for variations in temperature, and for other corrections, Mr. Siemens has been able to produce an instrument, which records the variations in the attraction, with sufficient nicety, to be able to estimate the depth of the sea which intervenes between the heavier portion of the globe and the position in which the bathometer is, so that by fixing one of those instruments on the keel of a ship, you can go down below and find out within a few fathoms, I believe, what is the depth of sea over which you are sailing. Whether an instrument of that kind can by any means supersede sounding is another question, but for merely preliminary surveys, with regard to laying cables, and for other surveys, I think it may prove to be an instrument of great value. I must not however detain you on this, which is not within the immediate subject of our communication. I will now only ask you to return your thanks to Captain Davis.

NAUTICAL MAGNETIC SURVEYS.

STAFF-COMMANDER CREAK, R.N. : The subject of my address to the Conference is that of marine magnetic surveys, which have now become an important branch of the science of hydrography, upon which you have just been addressed by the Hydrographer of the Admiralty—to whose department I am attached. We may define the term marine magnetic survey thus—given any area of the earth's surface, over which we require a knowledge of the three magnetic elements—variation, dip, and intensity—then on shore we select certain base stations where they are to be found absolutely, and at sea we multiply considerably our observations of the same elements. The values of the latter are however dependent upon those at the base stations, for, as yet, no instrument has been devised for obtaining absolute values at sea. After necessary reductions, the several observations are plotted on charts of the given area, and curves of equal values drawn.

I venture to remind you that the variation of the compass and its change of value by change of locality, was made generally known by Columbus in 1492. The dip, exactly 300 years ago, when the carefully balanced horizontal needle by Norman, on being magnetized, dipped in obedience to magnetic law. The earth's magnetic force or intensity as a quantity varying by change of locality, has only been generally known within the last 100 years, and the fact established by Lemanon, in the expedition of La Perouse in 1785-87.

Now when we consider the large proportion of sea surface to land in our planet, considerable importance attaches to magnetic surveys in ships at sea, and I therefore propose to lay before you how such surveys have been conducted in the past, and the system now adopted.

Among the early labourers in this branch of science were Humboldt, Erman, Lütke of the Russian navy in the " Siniavin "—de Freycinet and Duperrey of the French navy in the Uranie and La Coquille, all between the years 1798-1830. The sea observations of Humboldt, Erman, and Lütke were made with the shore dip circles principally in fine weather—the intensity being observed with the vibrations of a specially reserved dipping needle in its usual place in the dip circle.

Since their day there has been a zealous co-operation of sailors with their brethren on shore attended by most valuable results, prominent among which are those of Ross in the Antartic Regions, and it is a fact, that were it not for our knowledge of terrestrial magnetism, the great compass question in iron ships would still be in a very crude state compared with the precision which has been attained with that knowledge.

The instruments now exhibited in this room, and numbered 1145-46-47-49 in the catalogue, give a good idea of the form of instrument used in those days.

No. 1146 is a dip and intensity apparatus. This apparatus was made by Dolland, after a pattern described by Cavendish, in the Phil. Trans. Roy. Soc., Vol. LXVI. The vertical circle is graduated to twenty feet of arc and the horizontal circle to forty-five degrees. In 1776, Cavendish first introduced the existing method of resting the dipping needle on agate planes instead of the friction wheels previously used, but he retained Mitchell's cross. This cross was the invention of a clergyman, who sought to obviate irregularities in the balancing of the needle by fixing a light metal cross to one axle, with small adjustable weights screwed on the arms. In the circle we are now considering these crosses are dispensed with, but among its peculiarities are the provisions for finding the magnetic meridian. For this purpose an edge bar horizontal needle, with an agate cap is provided. To mount it, a metal bar, like the axles of the dipping needle, is fitted with a steel point and a pendulous balance. This bar rests on the agate planes, and the vertical circle is moved in azimuth, until the horizontal needle coincides with its plane. Of the three dipping needles two are flat, and one cylindrical and sharply pointed. The axles are made of gun metal, and one of the needles fitted with a brass sphere on the principle of Tobias Mayer. The sphere traverses on a small steel screw, fixed to the edge of the needle, its object being to alter the centre of gravity of the whole, and thus by giving the needle a momentum, in addition to that of magnetism, to overcome the effects of bad workmanship in the axles. The position of the needle under these circumstances is, therefore, not that of the true dip, but from four observations and a simple formula, this can, according to the inventor, be obtained with improved accuracy. For intensity observations the travelling box is fitted with two apertures, filled with glass and a torsion circle in the lid of the box. The vertical lines marked on these glasses were brought into the magnetic meridian with the aid of the horizontal bar needle, placed on a pivot in the bottom of the travelling box, and in the centre of a circle graduated to degrees.   Only one needle, for horizontal vibrations, remains with this apparatus, the others having been left abroad. It is flat, with rounded ends, and a gun metal pin for suspension is screwed into its centre. A knob on the top of this pin slips into a stirrup, suspended from the circle torsion by a thread of silk.   There is also a gun metal needle of the same form to eliminate torsion in the suspending silk.   There is every reason for believing that this apparatus was that used by Douglas during a journey to the North West Coast of America and the Sandwich Islands, where he died.   At Owybee, Mr. Douglas observed all three magnetic elements at a position, nearly 14,000 feet above the sea, but he found no appreciable difference between them, and others near the sea.

No. 1147 is also a land instrument, but for dip only. In respect to size and portability it is a remarkable change from the circle just described. For the convenience of stowing it in a small compass, the vertical circle can be detached from the base plate, and both laid flat in the travelling box.   The needles are six inches long, flat and pointed ; the axles of steel roll on agate planes in the centre of a graduated circle, and lenses are provided at the ends of a moveable arm for reading off.   The two vertical stems on which the agate planes are fixed, are pierced in order that readings may be obtained when the needle is vertical.   And it may be remarked that there is no other form of dip circle in which observations of both ends of the needle may be made for so large a portion of the circle.   In 1836 Major Estcourt, during his survey of the Euphrates made use of a dip circle of this pattern in conjunction with a Hansteen's apparatus, which I will next describe.

No. 1149 is that form of intensity apparatus first adopted by Hansteen in his magnetic survey of Norway and the Baltic shores in 1819—24, and afterwards in an extensive survey in European and Asiatic Russia in 1828—30. In this latter expedition he was joined by Due, and afterwards at St. Petersburg, by Erman, of Berlin ; their observations being all the more valuable from the constancy of the magnetism in the needle employed by Hansteen during a period of sixteen years. The construction of this instrument is very simple. The needles are solid cylinders, with conical ends of 2.65 inches in length by ·15 inches in diameter. They are suspended from the moveable pulley, at the head of the brass tube, by a fibre of silk secured to a brass strap and loop in the centre. By means of the pulley the needle can be adjusted to any required height in the vibration box above the graduated circle at the bottom. The latter is intended to show the arc of vibration. This instrument is very convenient when portability is an object, for the brass tube may be unscrewed into two parts, stowed in the vibration box, and the whole carried in a leather case, from a shoulder strap. The needles should be placed in the box, with their opposite poles adjacent. Observations with this instrument being only relative to stations where the absolute horizontal intensity is known, their value is dependent on the permanency in the magnetism of the needles—a property which it has hitherto been quite the exception to find in those carried into warm climates.

No. 1145 is a form of dip circle intermediate between that made by Nairn, of London, for Phipps, and the dip and intensity apparatus of Mr. R. W. Fox, of Falmouth. Phipps used Nairn's circle in his voyage towards the North Pole, in 1773, on board the "Racehorse," with her consort the "Carcase." The circle was suspended from a tripod by an universal joint to oppose the effects of the ship's motion. The needles were twelve inches long, with axles made of gold alloyed with copper, fitted with Mitchell's cross, and resting on friction wheels. As the needle was found to vibrate at sea, the graduated circle was adjusted by means of a screw at a given degree to the mean of the vibrations, and the angle between that degree and a pointer at the foot of the covering circle subtracted from ninety degrees, gave the dip. Phipps says that he found the dip increases in going north from England, but from observations in London there is no reason for believing in any secular change of that element. To return to No. 1145. It was constructed by Nairn and Blunt for observations at sea. It is suspended by an universal joint from a light wooden stand carrying one adjusting screw. The needle, nine inches long, with steel axles, vibrates within a circle graduated to twenty feet, and the ends of the axles are fitted to work in the agate holes of two adjustable screws in the vertical bars, which support the graduated circle and otherwise strengthen the instrument. The sliding pointers on the graduated circle are intended for adjustment to the mean position of the needle when the motion of the vessel causes it to vibrate on either side of the true dip.

The screw on the underside of the circle works the brass supports on which the needle is placed for introducing the axles into the agate holes. A thermometer graduated to minus thirty-eight degrees is placed inside. A point of interest in this circle is the arrangement for ascertaining when it is in the magnetic meridian. If the verticle circle be moved in azimuth there will be seen to turn with it a small brass horizontal plate graduated to degrees, and having a moveable double vernier with sights similar to those of a gun. On the two pins projecting from this plate is placed a wooden arm about two feet long, with a small compass gimballed at the end. The arm is moved in azimuth until the plane of the vertical circle coincides with the direction of the compass needle when it is removed and observations can be commenced.

The direction of the meridian being thus once obtained the small horizontal circle and its verniers give that direction for all azimuths as the ship turns round.

As an example of the instruments now considered necessary for a Marine magnetic survey, and the work to be accomplished with them, I will describe the equipment of the " Challenger," carried out under the directions of the Hydrographic Department.

Instruments.—For absolute observations at base stations.—1st. An unifilar magnetometer for horizontal intensity, constructed on the principles of Gauss and Weber, in which a very short magnet suspended in a copper box is deflected by a second and longer magnet—the second magnet being fixed on a bar in the same horizontal plane in the position end on to the first—at right angles to it—and at two or more distances from it. This process of deflection gives the ratio of the magnetic moment of the deflecting magnet to the earth's horizontal magnetic force. The deflecting magnet is now removed from the bar, suspended in the wooden vibration box, and the time of 100 vibrations observed. By this latter process the product of the magnetic moment of the deflecting magnet into the horizontal magnetic force of the earth is obtained. The values of the ratio and product of the two quantities being known, either may be obtained separately.

2nd. Dip or inclination.—The Barrow's circle, in which a reversible needle rests on agate planes protected by a glass case. Exterior, but in a plane parallel to the plane of action of the needle, is the graduated circle with microscopes and verniers for reading the observation off to minutes of arc. With a good circle of this kind the dip should not differ more than about twelve minutes in any of the eight positions in which the needle is observed. I may here remark that in ships going to high latitudes, where the horizontal force is weak, Barrow's dip circle may with great advantage be fitted with Dr. Lloyd's needles for ascertaining the total intensity.

3rd. Variation or declination.—When the unifilar magnetometer is not required for force observations, it may be used as a declinometer— the necessary additions to the instrument being generally supplied by makers as a part of it. One of the chief difficulties with a declinometer is the question of torsion in the suspending thread of the magnet when moving about from station to station. Therefore, as valuable auxiliaries to the declinometer, were added three compasses—a Kater, a Barrow, and the Admiralty Standard compass. Indeed, there are many occasions when, from their portability and simplicity of construction, they would be preferable. The principle of construction in the Kater's azimuth compass is that, having levelled the bowl (for which spirit levels are provided) a linear image of the sun is directed by means of a segment of a glass cylinder carried in a rectangular frame on to an ivory plate with a black zero line marked on it. The observation consists in bringing the linear image of the sun to coincide with this zero line, and the corresponding division of the card gives the sun's magnetic azimuth. The Barrow is a prismatic azimuth compass of superior workmanship. The Standard compass is that known as the fruit of the labours of the Admiralty Compass Committee in 1840. It consists of a bowl of stout copper carrying an azimuth circle graduated to minutes of arc. The observations are read off directly from the card by means of a prism fixed to the eye-piece. The cards are two in number, each fitted with four needles, or compound bars of laminæ of steel, fixed vertically and equi-distant on a light framework of brass, the centre needles being 7·3 inches long, and the external pair 5·3 inches.   The A card is fitted with a jewelled cap for traversing on pivots pointed with native alloy, and from its lightness and directive power is specially useful in fine weather under sail, and for observations on shore.   The J card is much heavier, with a speculum metal cap for working on a ruby headed pivot.   This card is used when the ship vibrates under steam power, and in fact does its work well under all circumstances.   I will now describe instruments for relative values at sea.

The Standard compass for variation, and Mr. Fox's apparatus for dip and intensity, principally for observations on board, but also on shore when required.   For instance, the Arctic Expedition, under Captain Nares, was supplied with three Fox circles, which, from the strength of their construction, might be carried about on the sledges far away from the base stations near the ships, under circumstances fatal to the more delicate absolute instruments.   Mr. Fox of Falmouth, the inventor, first brought these circles before the Cornwall Polytechnic Society in 1834.   The instrument consists of two parts, the base plate and the vertical circle.   The latter is fitted on one side with a glass door, on the other with a metal frame, carrying in its centre a moveable metal disc.   In the centre of this disc is fitted a jewelled hole, and in one segment of it the bracket for carrying the opposite jewelled hole, in another segment the arrangement for clamping the needle. The bracket and clamp terminate in the back of the disc with two large milled headed screws, by means of which the bracket may be moved when mounting the needle, and for clamping the latter.   At other times they are used in conjunction for turning the disc on its axis, and thus the jewelled holes may be moved so as to give a fresh bearing to the axles of the needle.   Immediately between these thumb-screws is a pointed projection which, on being rubbed with an ivory rubber, causes the needle to vibrate slightly, and thus remove the effects of friction on the axles as far as possible.   Concentric with the disc is the back circle engraved on the body of the instrument, with two verniers carrying also at right angles to them two arms, at the ends of which the deflectors are screwed.   The needles are seven inches long, flat, and tapering towards the ends.   On one axle a grooved wheel of brass is fixed for carrying the hooks and weights in intensity observations. The needle, when mounted in the jewelled holes, moves in a circle graduated to fifteen minutes, and in a plane parallel to this is a second circle graduated to thirty minutes. In reading off, similar divisions of both circles are made to coincide to avoid parallax. The dip is generally observed direct, and also with the aid of the north and south deflectors, each of which is screwed into the back circle at forty degrees from the dip, and the mean angle of the deflections produced gives two new values for the dip. The relative intensity to any given base station may be observed by four methods: 1st, by observing the angle of deflection of the needle produced by one or other of the deflectors; 2nd, with both deflectors combined; 3rd, by grain weights; 4th, by a combination of deflectors and weights. When deflectors are used a table of weight equivalents must be formed, as in that case the ratio of the intensity in different localities is inversely as the sines of the angles of deflection, and directly as the weights equivalent to the deflecting force of the deflector on the needle, at the respective angles. With weights alone the ratio of the intensity in different localities varies inversely as the sines of the angles of deflection with a constant weight. It is further necessary that temperature corrections be obtained for both needles and deflectors. Full directions for the use of all these dip and intensity apparatus are given in the article on Terrestrial Magnetism, by Sir Edward Sabine, published in the "Admiralty Manual of Scientific Enquiry." Like Hansteen's apparatus, the Fox circle has the defect of loss of magnetism in the needles and deflectors by time and high temperatures. This, however, may be successfully met by a frequent reference to base stations where there are absolute instruments, and in a lesser degree by observing with deflectors and also with weights as often as possible at the several land and sea stations.

All the absolute instruments and the Fox circles were examined at Kew Magnetic Observatory, and the necessary constants found and tabulated—the compasses at the Admiralty Compass Observatory at Deptford, where every compass supplied to Her Majesty's ships undergoes a rigorous examination. Some time before the departure of the ships, two officers were selected for conducting the survey, and, after instruction, made a complete series of observations of all kinds at Kew Observatory. As far as land stations are concerned, the expedition

A A

may now be considered as fairly started, and at every port visited the same observers with the same eye equation will carry on their labours. But not so on board the ship—before even she has quitted the port of departure or encountered any sea above a harbour-ripple, there is much to be done. Even in wooden ships a large amount of iron enters into their construction, producing errors in all three magnetic elements varying for every direction of the ship's head. To meet these sources of error a convenient spot is selected on the upper deck, where a small table, gimballed on four concentric rings, with a balancing weight hanging from it, is supported by an exterior table with four brass columns. The Fox apparatus is placed on this gimballed table (never in any other position in the ship), and observations of dip and intensity are made for every direction of the ship's head in the process called swinging, when the ship is secured with her head on sixteen points of the compass, beginning at north for example. The Standard compass is next placed on the table, and the ship swung again to ascertain its deviations at that position. The Fox apparatus is then observed with, on shore at the base station, and we are now able to compute the values of the horizontal and vertical components of the total disturbing force of the iron in the ship. The process just described must be repeated on any great change of dip and intensity (the oftener the better), and during a prolonged sea cruise a swinging or two is of great value.

For the registration of all classes of observations forms are provided, which are at once a guide to the observer and clear to an independent computer. On the principles just described are conducted the marine magnetic surveys in Her Majesty's navy.

The PRESIDENT: Is it your pleasure, gentlemen, to return thanks to Staff-Commander Creak for his interesting paper, in which no doubt he found it necessary to go into many details which probably rendered his communication not quite so interesting to those unacquainted with the details of the phenomena of terrestrial magnetism. I do not know whether any one wishes to make any remarks on this paper, or on the general subject of hydrography or deep-sea soundings which have been raised by the papers of those distinguished officers of Her Majesty's navy who have favoured the Conference with the remarks that we have just heard.

Captain BARON VON WRANGELL: With your permission, sir, I will say a few words on a new method of deep-sea sounding. Captain Davis has given all the methods used in the British navy with such excellent success, but still I think that an improvement can be made in some respects. Although great results have been obtained by the "Challenger" and other ships, still the amount of precision is perhaps not very great in great depths. As you know the moment when the bottom has been reached is ascertained by the difference in the periods of time which are needed for the descent of 100 fathoms of line; the observing officer takes the moment when a mark of each fifty fathoms passes into the water, and there is a difference in the period of descent when the bottom has been reached. This method of course gives good results, but only within a certain approximation. It is perhaps sufficient now, but it will not be sufficient after a few years, when a greater number of oceanic depths have been measured, and therefore I think a method which will give the depth, even if exceeding 1000 or 2000 fathoms, within a few fathoms, is important. My friend Professor Schneider, of St. Petersburgh, has devised a method of sounding by electricity, which I had the opportunity of trying in the Mediterranean some years ago, and though I could only make about eight soundings, as the ship was not sent out for that purpose, it seemed to work very well. The apparatus consists of an insulated line, a specimen of which will be sent over (I am sorry it has not yet arrived), and the whole apparatus, I believe, will be here in a few weeks. Some difficulty was found in the construction of this line; it is a copper line consisting of eight strands surrounded with a coating of india rubber, and an outer coating of hemp, which is covered with a similar compound to that which Captain Davis referred to. To the end of the copper wire is attached an apparatus by means of which the electric current is broken and again closed. It consists of an india rubber tube closed at the end with a stopper of copper; the tube is filled with oil, and inclosed in a cylinder of copper passing over the end of the line; this cylinder contains a strong spiral spring, one end of which is attached to the copper cylinder, and the other presses against the stopper. Two levers are attached to each side of the outer cylinder, the shorter ends of which press against the stopper, whilst the longer ones are connected with a clam such as has been described by Captain Davis. When the weight and clam are attached the levers keep the stopper against the copper end of the insulated sounding line, and a current is made with a small electric battery, the other pole being connected with the water. The moment the apparatus touches the bottom, contact is broken, and a galvanometer or other instrument gives you warning, showing that the bottom is reached, and then you begin immediately to haul in. It requires, of course, some force to drag the clam out of the bottom, and this force is sufficient to close again the current and give you a second sign, which shows the moment when you are pulling the clam out of the bottom; and by this means you know when the slack line is hauled in; but the moment the clam is out of the ground the current is again broken, which gives a third sign. We hope next year to do some work with this in the Black Sea, the depth of which does not exceed a thousand fathoms; but still that will be a good trial, I thought that it would be interesting to point out to you a new method which may perhaps be of use.

Captain DAVIS, R.N.: I should like to make one remark with regard to what Baron von Wrangell has been speaking of. I do not see that it would prove the accuracy of sounding by having this new method; because the great error which occurs in a deep-sea sounding is that of surface-currents. I do not believe in deep-water currents as far as concerns soundings, but the surface-current is quite as applicable to these machines as to any of those I have been describing. It is the effect of the draught of the ship, not anything at all to do with the apparatus you are using, that causes the error in deep water when using a marked line; and I think myself, when it came to the greater depths, we should have a great deal of difficulty with the electric line. Again, there is a compound used which is extremely awkward, namely, oil and india-rubber, because when the two are brought in contact, the india-rubber very speedily becomes rotten. India-rubber on board ship will go to pieces in six months under the action of the sea air; these accumulators which I showed you just now will only last by being constantly used. At the same time it is a very ingenious machine, and I hope next time the "Challenger" goes out the Baron will ship one of these electric machines on board and will go with it himself.

BARON VON WRANGELL : I am quite aware that the influence of the currents is one of the greatest sources of error, but then it is well to eliminate all other sources of error. If you cannot do without the surface-current it is a bad thing, but if you can do without other sources of error it is better. With the method now used you cannot with certainty fix within many fathoms the moment you have reached the bottom; and you cannot tell the moment when you have detached the weight, whilst Schneider's method gives you the moment when you detach the clam from the bottom, and you eliminate the error of the slack line. The india-rubber used is prepared in a particular way. At first it was as Captain Davis has said, but that has been improved, and the one which will be here shortly has been in use about eight years, and Captain Davis will see that it is in very good condition indeed.

The PRESIDENT : We must all return our thanks to Baron von Wrangell for having come forward a second time to give us this interesting communication. I will now call on Professor Roscoe, F.R.S., to give us an account of his

AUTOMATIC LIGHT REGISTERING APPARATUS.

Professor ROSCOE, F.R.S. : Mr. President, Ladies and Gentlemen, the few remarks which I shall have to make are with reference to an automatic apparatus for measuring the chemical intensity of total daylight and sunlight. It is acknowledged by all that the measurement of the total sun's effect at any given position is a matter of great importance, inasmuch as to a very great extent, the climatic relations of a country depend upon the amount of sunshine which it receives. Up to the present time we have had no apparatus at work for the purpose of measuring any one portion of the intensity of the various rays whether the heat-rays or the visible light-rays, or those rays which are termed the chemically active rays ; we have had no method which can be applied generally for the purposes of observatory work. The present method, the principles of which I shall very briefly indicate, is one which I think may come into such use, and which has in fact already been used at Kew, and is now at work at Chatham, under the care of my friend

Captain Abney. The variations of the intensity of sunlight at any given place is of course a matter of common observation; the intensity of the shadow cast in our climate and in any sunnier southern climate is of course a subjcet which has long attracted attention. Some time ago I was able to propose a method by means of which the intensity of these solar rays, not only of the direct sunlight, but of the total light of the heavens could be measured, and as showing the difference existing between our climate and the tropics I have here a curve showing the variation between the total intensity of the daylight as measured by the method I am about to describe, first at Kew, and then at a place called Pára on the Amazon, situated at about one degree of south latitude. The red curve at the bottom indicates the curve of daily intensity at Kew, beginning early in the morning on the 4th of April, 1866. It begins at seven o'clock in the morning, and shows it at every hour from sunrise to sunset; whilst the dark curve rising up to the top of the diagram indicates the like intensity of light at Pára, on the same day. By a very simple process of integration we can determine the value of the total sunlight cast on a horizontal surface at Kew under such circumstances as the one I have mentioned, and we find that on this particular day, with a unit of measure which I need not now explain, Kew was represented by 19·7, whilst the intensity at Pára was represented by 260. This indicates what a very large difference there is in the different situations. Similar variations take place at different times of the year, according to the altitude of the sun. Of course in the summer the intensity is greatest.

In order that this method should be regularly carried out, it is necessary that the process should be automatic—that is to say, that it should go on during the day without constant supervision. The method depends on the fact that it is possible to obtain a sensitive photographic paper which, exposed to a constant source of light, attains a uniform and certain fixed standard tint. It was found in some researches I made some years ago, in conjunction with my friend, Professor Bunsen of Heidelberg, that a certain law held good for the darkening produced when the product of the length of the exposure into the intensity of the light was constant, so that if we know the time of exposure, we are able to calculate the intensity having a certain tint given as the unit. The process then depends on obtaining a certain series, at regularly defined intervals, of portions of this paper exposed for a given length of time. It may be perhaps most intelligible if I begin with the beginning of the operation, and simply describe the mode in which the apparatus acts.

In the first place, then, we have here rolls of sensitive paper—that is to say, of paper which can be sensitized by dipping it into a solution of silver nitrate. This is pure photographic paper, which has been salted in a bath of common salt having a definite strength: here is a trough into which the nitrate of silver solution is put, then the strip of paper to be silvered, which is of sufficient length for twenty-four hours, is floated along in this trough, an operation which I need scarcely say must be done in the dark; it is then rolled on a little roller and allowed to dry in the dark. When it is thus dry it is fit to be placed on the apparatus in which it is to be exposed to the action of the sunlight. The apparatus consists essentially of three parts; first, of this small apparatus which is taken to the top of the house, or some situation where there is the best horizon, and where you can get the maximum amount of light. The paper still being in the dark, it is rolled on to this coil, and passes over a wheel; this machine, which is termed the insolater, is then covered by a dark metal shade, having a little hole at the top, so that the paper is only exposed to the light which passes through that hole against which it is pressed. It just receives a little round spot of sunlight, and to prevent the action of rain it is covered over with a semi-circular dome of glass. The apparatus being placed in this position, it is connected by means of copper wires with the second part of the apparatus, viz., a clock, which may be placed in the observatory, and that clock is also placed in electric communication with a battery below. By means of a little arrangement, into the mechanical details of which I need not enter, the clock is provided with one wheel connected with the battery, which wheel revolves once in every two minutes, and bears platinum points; whenever that point comes to the top of the wheel, electric communication is made which passes along the wire to the insolator at whatever distance that may be, and by means of a little magnet the densitised paper is moved on through a certain given distance. Each time that contact is made and unmade, the paper is moved forward, so that we have the means at any given time of moving it forward under the cover. Supposing we wish to make an observation every hour, this wheel would be made to revolve once every hour, and once every hour the paper would be exposed for a given interval, the interval depending on the distance between the platinum points on the wheel. As long as the contact is unbroken the paper remains exposed to the action of the sunlight. This, then, would be a very simple matter were the sunlight always of the same intensity, but the curves to which I have already referred, show that a very great difference occurs in the intensity of the light at sunrise and when the sun is in the meridian. Thus, for instance, taking it at seven o'clock at Pára, the action is represented by 0·16, whilst at twelve o'clock it is 1·16. Now in order to be able to estimate the degree of tint to which the paper arrives, or to recognise and to compare the tint which we have obtained with the normal tint, we cannot go beyond a certain degree of blackness. If we expose the paper in the early morning we only get perhaps a tint so slight that we cannot compare it, and if on the other hand we expose the paper for the same length of time which in the morning will not give us a dark enough tint, in the middle of the day we shall get too dark a tint, and shall not be able to ascertain the intensity of the light. Hence, then, you will observe it is necessary to ensure our having on our paper a tint lying between the lightest which we cannot read off, and the darkest; because if we do not get a tint laying between those extremes our measurement fails. In order to ensure that we do get such a result, it is necessary that the paper should be exposed for varying lengths of time, but for varying lengths of time which are known. Hence, then, this wheel to which I referred to in the clock bears a large number of platinum points placed on the periphery of the wheel at unequal distances from each other, so that each time this wheel revolves perhaps twelve or fourteen little platinum points come successively to the upper portion of the wheel in the whole of its revolution, and each time contact is made with our insulator; so that each time the wheel revolves instead of having one little spot we have ten or twelve spots each made by the exposure of a portion of this slip of paper to the light for varying but known lengths of time. If the wheel in the clock were constantly revolving, we should require an enormous length of paper, and hence it is necessary we should limit the number of observations and in the instrument we have on the table the observations by another mechanical arrangement are taken every hour; that is to say, every hour this wheel in the clock is made to revolve, and every hour we get on the paper a series of little more or less dark spots. The length of the paper is so arranged that the quantity on the wheel is sufficient for one day of twenty-four hours, so that if it be placed up at night and allowed to go on it goes by itself until this time to-morrow.

So far, then, the apparatus is an automatic one; and at the end of twenty-four hours we take out our paper. The next point is to be able to read off or determine the value of each of these little spots of darkened paper. For this purpose we take the paper off the reel. The object is to read off the value of some one or more of the spots at nine, ten, eleven o'clock, and so on throughout the day. In order to do this we make use of the third part of the apparatus which cannot be automatic, because it is one which must be worked by the eye. We cannot make any apparatus which will tell us the difference between two tints. The paper then having been thus exposed for twenty-four hours is brought into a darkened room, and there we have this little drum which carries a graduated strip of paper each tint of which has a known intensity and has a table attached to it so that we can determine the value of each of these different tints. All that is necessary then is for us to be able to say which tint at ten o'clock is exactly of the same shade as a particular portion —it matters not which—of this graduated strip. For that purpose a very simple mode of comparison is made use of. I have a little semi-circular punch, under which I place the strip, and punch out half of one of the spots. All this, of course, must be performed in the dark, or rather with the light of a soda flame. If you take a colourless flame, a mixture of common gas and air, and put into it a little bit of common carbonate of soda, attached to a platinum wire, that gives a flame of a beautiful yellow colour, and by this yellow light we can work, because it does not affect photographic paper. As you know, photographers always work behind a yellow screen, because that yellow light having all the chemically active rays cut out of it, does not produce any effect on the plate; so that you can work in a yellow light with perfect security that no darkening effect will be produced on the paper. We simply have then to place the little strip of paper on our drum, and to clamp it on by means of an apparatus for the purpose, and then to bring the drum together with the piece of paper in front of our lamp. We throw the light of the sodium flame on to the little drum, place the eye nearly on the same level, and turn the drum backwards and forwards until we have a tint which exactly corresponds with the portion that is cut out. In this way we can read the exact intensity of some of these various points, and in this way we get a set of readings for each particular hour.

Here we have tables showing the results of these measurements made at Kew. I have taken some from last year to indicate the kind of results obtained. For instance, on January 22, 1875, before nine o'clock the intensity of the light was so small we were unable to measure it. At ten and eleven it increased, and by twelve it reached its maximum or nearly so. It then went on until three, and after three there was not enough light for measurement. When the numbers in the table are tabulated they give the curve which you will see on the diagram. The next one I have taken is that on April 19, when the light could not be measured at six in the morning, but at seven it was sufficiently intense, and we got readings at eight, nine, and ten. At eleven it was too faint to be read off, probably from a passing cloud. Then it went on to one, two, three, and four, and at six we got the last reading, at seven it being again too faint to see. Of course it gives a much higher curve than in January, the total intensity in January being 11·1 and in April 33·6. The next table is on the 24th June, when the intensity at five o'clock in the morning was sufficient to give a reading. That went on throughout the day until seven in the evening, when the intensity become too small. Tabulating these results we arrive at the curve which you will see, the total intensity being represented by the number 65·6. This may serve to give an idea of the way in which the method can be used. I have here a long series of tables of similar observations, which have been made by Captain Abney at Chatham.

I need scarcely say that these measurements give only the intensity of those particular rays, which have the power of blackening the silvered paper. It does not pretend to be a method by which we can get an expression of the total effect produced by the sun. That, I suppose, can only be done by getting the total amount of heating effect which the sun produces by a method recently proposed by Professor Balfour Stewart. This, however, is a step in that direction which I think is an important one, and I venture to hope that this instrument and the method I have described may before long be so far perfected that we may see either it or some modification of it adopted by observatories throughout the kingdom and the rest of the world.

The PRESIDENT : Unless any one wishes to make any remarks on this interesting apparatus, I will beg to offer Professor Roscoe our hearty thanks, for having in so lucid a manner brought this registering apparatus before us, which promises to be a most valuable adjunct to the meteorological instruments which we have in use at various stations, and to afford what is much required, a record of the varying intensity of the sun, especially of those rays which are susceptible of acting on the silver salts. The ingenuity which has been lavished on this arrangement will have commended itself to all of you. The simplicity and ingenuity combined render it one of the most satisfactory instruments brought before us in connection with meteorology. I am very sorry to say that my explorations in search of the missing African traveller have been unsuccessful : but we have had, I hope, during the course of the afternoon a series of communications which have been of interest to those who have been present.

(The Conference then adjourned).

SECTION—PHYSICAL GEOGRAPHY, GEOLOGY, MINING, AND METEOROLOGY.

Friday, June, 2nd., 1876.

Mr. JOHN EVANS, F.R.S., in the Chair.

The PRESIDENT : Ladies and gentlemen, the Conference to-day will be devoted to geological, mineralogical, and crystallographical subjects, and we have on the list a series of papers which I am sure will be of great interest and value; not the least so will be that which is first on the list by Professor Ramsay, the Director of the Geological Survey of the United Kingdom, on the origin and progress of the survey. I think our survey is almost, if not quite, the first in date, and certainly deserves to be regarded as amongst the first, if not absolutely the first, in the perfection to which it has been carried. I will not detain you with any remarks of my own, but will ask Professor Ramsay to be good enough to give us his communication.

THE ORIGIN AND PROGRESS OF THE GEOLOGICAL SURVEY OF THE BRITISH ISLES, AND THE METHOD ON WHICH IT IS CONDUCTED.

Professor RAMSAY, LL.D., F.R.S. : In spite of Mr. Evans having expressed a hope that the subjects to-day shall all be very interesting, I doubt if that which I have chosen is likely to be interesting to a miscellaneous audience. When I selected this subject it was expected that a number of foreign gentlemen interested in geology, amongst other sciences, would be present at this Exhibition, and at these Conferences ; and, as at present, there are national geological surveys in progress in many continental countries, and as others are beginning in some of those regions to make a new start with geological surveys, I thought it might be of some interest to such persons to hear something of the history of the Geological Survey of Great Britain and Ireland, of its gradual progress, and the methods, in fact, by which it is now conducted.

Long ago when the ordnance maps of England began to be published, it was proposed that these maps should, in the long run, form a basis for statistical, antiquarian, and geological data ; but it was not until that survey had existed for a great number of years, that it occurred to the late Sir Henry De la Beche, then Mr. De la Beche, to propose to the government to institute a national Geological Survey, and at that time he undertook to commence the work single-handed, or with very little assistance, and chiefly at his own expense—but not quite, for the government made him the very small allowance of £300 a year towards the working of the national Geological Survey, all the rest of the expense to be borne by himself. He being a far-sighted man, thought that from small beginnings great things might in the long run grow, and as he well knew that the purely practical and possibly money-making part of the business would, at first, at all events, make most impression on the government and on the public, he decided to commence this work in the important districts of Cornwall and Devon, where mining enterprise, from Phœnician times downwards, had formed a large portion of the industry of that region. Accordingly, with such slight assistance as could be given by two officers, civilians attached to the Ordnance Survey, he commenced the arduous task of mapping the geological structure of Cornwall and Devonshire, and all the masses of granite, and the great sub-divisions of the strata, and all the igneous elvandykes and other igneous matters, and all the lodes in that country, and all the dips of the strata, or many of them, were wonderfully indicated at that early date, for his special commencement of the survey began as long ago as 1832. About this same time, or rather two years later, in 1834, Major Portlock, afterwards General Portlock, in Londonderry, commenced a geological survey of that district, and both of these surveys were, at that time, considered, and actually were, a branch of the Ordnance Survey, then under the command of Colonel Colby, afterwards General Colby. In that year, 1837, Major Portlock published a report in a large volume, on the geology of Londonderry ; and in the year 1839 Mr. De la Beche published his well known report on the geology and mining industries, and all other matters connected with the geology of Cornwall and Devon. Shortly after this there followed a work by the late Professor Phillips, of Oxford, giving an account and figures of all that was known at that time of the Devonian fossils of that region.

In the year 1835, Mr. De la Beche suggested the foundation of a geological museum, for the purpose of exhibiting specimens of all the valuable minerals of the metalliferous mines and other matters connected with the geology of Cornwall and Devon; and before that time a commission had been appointed to traverse England, for the purpose of determining what would be the best building stone with which to construct the new Houses of Parliament, now the great palace at Westminster. The late William Smith, though he can scarcely be called late now, but William Smith, that man who obtained the honourable name of the Father of English Geology, was one of those commissioners, Sir Charles Barry was another, and I do not remember the names of all the rest. They traversed the country, and obtained large numbers of specimens of rocks of all kinds, cut and carved into square blocks, by means of which to determine by the aid of experiment, what would be the best stone for the building of the Houses of Parliament, and this formed another nucleus, so to speak, of the collection of rocks in England in connection with building stones, and for all other purposes that might be useful to the architect in works connected with architecture. This museum was proposed, as I have said, by Mr. De la Beche in the year 1835, and in the year 1837, he obtained from the Government a moderate sized plain building in the retired region of Craig's Court, Charing Cross—a building without any front to the public street; but he saw it would be a good beginning. There was sufficient space to accommodate a great number of specimens, and to it, under Mr. De la Beche, there was attached a chemist—the late distinguished Mr. Richard Phillips, so well known for many years as one of the best chemists in this country.

Having finished the survey of Devon and Cornwall, Sir Henry wisely saw, that the best thing to do was to begin on another mining region, and accordingly he transferred himself, for that was all the staff in those days, to Glamorganshire, and immediately afterwards the Government, beginning to get a little more liberal, gave him two or three assistants, all young men inexperienced in geology, who were to be trained by him, and who afterwards became able geologists. Here he met with a gentleman previously unknown to him—the late Sir William Logan, who, being engaged in business as a copper smelter in Swansea, and having a great taste for geology, it occurred to him, that he would spend all his leisure time in constructing a geological map, as far as he was able, of the Glamorganshire coal field. Accordingly, when Mr. De la Beche arrived there, and made his acquaintance, he found that a large proportion of the great South Wales coal-field, in Carmarthenshire and Glamorganshire, had already been mapped by Logan, in a style of such beautiful detail, that no map of any coal-field that had been done before approached it in excellence, and none other in those earlier times came up to it, save that which was constructed by Mr. Thomas Sopwith of the Forest of Dean.

Logan, in the most generous manner, handed over all this work to Mr. De la Beche, and said he was welcome to it all if he found it was worthy to be adopted by the Geological Survey; and, as you may easily suppose, such a generous offer was willingly accepted. Not only had Mr., afterwards Sir William Logan laid down with systematic accuracy every crop of coal where it rises to the surface; but in addition to this he constructed geological sections on a true scale of six inches to the mile; in which the vertical measurements precisely correspond to the horizontal distances. In these, every bed of coal was laid down in its precise position, and all the faults and dislocations of the district indicated. But more than that, he had also constructed a number of vertical geological sections of these Coal-measures on a scale of forty feet to the inch, in which every bed of coal is not only indicated, but the precise nature of all the strata, shales, sandstones, fire-clays, and the like, that are interstratified with the beds of coal, of which from first to last there may be, of workable and unworkable about 100 in that coal-field. This was done with such excessive detail, that it enabled all interested in such subjects, not only to judge of the lie of the beds of coal with great accuracy, but to judge of the precise nature of the strata that they would have to sink through, in their attempts to seek for coal in regions where coal pits had not previously existed.

In the year 1841, I had the good fortune to be appointed one of the few assistant geologists on the Geological Survey. The survey had then progressed westwards into Pembrokeshire, and we were at work at Tenby and St. Davids, and the neighbourhood. I am not going to bore you with an account of all the doings of the Geological Survey; but I will now state in a more general way what has been the history of the survey since. There were then four assistants on the survey besides myself, and when the work in this portion of South Wales, Monmouthshire, Pembrokeshire, and Glamorganshire was finished, we began to work from the south-west to the north-east until the survey of Wales was completed. This took a long series of years, as you may well suppose, and while it was going on, various other members joined the Geological Survey—men of great skill, and who added to the efficiency of the survey, and the style of the work which was done in it. One of these men who joined as palæontologist —for it was then that the office of palæontologist to the geological survey was first formally founded—was a gentlemen well known, I dare say, to many in this room—I allude to the late Professor Edward Forbes, of Edinburgh, who, before he was transferred to the dignified position of professor of natural history in that university, for a long time was palæontologist to the Geological Survey; and as many of you know his skill, his judgment, his great amount of knowledge in all matters connected with the animals that live in existing or that lived in palæozoic seas was, so to speak, almost of an unexampled character. There are many also yet living, who remember his extreme amiability, kindliness, and mirthfulness of character, and altogether he was so endeared to his friends and associates, that to this day, although he has been dead for so many years, people mourn for him as if he had only lately died. Another person who joined the Geological Survey at this time, in 1844, was Captain James, now General Sir Henry James, who, long afterwards, became the distinguished director of the Ordnance Survey.

In 1845 the Geological Survey was transferred from the Board of Ordnance, and was placed under the charge of the Department of Woods and Forests. The Irish Geological Survey at the same time was transferred to the direction of Sir Henry De la Beche, and Captain James was appointed director for Ireland under him, whilst I was appointed as director under Sir Henry De la Beche, for the Geological Survey of Great Britain. There was then, therefore, a Director-General, Sir Henry De la Beche—a Director for Great Britain, myself, and a Director for Ireland, Captain James ; and each of these directors had a staff of five, six, or seven officers under him, who conducted the work under the superintendence of the directors, who, however—at all events as far as I was concerned—were not content with merely directing, for the survey was not so extensive in those days, but that the director could do a great deal of personal surveying in the field, besides superintending the work of others. It was in this year also, or about this time, that the publication of a systematic set of memoirs began to take place by the officers of the Geological Survey. The first were two volumes containing independent memoirs, one by Sir Henry De la Beche on the mode of formation of all the rocks of South Wales and Somersetshire, and one by myself on the denudation of those counties, one by Mr. Forbes on the geological origin of the existing floras and faunas of the British Isles, and other memoirs by Mr. Smyth, Robert Hunt of the mining record office, and by Dr. Playfair.

In going on with the survey by degrees, we progressed on through Shropshire and into North Wales, and the style of mapping, as knowledge increased, by degrees became more and more minute, so that when we came into North Wales, it was found not sufficient to separate the region into Cambrian, Upper Silurian, Lower Silurian, and Carboniferous rocks, but to separate out in the Lower Silurian, for example, all the different kinds of rock masses of which it was composed, and the result was we not only mapped the individual masses of felspathic and other intrusive rocks, but in doing so, it was discovered, that other igneous rocks, which previously had been considered in the main as intrusive, were distinctly divisible into a great series of felspathic lavas and ashes originating in volcanos undoubtedly of Lower Silurian age, and these rocks were all distinguished on the maps by peculiarities of colour corresponding to their lithological characters. You may well conceive this was a work of immense labour, where a large number of the mountains, as every one knows, attain a height of over 3,000 feet, being excessively craggy, and sometimes, to inexperienced persons, difficult of ascent. The result of some of this work, as shown on one of these sections on the wall is, that we have

B B

intrusive masses of porphyry in one place, at another intrusive masses of green-stone, at another beds of felspathic ashes and lavas that are perfectly interstratified among the common sedimentary rocks, and obeyed like the sedimentary rocks, all those forces that afterwards threw the country into those convolutions which are expressed in the sections; which sections, let me remark, like the others, are vertically and horizontally on a true scale of six inches to the mile. I should like you to know how such sections are constructed. The method of doing it is to level all across the country from a given point to another given point, and some of the sections levelled in this way are fifty, sixty, or 100 miles in length, and when all the data are collected in the note-books, then above a line representing the level of the sea, the surface undulations of the country are projected. A vast amount of data having been collected with regard to the various inclinations of all the strata, the lines of stratification are drawn on a true scale, with all their curves and undulations, and the section, when that is done, is completed. These are then engraved. In like manner the boundary lines of the formations are drawn upon plain sheets after the Geological Survey is completed, they are sent to the Ordnance Map Office, where they engrave our lines on the maps. Then they are returned to us, and coloured by our colourers, and are published by Her Majesty's Government, at the Stationery Office, for the benefit of the public.

The museum having for many years, up to 1851, remained in that obscure region of Craig's Court, Sir Henry De la Beche thought the time had come to make another attempt to induce the Government to erect a large building for the accommodation of the great amount of material that had been accumulated by the geological survey; not only a vast number of ores, minerals, and metals of all kinds, but a prodigious collection of fossils made under the superintendence of the various officers of the survey, which fossils he well saw would tend, if properly arranged in a museum, to throw a great deal of light on the geological structure of England, and on the history of palæontology in general, as regards the succession of forms of the different kinds of life which the formations contained. Accordingly, when Sir Robert Peel was in power, Sir Henry applied to him on the subject, and that enlightened statesman, who had always taken a great interest in all matters connected with the Geological Survey, saw at once the value of the suggestion, and, accordingly, a grant was made to erect that large building in Jermyn Street, known as the Museum of Practical Geology. In the year 1851 this museum was opened by Prince Albert in the presence of a làrge assembly, and attached to the museum, through the influence of Sir Robert Peel, the Royal School of Mines at the same time was founded, with a staff of Professors, whose duty it was to lecture there on geology, mineralogy, mining, physics, chemistry, and palæontology, and other matters, such as were considered necessary, at all events, for the commencement of the Royal School of Mines.

Shortly after this time, that is in 1853, the Geological Survey was transferred to the Department of Science and Art, in one of the halls in which we are now seated. During all this time the survey had been gradually increasing the number of its officers, and increasing in efficiency. In 1854 our old friend and colleague, Professor Forbes, died; and in 1855 the man who originated all of this, and carried it on in so admirable a manner—Sir Henry de la Beche—also died, and was succeeded in his office by the late Sir Roderick Murchison.

Now I must say a few words about the Ordnance Survey. When that was commenced it was considered, that all that was necessary, was to publish maps, with the hill shading, on a scale of one inch to a mile. But when the Ordnance Survey, or a branch of it, went to Ireland, that enterprising nation felt that a scale of one inch to a mile was far too small for Ireland, and accordingly the authorities were asked, and rightly so, to publish their maps on a larger scale; and a scale of six inches to a mile was fixed upon. Not only so, but the maps were not to be merely skeleton maps, with the hills shaded, but the different relative elevations of the country were marked by a series of horizontal contour lines, twenty-five, fifty, and by-and-by 100 feet apart, and the whole of Ireland was by degrees surveyed and published in this manner. About this time the Ordnance Survey in England had proceeded as far north as the southern borders of Lancashire on one side, and of Yorkshire on the other, and all south of that had been drawn and published, simply on the scale of one inch to a mile. But all in England who cared about such things, felt that it would not do to continue on this scale, and accordingly the Ordnance Survey of England was from this point north drawn on a scale of twenty-five inches to a mile, which scale was reduced and published with contoured lines on a scale of six inches to a mile, which scale was again reduced to one inch to a mile, and all the hills shaded in the ordinary way were put upon it. Therefore, when the maps of the Ordnance Survey were published for the North of England, we began, as the Irishmen had done before, to survey the ground on a scale of six inches to a mile, that is to say, to lay down all the work of the Geological Survey on that scale, and the result of this survey was that, minute as much of our work had been before in the mountainous regions of North Wales and elsewhere, when we came into the coal-fields of Lancashire and Yorkshire it was found possible to map these coal-fields in far greater detail even than had been attempted in the South Wales coal-fields by Logan and De la Beche. Here is an example of one of the maps of the coal-fields on a scale of six inches to a mile. All these white lines represent *faults*—some of them well known faults—in the mines; and all the beds of coal are accurately laid down as far as these things can be gathered by purely geological evidence, and also by the assistance of the miners of the district; and not only so, but all the individual beds of sandstones interstratified with beds of shale and coal, are laid down as accurately as it is possible to do it on these maps. The result is that an amount of valuable geological knowledge, both in a mining and scientific point of view, is laid down on the six inch maps, such as had never been attempted by us in our surveys in any other region before. The horizontal sections are also constructed across these regions on a scale of six inches to a mile also. Some of them are drawn to the level of the sea, some are drawn 1500 feet below that level, and they show the positions of all the well known beds of coal, whether they have been opened out in the regions where they are drawn, or whether they have not.

In the year 1856 the Geological Survey was first commenced in Scotland. Up to that period there had been no geological surveys, except of an amateur kind, in Scotland, because there were no large maps on which to lay down geological lines with accuracy: but by this time, several of the six inch maps having been engraved and published, the survey was commenced there, and I being the director at that time began it myself in the county of Haddington, and continued to superintend the work for the Geological Survey in Scotland and in England from 1856 to 1867. In that year the Department of Science and Art, having sometime before again largely increased the force of the survey, by the help of those liberal grants of money which governments are now and then in the habit of suddenly giving, it was found that it was impossible for one man to superintend the whole of the Geological Survey of England and Scotland, and accordingly it was split up, and there were two directorships, myself for England and Wales ; and my previous colleague, the present Professor Geikie, was appointed director in Scotland.

Here now I may as well shew you what is now the composition, so to speak, of the Geological Surveys of the United Kingdom.

In 1871 Sir Roderick Murchison died, and I was appointed to his place as Director-general. Before that time—and since that time—the constitution of the three surveys is as follows : There is the Director-general, who has the supreme command, so to speak, in all the three surveys, and his duty it is to see that the work is all done in harmony, and to settle any differences of opinion, should, unfortunately, any differences occur. He has three Directors under him—the Director for England and Wales, my friend Mr. Bristow, the Director in Scotland, Professor Geikie, and the Director in Ireland is Professor Hull, who succeeded the late Professor Jukes, who for a long time admirably conducted the survey in Ireland. Professor Jukes succeeded Professor Oldham, who was director in Ireland, but who was transferred to a much larger field of geological investigation, namely, to carry on, and almost to reorganise, the Geological Survey of India, which work he carried on for a long series of years with a small staff of assistants, and with the utmost efficiency, and which, on account of ill-health, which I hope will not last long, he has at length been compelled to resign, and has come home to spend his days in England. Under the Director in England, there are two officers next in command who are called District surveyors, and each of these district surveyors has the charge of a small staff of geologists and assistant-geologists. The same system, with smaller numbers, is followed in Ireland and Scotland. The geologists sometimes have the charge of two or three assistants, as the director Mr. Bristow, or any other director, may judge what is best for the progress of the survey. The staff in England being larger than in either of the sister kingdoms, the number of District surveyors and senior geologists is also larger. There is only one district surveyor in Scotland, Mr. James Geikie; and one in Ireland, Mr. Kinahan; and they each have a staff of men under them. It is hard for me to count how many men are employed on the Geological Survey; but I think not less than twenty-six or twenty-eight are now in the field in England alone; eleven in Ireland, and nine or ten in Scotland; so that you see by slow degrees, and without any pressure being put on the government, the survey has increased in numbers and acquired an amount of popularity simply, I presume, from its utility. Unsolicited, the government has gone on increasing it in numbers, that, as is supposed, its efficiency might be still further increased, and that, if possible, its labours might be brought to a close in a reasonably short period of time. I need say nothing more about the method of constructing six inch maps. It is much the same as that of the one inch scale, or of the method of constructing the sections, except this, that the six inch maps being all covered with contour lines, fifteen, twenty, twenty-five or fifty feet apart, all that you have to do in constructing the section on a true scale, is to measure off certain places on a horizontal line, and mark the points where the ground is ten feet, twenty feet, thirty feet, forty or fifty feet above the level of the sea, and so on; simply draw this line on a true scale for the distance, and the whole thing is done with an immense saving of trouble and expense; and after that the geology is laid down upon it with sufficient facility.

Now there is one thing I should like to draw your attention to—I have missed a great deal, but I am merciful to you, and I will be as short as possible; and will only mention this, as somewhat connected with the Geological Survey. It happened that there was a great outcry some time ago in the Houses of Parliament, about the near approach of the working out of all the beds of coal in the country, when Britain would be obliged either to import its coals, or to begin to plant forests as the Prussians do, for the sake of supporting the fires even on domestic hearths. A great deal of controversy took place in consequence of these representations, and accordingly the Royal Coal Commission was appointed, to enquire into the quantity of coal still left available in our native coal-fields, and if it were possible to find out the quantity of coal that might be available, not only in the visible coal-fields, but in those other regions covered by secondary strata, under which it might be presumed that the Coal-measures in some places here and there might still be found, and a number of gentlemen were appointed as commissioners. They broke themselves up into several committees, so many taking the charge of one coal-field and another taking the charge of another coal-field, and so on. Mr. Vivian, M.P., and Mr. Clarke, of Dowlais, undertook to work in the great South Wales coal-field, and Professor Prestwich and other gentlemen undertook to work in the Bristol coal-field, the Staffordshire coal-field, the Lancashire, the North Wales, and so on. One of the results is this : it was shown that there is a prodigious amount of coal left untouched, sufficient to serve the country for thousands of years. And another important point, to me at all events, and to those connected with the Geological Survey, was that most of those gentlemen who investigated the quantity of coal left in each coal-field stated, not only publicly, but in conversation and in board rooms, that they could not have carried on their work with half so much efficiency if it it had not been that the geological maps and sections of the coal-fields had been previously published.

There is one other point I should like to draw your attention to, which is this. In this section on the wall you have a large extent of country all plotted on a true scale. Here are the Upper Silurian rocks of the Wenlock region and the Colebrook Dale coal-field. Here is the South Staffordshire coal-field, and here is the Yorkshire coal-field, and in the margin of this coal-field you observe drawn Permian rocks, New Red Sandstones, and Marles, and so on. What I wish to insist upon is the value of such geological sections being actually drawn and published on a true scale. I will not go into all the proofs connected with the continuation of the coal-fields under the ground, but they are numerous. For example, the various beds of coal in this Colebrook Dale field, with the intervening beds of ironstone, can be identified with the various beds of coal and ironstone of South Staffordshire, and the various thicknesses of the sub-formations of New Red Sandstone, and its various subdivisions, are laid down on a true scale. They all dip to the east, and they are intercepted by a fault, the

magnitude of which is inferred with, I believe, a very near approach
to accuracy.   Then we have underneath that the Permian strata, and
underneath the Permian strata lie the Coal-measures; and there is
every reason to believe, after reasoning out the subject, with all the
skill and judgment that man could exercise on it, that this coal-field
is continuous underground, in the manner shown in the section, in a
long broken curve, the beds cropping up again to the surface in the
South Staffordshire coal-field, which coal-field was itself once covered
with New Red Sandstone and Permian rocks which have been
removed by denudation.   If that be true, you are able to determine
at what depth the Coal-measures will be found underneath the surface
all the way along that line.   In some places it will be within an acces-
sible depth, but in other parts, the depth before you touch the Coal-
measures will be 3,000 feet from the surface, and in other sections,
prepared in the same manner as this, it can be shown that underneath
similar rocks the Coal-measures cannot be less than five or 6,000 feet
from the surface.   Those persons who are ignorant, every now and
then rashly sink pits in search of coal, and then give up the undertaking when all their money is expended; but if they will take the
trouble to consult such sections as these, which are all published in
the Report of the Coal Commission, and the written Reports of the
Coal Commissioners themselves, they will get a large amount of data
to enable persons to judge when they may successfully attempt such
investigations, and when they had better let it alone.

Now I think I have said enough; indeed, I see by the clock that I
have said a great deal too much; but there is one other point I will
mention, and then I have done.   The Geological Survey of the British
Islands has done a great deal more than carry on its own work, so to
speak, for it has given rise to a number of other geological surveys,
some of which are almost equal in importance to that which has been
conducted in Great Britain.   For example: in consequence of Sir
William Logan's connection with the geological survey of Great
Britain, for he always considered himself an officer of that survey,
(although he merely worked as an amateur, and without any pay), he
was appointed to the directorship of the Geological Survey of Canada,
when it was determined wisely by the Canadian Government to found
a Geological Survey in that colony; and the great success with whic

he carried it on, and the very disinterested manner in which he did it, with a staff of assistants, expending himself more money than he got, led to the production of that admirable geological map of Canada which you may see in one of the other rooms. After his retirement from age and increasing infirmity, Mr. Selwyn was appointed to the direction of the Geological Survey of Canada, and not only of Canada, but of all the British dominions from the Atlantic to the Pacific ocean. Previous to that, Mr. Selwyn was long one of the most skilful officers of the Geological Survey of England and Wales, and had been transferred to the colony of Victoria in South Australia, and began to carry on a geological survey there. This he did with all the skill and judgment that such a man might be expected to show; but the Government there, seemed not to be so permeated with scientific ideas as the Government of Canada, for after a time they began to desire that he would do nothing but go about prospecting for gold, which did not come up to his ideas of what a Geological Survey ought to be, and accordingly he resigned his post. By great good luck it was at the very moment when Sir William Logan had retired from the Geological Survey of Canada, and, upon my speaking to him about it, he saw that here was the right man to succeed him, and Selwyn was appointed Sir William Logan's successor.

And not only that, but we have springing from the surveys here, the geological survey of Queensland, commenced under Mr. Daintree, who previously, in Victoria, had been Mr. Selwyn's best officer; and, by far the most important of all those surveys, the Geological Survey of India entirely sprung out of the existence of the Geological Survey of Great Britain. It originated merely from the wish to look for beds of coal, and one of our officers, Mr. Williams, was sent to India to examine into the coal-fields. Then Mr. Williams died, and they saw they must do something much better than that, and accordingly I was applied to, to go out and to found a Geological Survey of India. But having that very year been appointed director over Great Britain I would not go, and accordingly Professor Oldham went in my place. From that has grown the grand Geological Survey of India, conducted on a great scale, but without all those appliances and perfect maps which we have. But still it is wonderful the amount of admirable work that has been done in consequence of the appointment of Professor Oldham, the judicious selection of his staff, and their zeal in the service. There have also been Geological Surveys of Jamaica, partly by men connected with us, or who have sprung from the School of Mines; another of Trinidad; another of British Guiana, executed by Mr. Brown; and the Geological Survey of New Zealand, which is now in active progress, under the direction of Mr. Hector. All of these surveys sprung from the Geological Survey of Great Britain, and are more or less conducted in the same manner as we do our work here.

The PRESIDENT: I am sure we must all feel that Mr. Ramsay ought to withdraw the words with which he commenced his address, when he said that it was likely to be wanting in interest, for I think you will all agree that he has given us a most excellent account of the situation, the rise and progress of the Geological Survey of this country, and of the other surveys which have arisen out of it; and not only so, but he has also brought before us a series of personal reminiscences which have given great interest to his communication, apart from its scientific value. I therefore beg to offer our best thanks to Professor Ramsay for this communication. I am glad to think that we have —although not, perhaps, strictly speaking, representatives of foreign surveys here—a number of foreign geologists of distinction, and we shall be glad to hear any remarks either from them or from any other gentleman present who wishes for further information, or to supplement any further point which Professor Ramsay may have left untouched.

M. DAUBRÉE asked permission to say one word expressive of his high appreciation of the value of the Geological Survey which had been described by Professor Ramsay, and which had excited so much interest both in England and on the continent. The survey was admirably commenced by Sir Henry De la Beche, continued in the most able manner by Sir Roderick Murchison, and it had since been followed up in an equally scientific spirit by Professor Ramsay. He only wished to say that this admirable work was, in his opinion, of the greatest value, and was calculated to become still more so in the future.

M. le Professor RÉNARD : The remarks just made by M. Daubrée on the work published by the Geological Survey, the organisation and development of which have been so ably expounded by Professor

Ramsay, scarcely leave me further scope for discussion. If I make any remark at all it is simply to express my congratulation to the Director-general of the survey upon the work done under his superintendence ; and after several weeks study of certain portions of Wales which have been mapped individually by him, I have step by step convinced myself of the accuracy with which the country is mapped, and the labour which it must necessarily have entailed. I will, however, ask one question. Would it not be better to represent actual exposures of rock upon the geological maps, rather than, in places, to prolong the boundary lines hypothetically between the visible outcrops. I cannot say that this remark applies generally to the work of the Geological Survey, but judging from those maps which I have used in the field, notably those of Shropshire, I have frequently experienced difficulty in finding any exposures of rock to prove the validity of the published boundary lines, and hence I have inferred that the drawing of those lines has sometimes been a purely hypothetical matter. This, however, may be regarded as having been a necessary evil, which existed in the early days of geological mapping, and in no way diminishes the general value which I attach to these admirably constructed maps.

Professor RAMSAY : I am very glad that my friend M. Rénard has mentioned these matters, which are very easily explained. When these maps of Shropshire of the Lower Silurian Regions were done, it was at an early period of the history, comparatively, of the Geological Survey, and they were, of course, simply mapped on the scale of one inch to the mile, and there very often happens, as every one knows who has mapped a great deal, that in tracing a geological formation you are able to follow it for some distance along a boundary line with great accuracy, and then by-and-by there is a portion where everything is concealed, perhaps by vegetation, or sometimes by Boulder clay, or any of those superficial gravels and formations that are sometimes called Quaternary. Then by-and-by you find it again in the same line of strike, and the same kind of lithological divisions between two sets of rocks. In old days people had no hesitation in joining up these in a bold manner one to the other, and if it was no great distance, perhaps not one-eighth of a mile, or a quarter of a mile, or something of that kind, being on a map of a scale one inch to the mile it would hardly be noticed ; but now we have maps on a scale of six inches to the mile we have more space to draw upon, and then if we have a line going along clearly in the manner I have just stated, and breaking off, instead of drawing it continuously, we put it in with a dotted line, to indicate that it is not quite certain, though it is most probably there. That indicates that it is not quite certain, but is in some slight degree hypothetical. There is another point I forgot to mention, and I am obliged to M. Rénard for recalling it to my memory ; that is, about the Quaternary formations, as they are called on the continent, and sometimes in England, that is the superficial formations which do not consist of solid rocks, but of Boulder clay, river gravels, and such like. We now map all these, and we divide them into a great variety of different kinds, Lower Boulder clay, Upper Boulder clay, river gravels, terraces, and so on. These are all mapped ; and here is a map of a portion of the London district, in which all these Quaternary formations are actually laid down. The London Clay and the other formations are marked upon the map, where they actually form the surface of the soil, and where the plough share would sink into them if it passed over that region. Here is a map of London, on a larger scale, which shows London itself and the neighbourhood in the same way. We have been ordered to do this now for the whole of Great Britain, so that, I think, the termination of the Geological Survey seems very far distant. I shall not see the day, but I hope better men will.

The PRESIDENT : I am sure we shall all join in the hope that the Geological Survey may be yet some time before it is completed. We will now take a paper in some degree connected with the same subject, namely, the one by Mr. Topley on the Sub-Wealden Boring. Unfortunately, Mr. Topley is not here, but Mr. Bauerman has kindly undertaken to read the paper for him.

ON THE HISTORY AND RESULTS OF THE SUB-WEALDEN EXPLORATION.

Mr. WILLIAM TOPLEY, F.G.S., Assoc. Inst. C.E., Geological Survey of England and Wales : So much has been written lately concerning the Sub-Wealden Exploration, and the chief results obtained are so fully illustrated in the Exhibition and detailed in the catalogue (No. 3270), that it will not be necessary to enter largely into detail concerning it.

The main object of the undertaking has been to reach if possible the palæozoic rocks which are believed to lie beneath the secondary strata in the south-east of England. These palæozoic rocks (with included coal measures) occur with an east and west strike in South Wales and the West of England ; they pass under the oolitic rocks to the east, beneath which coal is worked. In Belgium and the north of France these rocks again occur, stretching on the one hand far into central Europe, on the other hand ranging westwards and passing under the cretaceous rocks, beneath which again coal is worked.

In the Bas Boulonnais the palæozoic rocks are exposed at the surface for a small space ; here they are overlain unconformably by oolitic and cretaceous rocks. Here also coal occurs, which was formerly thought to lie *in* the carboniferous limestone, in the same way that such beds occur in the limestones of Scotland and Northumberland ; but Professor Gosselet, M. Rigaux and others, now believe that these beds are true coal measures brought in amongst the carboniferous limestone by faults and inversions. In general range and character, and even in their faults and disturbances, the carboniferous rocks of Belgium closely resemble those of the West of England. Reasoning upon all these facts Mr. Godwin-Austen long ago concluded that beneath the South-east of England coal measures occur at a workable depth.[*] Although the Sub-Wealden boring has failed to reach them, I believe that nothing has been here discovered which affects this argument. Mr. Godwin-Austen's reasonings were remarkably confirmed, immediately after the publication of his paper, by the discovery of carboniferous rocks beneath the cretaceous strata at Harwich ; and of strata, believed by many to be old red sandstone, in a like position at Kentish Town.[†] Palæozoic rocks beneath the cretaceous series have also been discovered at Calais, Dunkirk, and Ostend. Had the sole object of the exploration been to seek for coal, or even for palæozoic rocks, the boring would have been started near the Thames, or at any

---

[*] Quart. Journ. Geol. Soc., vol. xii. p. 38, 1856.
[†] At the boring at Messrs Meux's Brewery, Tottenham Court Road, *Devonian rocks,* with characteristic fossils, were found beneath a small thickness of lower green sand.

rate on the northern side of the North Downs, as there is reason to believe that the main line of coal measures will probably take that direc_tion.   There is, however, a possibility that coal may be found furthei south, because the carboniferous limestone of the Boulonnais, amongst which the coal is faulted, dips south, and if this dip continues coal must occur beneath the oolitic rocks further south.   This line, if prolonged westwards, would pass under the Weald.

The boring owes its origin to Mr. H. Willett, of Brighton, and by him it has been superintended throughout.   It was commenced in Sussex, in order to commemorate the visit of the British Association to that county at its Brighton meeting in 1872.

The site of the boring was chosen because at that point the lowest beds of the Weald are exposed, and because, lying on the anticlinal, the beds are there horizontal.

Reasoning from what was then known as to the thickness of the oolitic rocks in the Boulonnais and in Dorsetshire, it was hoped that within about 1700 feet of the surface the palæozoic rocks would be reached.   In this we have been mistaken, for at that depth the beds are Kimeridge or Coralline oolite.   But the unexpected development of the Kimeridge clay has enabled us the better to attain the second object of the exploration, which was to ascertain the thickness and nature of the secondary rocks which underlie the Weald.

Although unexpected by most geologists the great development of the oolitic rocks here was anticipated by some ; Mr. S. V. Wood, junr., Mr. Kinahan and Mr. Bauerman having stated to me their belief that they would here be found to be thick.

The beds passed through may perhaps be arranged as follows:—

Purbeck beds	...	...	200
Portland ...	...	...	57
Kimeridge	...	...	1512
Coralline oolite	...	...	17
? Oxford clay ?	...	...	117

1903

The diagrams exhibited give, in sufficient detail, the chief facts ascertained as to the lithological character of the beds passed through, it is therefore unnecessary to enumerate them here ; but a few words may

be added as to the general results. One point occurred to me frequently whilst examining the cores; that was—the very gradual
transition from one kind of rock to another. The long diagram gives
apparently somewhat abrupt changes, and it is almost impossible to
note the beds in such a manner without giving this idea. The section
is founded upon the daily reports of the foreman at the works, supple-
mented by the notes of Mr. Willett and myself. Shale passed
gradually into sandy shale and shaly sandstone, sometimes becoming
a true sandstone. On the other hand shale gradually became calcareous, passing into cement stone and sometimes into a tolerably
good limestone. The chief cases of fairly abrupt change were at the
occurrence of some highly fossiliferous bands, when it would seem that
forms of life, chiefly oysters, suddenly migrated to the area and as
suddenly departed.

Whatever may be the cause, it is certain that lithological changes
are generally more abrupt in exposed sections than they are in this
deep section, where the beds seem to have been but slightly altered,
except by pressure, since their deposition. It is also certain that
the gradual changes here mentioned accord best with what we know
to be the conditions of deposit in existing seas.

Again, as regards the formations passed through, there is no well-
marked and unmistakable line between any two of them. The classification suggested on the section, and illustrated in the case, may
possibly require some modification in the future. As regards the
Portland beds, the change, if any, will not be of great moment; but
as regards the lower beds of the Kimeridge clay, it may turn out to be
considerable.

It appears probable that there is no sufficient ground for dividing
the oolites into middle and upper at this point. The beds from the
Portland to the Oxford clay constitute one continuous series; evidence
to the contrary being only of local value. According to Dr. Waagen,
beds in different parts of Europe, which are classed as Corallian by
various authors, are really referable to one or other of the stages into
which he divides the Kimeridge clay. In one case a so-called Corallian bed is classed by him with the Portland.

Mr. Blake and Mr. Hudleston are now engaged in studying the
Corallian series of Weymouth, and the relation of this series to the

beds above and below. Their results are as yet unpublished ;* but
Mr. Hudleston, who has examined the section and cores, suggests that
many of the lower beds here marked as Kimeridge, show the survival
of Corallian conditions. The lowest beds, here marked as Oxford clay,
contain oolitic bands resembling those of the Coralline oolite.

Mr. Blake estimates the Kimeridge clay of England at 1,050 feet
thick, this total thickness however not occurring at any one place ; the
lower beds being most developed in one locality, the higher beds in
another.† It is probable that the lowest estimate which can be given
of the Kimeridge in the Sub-Wealden Boring will much exceed this.
At present, reckoning all to be Kimeridge down to the true Coralline
oolite (the *typical* upper Calcareous grit not occurring here) we have
a thickness of over 1,500 feet.

Mr. Etheridge has given his valuable aid in naming and classifying
the fossils; those shown in the case—some forty-five species—are
mostly named on his authority. Mr. H. Woodward has described a
new species of crustacea (*Callianassa isochela*) from the Kimeridge
clay of the boring. There appear to be new species of *Arca* and
Astarte, and many of the other forms, if not absolutely new, require
further study.

There have been two borings on this spot. The first, began by Mr.
Bosworth, and continued by the Diamond Rock Boring Company,
was abandoned at a depth of 1,030 feet. The second boring has been
wholly done by the diamond system. The advantages of this method
for scientific investigation are evident from an inspection of the cores.
It is more expensive than other systems ; but when in full operation it
does the work in less time, and hence, where time is money, it is
cheaper in the end—even for commercial purposes. Dr. Hartig,
classifying the performances of various rock-drills for boring, concludes that the diamond system involves a greater expense of 1·75
per cent., but gives a saving in time of 30 per cent.‡

The Sub-Wealden Exploration is remarkable as being the only boring of the kind executed solely for scientific purposes, with no hope ot
gain to any ot the promoters. As regards depth, it is exceeded by

---

* Since published in Quart. Journ., Geol. Soc., vol· xxxiii., p. 260, 1877.
† Quart. Journ. Geol. Soc., vol. xxxi., p. 216.
‡ *Civilingenieur* vol. xxi., part 2, (See an abstract in *Prol. Inst. Civ. Eng.*, vol. xli.,
p. 264.

several holes. Mr. Judd informs me that some borings are now in progress in Hungary, remarkable not only for their depth, but also for the great thickness of certain strata, which they have explored. At Buda Pesth a hole has been carried 2,300 feet through Oligocene strata without reaching the Nummulitie series. At Petrozseny, just on the Turkish frontier, another boring has proved 2,238 feet of Neogene (Miocene and Pliocene) rocks without touching the Oligocene.

The boring at the Insane Asylum, St. Louis County, Missouri, is 3,843 feet deep. The deepest boring in the world however is that at Sperenberg, near Berlin, which is 4,172 feet deep. This is remarkable for the enormous thickness of rock-salt which it has explored. So great a thickness of so soluble a rock could have been proved in no other way.

It is most important that temperature observations should be taken at the Sub-Wealden Boring ; it may be many years before another such opportunity occurs in the south-east of England. It is desirable to have records of temperature at many localities, and when these are obtained in deep holes the information is of great value. The results obtained at the Sperenberg boring are of high interest. It has been shown by Professor Mohr that the temperature increases at a *diminishing* rate.* Observations in deep holes in other localities must be made before this result can be accepted as fully established. Perhaps the Sub-Wealden hole is not deep enough to be of much value here, still it is probably deep enough to test the point. In the boring at La Chapelle, 1,968 feet deep, the temperature below 1,312 feet was found to increase at a diminishing rate.†

The bearing of this question on current theories of ' vulcanicity ' is obvious. If these results should be confirmed by further researches we shall be furnished with an important argument in favour of Mr. Mallet's theory, that volcanic phenomena are due to actions going on within the superficial crust of the earth, and not directly to central heat.

The PRESIDENT : I am sure you will return your thanks to Mr. Topley for this communication. With regard to this set of borings those who have examined the case at the end of the gallery will have

---

* *Neues Jahrbuch,* 1875, 4th Part, p. 371. See also *Nature,* vol. xiii., p.
† *Rep. Brit. Assoc.* for 1873, p. 254.

seen one very remarkable feature in connection with it, namely, the beautiful nature of the cores or columns of rock which have been extracted from such an enormous depth of earth. When you think that some have been brought up from a depth of between 600 and 700 yards from the surface, you will see to what a pitch of perfection boring apparatus has now been carried. I am glad that we have here Major Beaumont, M.P., who is intimately connected with the boring process; and I hope he will favour the Conference with a few words as to the method by which these results have been attained.

Major BEAUMONT, M.P. : I shall be very glad to avail myself of this opportunity of saying a word or two, but as I am warned that I have only six minutes at my disposal, it is rather difficult to compress into that anything like an explanation for the process, and consequently I shall dismiss that portion of the subject, with the remark that it has been entirely done by the use of the black diamond, which is a substance exceptional in nature for its hardness. There is no comparison between the relative hardness of the diamond and of the rock that it is called upon to cut, and consequently you are able, profiting by that, to introduce a description of machinery for the purpose of rock boring that is not possible where you are using percussion ; that is to say, you are able to scrape the rock instead of striking a blow, and it is entirely owing to taking advantage of that principle that the present amount of progress has been attained.

The results obtained in one or two other instances which were mentioned, I think it will be interesting to the Conference to know something about. First, with reference to this boring, the absolute speed at which this work can be done is the principal point of importance, inasmuch as people who were not likely to wait for four or five years to obtain a result, as they would be obliged under the old system, will very willingly spend their money, and encounter a delay of two or three months, and consequently, I venture to predict, that the time has now come when no person will ever think of putting down a shaft, without first of all exploring the ground by boring. This bore hole of the Sub-Wealden was commenced eight inches in diameter on the 11th of February. On the 18th May, that is a little over four months, the depth reached was 1137 feet ; on the 28th August, a little over six months, the depth reached was 1905 feet, close upon 2000, and at that point the boring was stopped, owing, unfortunately, to the inability of the committee to find the funds necessary to continue it. The boring was completed 1812 feet when the collapse took place, but the Diamond Boring Company which I represent, very generously, I think, agreed to attempt to continue the boring, the difficulty being not that which was inherent in the boring, but that which was inherent in the necessity for lining the hole, because it is under ordinary circumstances, impossible to continue a boring in running strata, unless the holes are lined from top to bottom. There is no difficulty whatever in the lining, but it is a matter which takes a certain amount of time, and what is of more importance, of money. Our attempts succeeded to a certain extent, because we continued the boring from 1812 down to 1905 feet ; at that point, where the boring rests now, the diameter is two inches, and we cannot continue it owing to the sides of the hole falling in. It is extremely desirable, as has been pointed out to you, that this scientific exploration should not stop, and as the contractor doing the work, I may say there is not the slightest difficulty in continuing it, provided only the necessary funds are forthcoming. I hope, therefore, the gentlemen of the Press, if they notice my remarks at all, will be good enough to give prominence to this particular point of the desirability which exists for further funds being forthcoming to continue this important scientific experiment, and further that there is not the slightest difficulty if the necessary funds are forthcoming. Mr. Willett, who was the honorary secretary of the committee, finding the work too heavy for him, has given up his position, and I have taken that most undesirable position of being honorary secretary, who generally has a great deal of work to do, and gets no thanks for it. The position in which I find myself is this :— We have a hole, bored close upon 2000 feet, and a small balance, or rather no balance, I think, at the bank, with promises to the extent of £150, and with conditional promises of £800 more, that is, £400 from Mr. Willett and £400 from Mr. Walker, making a total of £950, but, unfortunately, that is conditional on the remainder of the money being subscribed. The total amount of money required is £1500 for the actual boring, which represents 500 feet at £3 a foot, and £600 for lining tube, being a total of £2100. Consequently, we are stuck fast for £1100 or £1200. I cannot help thinking that when we see what the object is in view, the determination of what underlies the Wealden, which is not only a matter of interest from a scientific point of view, but also from a commercial point of view that the money will be forthcoming.

With reference again to two other borings which we have made. One is at Ham, in Westphalio ; in fourteen days we went 500 feet, namely from a depth of 1500 feet to 2000 feet, being 500 feet absolutely bored in fourteen days, which is more like a tale of fairyland than a matter of reality. Then a very interesting boring was conducted at Reinfelden, near Berne. The Swiss were exceedingly anxious to find coal on their property, both for strategical as well as economical reasons, because in the late Prussian war they found themselves exceedingly inconvenienced from having no coal. A society was formed to make an exploration, and the first effort was at Reinfelden. That was done by my company. We commenced boring on the 14th of August, and we ended on the 15th of October—only two months and a day—and in that time we had bored a distance of 1400 feet, the diameter being three inches, and it gave a complete solution to the question ; the geological point being whether the coal formation was to be found there or not. Unfortunately, we got through into the older rocks, so that the answer there was in the negative, but the amount of boring which was done in the two months was remarkable, and bore a very small proportion to the time it took to make the preliminary arrangements. If I remember rightly I was in communication with the Swiss Committee something like a year, but when we got to work the boring was effected in two months, and I have little doubt that if the money is forthcoming for the Wealden boring, that the answer to that important question of what underlies it, will be forthcoming in something like two months.

The PRESIDENT : If no one else has anything to say, or is ready to come forward with a liberal offer to complete this undertaking, I can only add my testimony to that of Major Beamount, as to the amount of intelligence and ability which has been devoted to this boring by Mr. Willett, of Brighton, with whom the idea originated, and whatever may be the future of the boring we cannot but be grateful to him for the scientific results which we have already attained. I again ask you to return your thanks to Mr. Topley and to Major Beaumont, and I will next call on Mr. C. E. de Rance for his paper on the geology of the known Arctic regions.

SKETCH OF THE GEOLOGY OF THE KNOWN ARCTIC REGIONS.

Mr. C. E. DE RANCE, F.G.S., of Her Majesty's Geological Survey. When engaged in the spring of last year, with Captain Feilden, R.A., the naturalist of the senior ship of the British Arctic Expedition, in working up what has already been written on Arctic geology, I was often asked the questions—Are there any rocks to be seen in the Polar regions? Is not Greenland and the islands of the Archipelego covered with a perpetual mer de glace—breaking off in bergs and filling the frozen sea?

Investigation proved, however, that so far from this being the case, not only does the immense extent of coast section of the lands of the Arctic regions, though here and there concealed by ice, afford a ready key to the geology of the country, but this key has been already used by naval officers, who, though lacking regular geological training, have brought back observations and specimens so numerous, that in the hands of such palæontologists as the late Mr. Salter and Dr. Haughton, not only has a long sequence of formations been made out, but Dr. Haughton was able to add a geological sketch map of the Arctic Archipelego to the voyage of the "Fox." This map I have taken as a basis for a portion of that which, at the suggestion of the chairman of this section, I have now the honour to briefly describe.

PALÆOZOIC ROCKS.

The crystalline, granitic, and gneissic rocks, described by Sir John Richardson as occupying the central and eastern countries of the Hudson's Bay Territory, were believed by Sir Roderick Murchison to belong to the Laurentian system, and he considered that the whole of the country north of the Laurentian mountains was dry land during the deposition of the Lower Silurian rocks, for no Lower Silurian rocks or fossils had then, or have been now found, in the whole known north Polar regions. And even in Upper Silurian times, the sea in the Arctic regions, as in Europe, was evidently a shallow one, as evidenced by the presence of a profusion of corals of Wenlock and Dudley type, the shell *Penta-merus oblongus*, and a trilobite—*Encrinurus punctatus*. From the rock specimens brought back by Dr. Rae from Melville Peninsula, and from specimens collected by various observers along the entire northern coast of Arctic America, there is no doubt that the Lauren-tian rocks form the floor on which rests unconformably the Upper Silurians of the Archipelego. With the Laurentian rocks of the northern coast of America may probably be classed the granitic, crys-talline, and gneissic rocks, forming the greater portion of the bleak coast-line of Greenland. On the southern coast Professor G. C. Laube collected more than 200 rock specimens, which were examined micro-scopically by Dr. Karl Vrba, who found them to consist of granite, gneiss, eurite, syenite, orthoclase porphyry, diorite, diabase, gabbro, and serpentine. From these crystalline rocks of the west coast the Greenlanders derive the steatite, from which they make their lamps and other utensils.

The curious fluoride of sodium and aluminium—cryolite, which was discovered by Danish missionaries taking weights to Copenhagen composed of it, used in the capelin and cod fishery by the Green-landers, occurs in veins eighty feet in width in the Gneiss of Evigtok, on the west coast, which (with the exception of Miask, in Siberia), is the only known locality for the mineral. The Greenlanders especially value the white variety, which, from soft greasy appearance, they call orksoksiksæt, from *orsok*, blubber, which they pound up with tobacco leaves, and use as snuff. Some 5,000 tons of cryolite per year are imported into Denmark, and used in the manufacture of soda—100 tons of cryolite yielding forty-four of caustic soda. Mixed with silica cryolite produces a beautiful glass, and for a time aluminium wa extracted from it, but its use has been given up in favour of bauxite.

Dr. Sutherland describes the coast of north-west Greenland, Cape York, Wolstenholme Sound, to Cape Hatherton, as trap. At Whale Sound, horizontal beds of sandstone occur, but on the opposite side of Smith's Sound the cliffs are high, rugged, and inaccessible.

Dr. Conybeare, in his Report on Geology to the British Association in 1832, was perhaps the first to notice the similarity of the fossils from the Arctic regions to the English Upper Silurian series; and Dr. Haughton, from his examination of the fossils collected by Sir Leopold

M'Clintock between 1849 and 1859, was definitely able to refer, the limestone of which nearly all the islands of the Arctic Archipelego consist, south of Lancaster and Melville Sounds, including Prince Albert, Prince of Wales, Banks Island, &c., to the Upper Silurian system.

At a late date Sir Roderick Murchison, commenting on the results of the collections, brought back by Parry, Franklin, Ross, Back, Austin, Ommaney, and the private expedition of Lady Franklin, par-- ticularly those of Penny and Inglefield, and by the expedition under Sir E. Belcher, endorsed the opinion of Mr. Salter, that the Upper Silurian' Arctic species, though having many in common with those from Wenlock, Dudley, and Gothland, have on the whole an American facies.

At Dundas Island, in latitude 76⁰ 15, one of Captain Penny's crew found a Silurian trilobite, and from the cliffs of Lady Franklin Bay, where the "Alert" is to winter, Dr. Hayes obtained (?) thirteen species of fossils, which Professor Meek identified as upper silurian species belonging to the fauna of the New York Catskill Shaley limestone of the Lower Helderberg groop, so there is little doubt that the strike of the Carboniferous rocks, with their outlying Liassic patches, is cut off to the east, and the Upper Silurian rocks entirely surround them, an E.N.E. synclinal running through Prince Patrick's Island towards Hayes Sound. Detailed examinations however of these coasts by the present expedition will prove whether another roll, brings in the Carboniferous beds, and if so, in what manner they rest on the Silurians. Should coal-bearing be discovered, they would be of great value to the expedition.*

CARBONIFEROUS ROCKS.—The Lower Carboniferous close grained sandstones, over-lie the Silurian limestone, and strike south-west and north-east from Baring Island or Banks Land, through Melville Island to Bathurst Island, where they disappear under the Mountain Limestone, which strikes about east and west, and probably lies in a synclinal fold in the Lower Sandstone series, which contains the coal seams, which have been described by so many observers, and

---

* True bituminous coal was discovered by the Expedition at Lady Franklin's Inlet, in beds of Miocene age, resting directly on rocks of probably Laurentian times, the Carboniferous and Silurian rocks being absent.—C. E. R.

first discovered by Parry in the islands named after him, and afterwards by Austin and Belcher in Melville Island and Bathurst Island.

The coal burns with a bright flame with much smoke, and resembles some of the gas coals of Scotland. In Byam Martin Island the
coal occurs at a height of 350 feet above the sea, and is rather a
lignite than a coal.

Professor Heer states that not a single species of the Bear Island
Ursa stage flora, collected by Professors Nordenskjöld and Malmgren,
exists in the Upper Devonian *Cypris* shales of Saalfeld, in Thuringia,
species belonging to the true coal measures are equally absent, and
he therefore correlates the Ursa stage with the Kiltorkan beds in
Ireland. The Greywacke of the Vosges and Southern Black Forest,
and the *Spirifera Verneuilii* shales of Aix, and the coal-bearing sandstones of Parry and Melville Islands, the whole of which he considers
to be of Lower Carboniferous age.

The Bear Island flora is characterized by *Calamites radiatus, Lepi-
dodendron Veltheimianum, Stigmaria ficoides, and Knorria acicul-
laris,* all of which occur in the yellow sandstones of Kiltorkan, and
the last named species is also found in the islands of the Arctic Archipelego.

This flora, the first rich one of which we find evidence in the earth's
history, can be traced in the northern hemisphere, both in the old
world and the new, from 47° of north latitude to 74° and 76°, and
affords abundant proofs of a wide spread continent, occupying much
of the Arctic, as well as of the temperate zone, traversed by rivers
tenanted by the freshwater mussel (*Anodonta*) and neuropterous
insects.

From the fact that the Kiltorkan sandstones contain *Asterolepis,*
and other genera fish found in the Old Red Sandstone of Scotland, Sir
Charles Lyell considered these deposits to be of Old Red age, but having regard to the persistence in time of freshwater forms, it has been
suggested that is not remarkable that Old Red genera should live on, in
the later sandstones of Kiltorkan.

The Lower Carboniferous coal measures of the eastern United States
of America lie unconformably on the Devonian, but in Ohio, a transition of the two floras occurs, and Principal Dawson suggests a similar
blending in Bear Island. The subsidence and other changes, which

brought into existence extensive coral reefs, and the formation of the Mountain Limestone of Europe, equally affected the Arctic zone in both the old and new world, equally also is the return of continental conditions expressed by the European Millstone Grit, represented by the siliceous schists of Bear Island, which overlies the *Productus* and *Spirifer* limestones, which rest upon the Ursa stage.

MESOZOIC ROCKS.

In Spitzbergen black bituminous shale, stratified hyperite, limestone coprolite beds and sandstones 1,500 feet thick, have been referred to the trias, but as they contain, or at least the upper portion of the series—saurians, nautili, and ammonites, it is more probable that they may be referred to the Rhœtic or Lias. Amongst the fossils are *Icthyosaurus Polaris*, *I. Nordenskiœldii*, and *Hybodus Spitzbergensis*.

Though liassic fossils were discovered by Lieutenant Anjou of the Russian navy and described by Wrangel, from new Siberia in Asia, in 74 north latitude, their presence in these high latitudes remained unnoticed until Belcher's discovery of a large number of Liassic fossils at the top of Exmouth Island, which consists on the base of soft sandstone, overlaid by Carboniferous Limestone, dipping to the west 7°, containing *Zaphrentis*, *Spirifer Keülhavii*, at the top occurs the Lias, the fossils from which were determined by Mr. Salter and Professor Owen, who described the vertebræ and ribs of an *Icthyosaurus* near to *acutus* of the Whitby Lias. Other fossils were afterwards brought from the same locality by Sherard Osborne, and Sir Leopold M'Clintock, amongst others the ammonite named after the latter by Haughton.

Liassic patches also occur at Table and Princess Islands, at Depot Point in North Devon, and a remarkably fossiliferous patch at Point Wilkie in Prince Patrick Island.

Dr. Toula reporting on the specimens collected by Lieutenant Payer and Dr. Copeland, describes an extensive series of Jurassic shoals and sandstones on the east and south side of Kuhn Island with a fauna nearly allied to the Russian Jurassics. *Aucella concentrica* in five varieties connected by intermediate forms is frequent on the east coast of the island. The genus is frequent in all the Russian deposits of age from the Lower Volga northward to the mouth of the Petschora—and it occurs in the Spitzbergen Jurassics. On the south side of Kuhn

Island the jurassics contain seams of fossil bituminous coals, and may possibly be referred to the carbonaceous Jurassics of Brora, Mull and Skye (Scotland).

These rocks overlie Gneiss, and crystalline rocks forming the massive of the country as well as that of Frantz Joseph Land.

On the west coast of Greenland, in the island of Disco, and the peninsula of Noursoak, occurs the remarkable series of Cretaceous, and Miocene Tertiary strata, the whole of which appear to have been deposited in fresh water, around which grew leafy trees, including nine species of oak, of which two were evergreen, like the Italian oak, two beeches, two planes, a walnut, hazel, sumach, buckthorne, holly, and a Guelder rose, in all numbering, according to Professor Oswald Heer, no less than 167 species; while from the entire Arctic regions he has described and determined no less than 353 species of plants.

The Cretaceous strata of the north coast of Disco were found by Professor Nordenskjöld to consist of two series: an upper, called by him the *Atane strata,* and a lower, called the *Koma strata.* The latter are about 1,000 feet, and lie either horizontally, or dipping slightly towards the Noursoak Peninsula, in hollows and depressions in the old gneissic rocks.

It is in these Lower Cretaceous strata that the most numerous seams of thick coal occur, some of which consist of a mass of felted leaves, forming a flexible substance, resembling the vegetable parchment produced by the action of sulphuric acid on lignite. It is probable that the plants brought back by Giesecke and Rink were derived from these lower strata.

The thick coal of Atane belongs to the Atane, or upper division, which occurs on the southern side of the Noursoak Peninsula—the Ritenberg coal mine at Kudiliset, the retinite beds of Hare Island, all probably belong to this upper series. Dicotylednous leaves occur in it, one being nearly allied to *Magnolia alternans,* Heer, from the Upper Cretaceous of Nebraska.

CAINOZOIC ROCKS.

Resting on these cretaceous strata are the Miocene rocks, first described in 1821 by Sir Charles Giesecke, F.R.S., and since worked out in detail by Nordenskjöld and Dr. Brown. The strike of the strata at the creek of Atanekerdluk is east-north-east, and plant impressions are very numerous at 1,000 to 1,200 feet above the sea. No valuable coals occur in these lower Miocenes, which are separated from the middle Miocene of Ifsorisok and Assakak by several thousand feet of basalt. The flora of these middle Miocenes is however similar to the lower beds. These Miocene sedimentary deposits, with their associated basaltic tuffs, lava flows, and dykes, are probably coterminous across Greenland with the deposits of similar age discovered by Scoresby, on the coast of East Greenland, from which the second German expedition has brought back a large number of specimens.

The Greenland meteoric iron has been known since 1819, when Sir Edward Sabine described certain Esquimaux knives obtained by him, and stated that they procured the iron from a dark rock in 76° 10′ lat. and 64° 75′ long.—called " Sowallick " from *sowil*, iron. Since then, Nordenskjöld has discovered this meteoric iron over an area of 200 square miles in the south-west corner of Disco Island, often occupying cracks in the basaltic tuff, which he considers to be consolidated volcanic ash, and believes the meteorites to have fallen in Miocene times ; the largest block observed probably weighed 21,000 kilogrammes, that now in the British Museum weighing eighty-seven. M. Daubrée, though considering these meteorites to differ from all other known types, considers the presence of nickeliferous iron and schreiberite to prove their meteoric origin, in spite of the combination of the iron with oxygen, and the abundance of carbon and the larger proportion of soluble salts, considering that the preservation of the latter may be due to the feeble tension of the vapour of the northern regions.

These Miocene coals and plant-beds spread over a large area, for many species are in common with those found in Spitzbergen, the Baltic coasts, France, Italy, and Greece, with the potters' clay and lignite beds of the banks of the Mackenzie, described by Sir John Richardson, and no less than four Greenland species, including *Sequoia Couttsiæ* so common in Bovey Tracey, occur in Britain. It is worthy of note that palms, whether of the European or American types (*Chamærops* and *Sabal*), and other exotic forms, present in the upper Miocenes of Eningen, in which occur many North American types, are absent in the Miocenes of the more northern area, proving that though the Arctic regions did not experience then an Arctic climate, yet the temperature was even then cooler than the middle and south of Europe. There is perhaps no problem of greater interest to the palæontologist, amongst the larger number that may be solved by the British Arctic Expedition, than to see how far southern species die out in advancing to the present pole, and what minimum of cold the surviving preserved species appear to indicate.

The former continuity of land with Greenland is borne out by the presence in the miocenes of that country of the Japanese genera, *Glyptostrobus* and *Thujopsis*, which last one, *Europæus*, occurs in Europe in amber, and at Armissan (Narbone); associated with it are American forms which characterise the flora of the Baltic amber pine district, according to Professor Göppert.

Another link of evidence occurs in the presence of the so-called wood hills discovered by Sirowatskoi, on the south side of the island of New Siberia, stated by Wrangel to consist of horizontal sandstones, with vertical bituminized trunks of trees.

Of the 169 miocene species in West Greenland, sixty-nine occur in Europe and forty-two in Switzerland, of these thirty-five occur in the lower and twenty-four in the upper molasse; on the whole they accord, like those of Spitzbergen, with the Swiss Lower Miocene.

In East Greenland, MM. Payes and Copeland found remains of several plants at Sabine Island, including *Populus Arctica*, all of which occur on the west coast; and Professor Heer considers it probable that the whole of the thirty-five species of plants common to Spitzbergen and West Greenland existed throughout the whole of the intermediate area.

From Spitzbergen Heer records no less than 179 species, of which forty occur in the European Lower Miocene. Both in Greenland and Spitzbergen, the swamp cypress, poplar. hazel, and plane, common to both, are such species as *Quercus Greenlandica, sali varians, Fagus deucationes*. Professor Heer states that no less than fifty-four per cent. of the brown coal flora of the Baltic occurs in the Arctic regions.

The *Lower Carboniferous flora* agreeing as it does with that of southern latitudes and in its profuse vegetation, testifies to the existence in the Arctic regions of a far greater degree of light than that area now obtains, while the saurians and ammonites of the Jurassic period testify to the former existence of far greater heat, and the presence of a comparatively warm ocean around Spitzbergen and Exmouth Island.

In the Arctic *Lower Cretaceous*, also, the character of the flora, according to Professor Heer, point to climatal conditions similar to that now obtaining in Egypt and the Canary Isles.

In the *Upper Cretaceous* period, though the climate was cooler, deciduous trees, dicotyledons, flourished in Noursoak, as well as a fig, and two species of *Magnolia*.

The *Miocene* flora also indicates a temperature in North Greenland at least 30° Fahrenheit warmer than at present, and planes and beeches occur in Spitzbergen 8° further north, so that assuming these to have reached their northern limit, there is nothing to have prevented poplars and firs to have grown as far north as land might extend, for they reach 15° north of the artificial limit of the plane.

GLACIAL PHENOMENA.

The entire western coast of Greenland is surrounded by a circlet of bleak bare islets, often 2,000 feet in height, separated from each other by deep glens or fjords, through which pass the overflows of the great *mer de glace*, which covers the country to an unknown depth, and which Professor Nordenskjöld found thirty miles inland to rise to a height of 2,000 feet above the sea; the surface of this ice-field rises so gradually, that effect of travelling upon it is compared to that of sailing out to sea when the land gradually fades away. In summer the snow covering of this great ice desert melts away, and rivers of icey-cold water flow over the surface, and lost in crevasses which appear to increase in width and number in advancing inland. The overflow ice, the result of eight months additional snow, is carried off by larger affluent glaciers, some of which, like the Humboldt glacier in Smith's Sound, are sixty miles across. Where no fjords occur, the ice bodily pours over the cliffs, and masses hang in mid air, until gravity overcomes cohesiveness.

The rain and snowfall of Greenland is estimated by Dr. Rink at twelve inches per year, while the amount carried off in the form of glacier ice is not more than two inches, so that evaporation being unimportant, a very large amount must flow away in the sub-glacial rivers beneath the *mer de glace*. The result of the trituration and gigantic sawing of the rocks, by the slow movement of this enormous mass of ice, must have covered the surface of the rock with a thick *moraine profonde*, which will well account for the muddy water issuing from beneath the ice at the entrance of all the ice-fjörd, and it is easy to understand that such materials being sifted by the sea, and distributed by currents, would produce that curious and recurrent series of clays, gravels, sands, and finely laminated silts, associated with glaciated pebbles and sea shells, which characterise the glacial drift of the north of England.

Both in the islands of the Arctic Archipelego, and in the whole coast line of Western Greenland, and in Smith's Sound, there are distinct traces of marine terraces at various levels above the present high-water mark, in the hollows of which occur clays containing such marine groups as various echinodermata, crustacea, and mollusca, which are found at all heights up to 500 feet, and in " Polaris Bay " up to 1,800 feet above the sea; were another elevation to take place, similar deposits, containing identical organisms, resting on glaciated rock surfaces, would be raised above the sea level. How closely this sequence resembles that found in the marine drift deposits of the midland and north-western counties of England, where shell-bearing sands and clays occur at successive horizons, and rest on glaciated rock surfaces, believed to be the result of the action of land ice.

Dr. Sutherland states that the sea-bottom of Davis Straits, from Cape Farewell to Smith's Sound, swarms with brittle starfish and other echinodermata. In Melville Bay immense crops of laminaria are now and again mown away by the grazing of iceberg, and it is obvious that fragments of the shells of *Mya* and *Saxicava*, entangled in their leafy fronds, would be carried with the seaweed long distances by currents, which may account for the occasional occurrence of marine shells, in isolated areas, in the English Boulder Clay. Again, the stranding of bergs, earth and boulder laden, causes submarine banks to be thrown up which constantly increase in size, through the stranding of small bergs upon them, and form the favourite haunt of shoals of halibut, cod, and myriads of sharks. A section through one of these banks would disclose in the centre a tumultuous mass of clay, with included stones and boulders, some weighing as much as 100 tons, wrapped round at the sides and overlaid by unconformable deposits of ordinary marine sediments, which is precisely the section observable in the cliffs of Blackpool in the north-west of England. Stratified sands and gravels rest unconformably on Lower Boulder Clay, if this theory of the formation of the latter be correct, and the light thrown upon it by the observations of the Arctic explorers gives much support to it. It is easy to understand the occasional stratification observable in these clays, and the occurrence now and again of marine shells in the glacial drifts.

The sea-water of Davis Straits has a specific gravity of 1·0263, and freezes at 28½ Fahrenheit, when the salt amounting to 5¾ ounces per gallon is precipitated, and the "bay-ice" of the whalers formed. This eventually becomes a flock, and, still later, pack ice. In summer the ice at the sides melts quicker than that at the centre, which forms the "middle ice" of the whalers. The ice between the land and the open water hangs to the cliffs, and receives the various debris, landslips, and boulders falling from them. As the summer advances, this belt of ice, "iis fod" of the Greenlend Danes, becomes detached, and floats off with its stoney freight. Comparing these facts with the phenomena observable in the glacial drifts of the north-west and midland counties of England, it is in the highest degree probable that floating portions of the ice-foot which surrounded the mountains of the Lake District, carried southward by currents, were the transporting agents by which the erratic fragments derived from the volcanic rocks of the Lake District were carried over the plains of Lancashire, Cheshire, Stafford-shire, and Warwickshire, and that their stoney freights were deposited in clay, the result of various denudations of the older secondary rocks. The water-worn appearance of many of these erratic fragments, which is evidently of later date than the glacial striæ observable on nearly all these stones, may well have come into existence in tidal movements of the ice-foot, or by the beating of the waves against it, which must have caused continually grinding of pebbles and rocks which had sunk to the bottom of the ice-foot and become entangled in the ice; and it is easy to understand the origin of the nests, or pockets of sand, so often found in the English Boulder Clay.

In 1828 a Danish expedition, under Graah, examined the west coast of Greenland in search of the lost Icelandic colonies. Dr. Pingel, the geologist attached to it, discovered in latitude 61° north a series of red sandstones at Igalliko and the fjord of Tunnudleorbik, consisting of hard compact rock, with fused quartz particles. No fossils have been discovered in it, but it is believed to be of Devonian age; the rock was re-discovered by Laube, of the second German Polar Expedition.

Dr. Pingel, in 1835, described various phenomena proving the western coast of Greenland is being gradually submerged; amongst others, that the Moravian village of Lichtenfield, founded in 1758, had to be moved forty years later, and the poles to which omiaks (women's boats) were tied still remain uncovered at every low tide.

The PRESIDENT : I must now ask you to return your thanks to Mr. De Rance for his interesting communication. It opens up a considerable number of questions with regard to the causes which may have led to the existence of such a climate as permitted the growth of these plants, some questions would also arise with regard to the ammonites which have been found there of such large size and in such remarkable profusion. We shall be glad to hear any remark on either of these subjects or on any other raised by the paper.

Mr. R. H. SCOTT : I have listened with great pleasure to Mr. De Rance's paper, and I am very glad indeed to find anyone who has taken the trouble to put together the scattered notices about Arctic geology which are in existence. For the last five years Dr. Rae and I have had money in our hands and the order to get fossils from the Mackenzie River to see what these lignite beds are, but no one will send a bit of stone here. We hope, whenever we do get them, we shall be able to trace the connection across the whole way from Spitzbergen through Greenland and up the Mackenzie River and out to Alaska, where we have already found these Miocene fossils.

The PRESIDENT : If no one else wishes to say anything I would ask you to return your thanks to Mr. De Rance, and I will now call upon Mr. Galloway to read his paper on explosions in coal mines.

COLLIERY EXPLOSIONS.

Mr. W. GALLOWAY : The only noxious gases found in appreciable quantities in mines are fire-damp and choke-damp : the former exists in the seams of coal in a state of tension, while the latter appears to be mostly produced in the workings or in the open fissures into which air can penetrate.

Fire-damp is an inflammable gas, that is to say, when a certain proportion of it is mixed with a given volume of air the mixture can be ignited by the application of a flame, and continues to burn until the whole of its combustible elements have been consumed. This process of combustion is accompanied by flame, and by a sudden expansion of volume, and when it occurs in the workings of a mine it constitutes what is called a colliery explosion.

As the workings of a mine progress through a coal seam, they intersect innumerable small fissures, through which the hitherto pent-up fire-damp escapes into the space from which coal has been removed and mixes with the air contained in it ; and if this air were not continually being renewed by the process called ventilation it would soon contain the requisite quantity of fire-damp to render it explosive. It is therefore the obvious duty of those who are responsible for the safety of mines to see that this renewal of the air is taking place with regularity, so that the workmen may never be exposed to the danger of working in or near an accumulation of explosive gas.

It may be said, in a general way, that every mine might be made absolutely safe from liability to the occurence of explosions, if the quantity of fresh air supplied to the workings were sufficiently great and sufficiently well distributed. There is, however, in many cases considerable difficulty in the way of attaining such a desirable consummation, consisting chiefly in the expense which would be incurred in making and maintaining large enough air passages between the working places and the surface. A kind of compromise is therefore usually adopted: the airways are made large enough for present necessities, that is with a larger or smaller margin on the side of safety, and the workmen are supplied with safety lamps, or not, according to the smaller or greater value of this margin. The various causes which tend to disturb this normal state of affairs, which may be defined as the condition under which work can be carried on without incurring the risk of explosion, may be described as follows :—

1. Internal derangements: Falls of roof which block the air-passages or destroy partitions which conduct the air ; the gradual collapse of airways ; pushing forward particular parts of the workings beyond the point at which they can be efficiently ventilated ; neglect of, or accidents to, the ventilating apparatus. All of these have the effect of

reducing the quantity of fresh air supplied to particular localities or to the whole workings, and as a consequence they favour the formation of explosive accumulations where the ventilation is least effective or the outflow of fire-damp most plentiful. They are all, however, under the direct supervision and control of the management, and are to be provided against by the establishment of a proper system of discipline.

2. Changes of weather : A sudden decrease of the pressure of the atmosphere indicated by a fall of the barometer ; an abnormally high or low temperature at the surface. A fall of the barometer has the effect of temporarily increasing the rate at which the fire-damp is given off by the coal, just as the opposite phenomenon retards it for a time. The aggregate quantity of gas emitted is the same, if we consider a lengthened period of time, but the rate at which it is emitted may be temporarily accelerated or retarded by variations of atmospheric pressure. A high temperature of the air at the surface reduces the quantity passing through the workings, and its effects are most apparent where the ventilating appliances are unimportant. These changes of weather, although perhaps insufficient of themselves alone to bring about a critical state in tolerably well ventilated mines, are nevertheless certain in their action, and instantly produce accumulations of explosive gas, or increase the volume of previously existing ones, at what may be called the sensitive points of the workings, that is to say, in localities which are already in a critical state owing to one or other of the causes mentioned under the first head.

It had been observed many years ago that explosions occurred most frequently when the barometer fell, and several papers have been recently written to show how these changes of weather act, and to correlate them with the actual occurrence of explosions by means of diagrams. Mr. Scott, of the Meteorological Office, has contributed several in conjunction with myself.

An abnormally low temperature at the surface has the effect of drying the coal-dust in those mines in which it exists ; but it has not yet been positively ascertained whether a preliminary process of desiccation is necessary in the case of explosions in which coal-dust acts a part.

3. The sudden influx of a large quantity of fire-damp causing the air to become explosive in the immediate neighbourhood, and sometimes throughout a considerable extent of the workings. It is hardly necessary

to refer to the precautions which are generally adopted, or at least should be, in cases in which occurrences of this kind are likely to take place : it is sufficient to say that they resemble those which are observed when water is expected to be encountered.

4. Shot-firing : This term is applied to the operation of blasting, by means of which hard rocks and coal are broken down with infinitely more ease than they could by any other known means. Blasting may be carried on with perfect safety when the air is free from fire-damp and coal dust. In the event of an explosive accumulation existing in the neighbourhood of the intended shot, it would be highly dangerous to have one of the safety lamps placed in it for any purpose whatever, since the flame would in all probability be driven through the meshes of the wire-gauze at the instant it was traversed by the sound wave ; and thus an explosion would be initiated.

5. Coal-dust : It has been recently shown that if a cloud of coal-dust be raised in air containing a small proportion of fire-damp, an explosive mixture is formed even when the quantity of fire-damp is less than one seventh of what would be required were it mixed with air alone.

It so happens, however, that in the working places and airways of many mines in which there is plenty of dry coal-dust on the floor, the air contains a larger proportion of fire-damp than that mentioned above. In order then to fill these workings with an explosive mixture it is only necessary that the coal-dust be raised up by the puff of a slight explosion of fire-damp or of a heavily charged shot ; and when this has been effected and ignition communicated to the new mixture by the flame of the first one, the explosion may be said to become self-sustaining, and may then be conceived to be in a position to propagate itself as far as the same conditions continue.

In concluding this short and necessarily imperfect account of the various causes which tend to produce explosions, I will take the present opportunity of entering a protest against the undiscriminating demands which are reiterated from time to time to have the use of gunpowder entirely prohibited in mines in which safety lamps are used. Such a prohibition would deprive these mines of the advantages of a powerful auxiliary : and, although it would no doubt prevent the occurrence of some explosions *if everything else were allowed to remain as at present,* it would be utterly valueless in many other cases.

Safety lamps should be quite unnecessary in any mine except as a means of detecting the presence of small accumulations of explosive gas and guarding against their accidental ignition should they be formed through the operation of any of the causes already specified. It is now well known that they have no other legitimate use, and yet, at the present day, we find them employed in very many mines in which an increase of ventilation would render their use superfluous.

Finally, then, I would say, that the majority of great explosions are probably due, not so much to the use of gunpowder or to recklessness on the part of the workmen, as to the pernicious system of stinting the supply of air, and the partial or entire neglect of a few obvious and necessary precautions.

The PRESIDENT: I am sure you will all return your thanks to Mr. Galloway for this interesting communication. I do not know whether any one wishes to make any remarks, but I should like to call attention to the concluding observations of Mr. Galloway, in which he points out how desirable it is that instead of taking precautions to prevent the explosion of fire-damp, far simpler means should be adopted to ventilate the mines thoroughly, so as to render explosions impossible. It is just one of those simple principles which it is so often desirable to adopt where danger of different kinds exists. It is the same thing in cases of health where, however valuable a medicine may be, it is far better to obviate the causes of disease than to introduce a medicine. In the same manner these precautions, which may be adopted as suggested by Mr. Galloway, appear to me to be worthy of the very highest consideration. The various causes which he has pointed out which give rise to explosions are to a great extent already known; but I think that to which he has called attention, the danger of mixing of the inflammable dust with air containing only a small proportion of inflammable gas, is one which, if fully appreciated, may lead to the number of explosions being greatly reduced. At all events, we must all feel that this paper is of a very practical character, and that it is deserving of great consideration.

(The Conference then adjourned for luncheon.)

The PRESIDENT: The first paper this afternoon is by Mr. W. Stephen Mitchell,

On the Manuscript Table and Geological Maps of William Smith.

Mr. MITCHELL: During the Conferences, there have been several references made to instruments which no longer have any practical use, but which still are respected for their high historical value. Perhaps one of the most interesting features of these Conferences and this Exhibition has been the way in which memorials of the dawn of particular sciences are brought in contrast with their latest developement. You have this morning seen on these screens some of the splendid maps and sections exhibited as the latest triumphs of geological mapping by the Geological Survey of Great Britain. By way of contrast the Geological Society of London has sent you the very first geological map which ever was made—a map of the neighbourhood of Bath ; the first geologically coloured map of England, and the very first known list of the strata of England ever made.* They have also sent what are termed "index sheets," that is to say, sheets shaded of different tints, showing the progress of the way in which the division of strata have been recognised. There is not much fear that the splendid work of the geological survey will be overlooked, but I think there is some fear that these small frames may not have that attention paid to them they deserve, unless their importance is specially pointed out. What I propose to do, therefore, in the few minutes allotted to me, is to tell you something of the history of these three manuscript maps and list of strata, and to very briefly tell you the way in which discovery of the sequence of strata was made.

Rather more than 100 years ago there was born at the little village of Churchill, in Oxfordshire, William Smith. The only thing I need mention about his early days is that he received his education at a dame's school, running away as much as he could, and spending his time in the country ; his principal amusement being collecting the "pundibs" and "quoit stones" or fossils of the neighbourhood. His father died when he was eight years old, and he was transferred to an uncle who was a farmer, and who was very anxious that he should take to his

---

* In saying this, I have not forgotten Michell's account of the strata of a portion of England.

property, and be a farmer also.   This uncle lived at Chipping Norton, and it is worthy of notice that at Chipping Norton there is a junction of the lias and the oolite.   The valleys there are deep, and there are in consequence, in a very short distance, very marked alternations of strata.   This young lad, with his quick eye following the plough, could not fail to notice these differences in strata; and his wish was not to be a farmer, but to be a surveyor, and it so happened that his wish was in an unexpected way fulfilled.   The "common land" of the village, to which he belonged, was being surveyed for the purpose of enclosure, and the surveyor who was sent to do it happened to take a fancy to him, and removed him to his own place as an assistant, where he speedily became a very experienced surveyor, and in the course of a few years was sent to different parts of the neighbouring counties, and even to Hampshire and Somersetshire and more distant parts. Wherever he went he made a record of the kind of soils that he saw. What his object was at the time we do not know, but it was afterwards of great practical value to him.   The first survey he undertook on his own account was in Somersetshire, not very far from Bath, at the spot where you see a coal-field marked on this map. [High Littleton].   He was so much respected that the coal owners wanted him to make an underground survey.   At that time there were no steam engines to work the coal pits, and such a thing as a regular survey of a pit was not at all thought of.   His first underground journey very much astonished him, for he was not at all prepared to see the regular alternation of coal beds and shale, pennant, and so on, which he met with.   The colliers told him, further, something which very much astonished him, namely, that all the coal pits in the district had the beds in the same sequence, and when they were sinking a shaft they knew beforehand exactly what they were coming to. He thought over this, and it occurred to him if there is this regular order underneath the ground, why is there not the same order above ground?   The colliers told him that was absurd, there was no such thing; you had red earth in one place, lias in another, and oolite in another; but there was no sequence in those different materials as there was in the coal pits.   However, he was a very independent young man, and he set to work to think for himself; and from this point it is interesting to see how, whenever he wanted any explanation, circumstances seemed to put into his hands just what he wanted. At this time the coal owners of Somersetshire wanted to have a canal made to convey their coal away, and it was decided to run one up to Bath, to join another which was to be made to London, and Smith was appointed to take some of the measurements. The whole valley was well worked, the lias was quarried, the red earth was quarried, the oolites were quarried; and so he became well acquainted with what there was in the neighbouring fields. Having an idea in his mind that possibly there was a sequence in the beds above the coal, it happened that, having to make a survey for this canal, the facts he wanted came before him in a way in which he could hardly help recognising them. The district is a hilly one; there (referring to map and section) is a hill of red earth, and following on that you come to the lias, and you go still further on to the east, and you come upon the inferior oolite, then upon fuller's earth, and then upon the great oolite. The canal line went all through these different beds, and he recognised at once that there was a sequence in the beds, and he also learned that they had a general dip to the south-east, because he found they dipped down below the artificial line he had to run for his canal. Having satisfied his mind on that point, the next thing which occurred to him was whether that would hold good for the rest of the country; and, fortunately, the very thing he wanted was offered him. He was sent with two members of the committee connected with the canal to make a journey as far north as Newcastle, to see how canals were made, and they went up the eastern counties and came down through the middle of the country. He there found that the beds followed one another in the same sequence in which he found them in the district he had examined and extended his knowledge to other beds. He then thought it time to put some of his information into writing, or map it in some way, and this map of the district of Bath shows what he was able to do as early as 1795, whilst this map of England shows the information he had gained at the same time. I wish to draw special attention to this, that though this map of England is dated really in 1801, yet the northern part therefore is based entirely on the information he gained in that journey through the country, for we know he did not for many years go north again. At that time he had arrived at the notion of the sequence of beds, but had not

arrived at the theory with which his name is so thoroughly connected
—that of the identification of strata by the fossil remains; because
it was not until after the survey for the canal was completed,
until the making of it was commenced, that he found the necessity
of endeavouring to identify beds by looking at their organic remains; consequently, this map, although generally assigned to
another date, I feel certain really represents the information he gained
through thus travelling through the country, without at the time having
any knowledge whatever of the identification of strata by their fossils.
This M.S. table (holding up the framed sheet), although drawn up from
Smith's dictation as late as 1799 by the Rev. Benjamin Richardson,*
at the house† of the Rev. Joseph Townsend, the author of that curious
book, "The Vindication of Moses"—records the information really
gained in 1794. It would take up too much time to speak of the
theory which he originated after he commenced cutting the canal—
that of the identification of strata by their fossils; but he states that
he was obliged, owing to the similarity of some of the beds, the lias
and limestones, and fuller's earth, to make a careful examination, and
that he was guided more by the fossils than by the mineral characters
of the rocks.

Without going away from the immediate subject, the only other
point to which I wish to draw attention is this, that these maps and
this table of strata take precedence by many years over the work of
Cuvier, and over the work of Werner, and, consequently, the very first
recognition of sequence of strata is due to an Englishman.

The PRESIDENT: I am sure you will return your thanks to Mr.
Mitchell for this communication. It is very interesting to have this
tribute of respect paid to William Smith, who was the father of stratographical geology throughout the world. These records which have
been sent by the Geographical Society are of very great historical
interest; but there is one other map in the gallery (referring to large
map of England, 1815) which I think will also show the great sagacity
and acumen of Smith even more than these show his general grasp
of mind, and the manner in which he was able to conceive his great
idea as to the stratographical succession.

* See Proceedings.
† 29, Pulteney Street, Bath.

As an illustration of his powers of observation, you will see a survey undertaken by him, in which the various strata are coloured; and although the map which acted as the basis of his survey was deficient in geographical accuracy, yet when you compare it with the results of the more recent geographical survey which are exhibited below, one cannot help being struck with the remarkable accuracy of details with which the map of William Smith is furnished. It is a curious instance of how a man making use of the eyes with which every one is endowed, but which so few know how to use, was able to place himself in advance of those who lived at the same time with him.

I do not think there is anything to call for discussion on this paper of Mr. Mitchell's, but if anyone wishes to add anything to it we shall be glad to hear it. If not, I will now call on Professor Baron von Ettingshausen for his paper. You will have noticed the magnificent collection of fossils which is in the gallery beyond, exhibited by Baron von Ettingshausen, and you will, I have no doubt, be glad to hear what he has to say on the subject. Unfortunately, he prefers addressing us in German; but his paper will be followed by an abstract giving the general results in a language with which you are more familiar.

Professor Baron VON ETTINGSHAUSEN then read in German a paper on the Tertiary origin of the existing floras. The Tertiary flora contains representatives of all the recent floras of the globe. This theory, said the Baron, I have put forth for twenty-five years in my publications on the Austrian Tertiary flora. During this time this theory has become more and more self-evident. The Loan Exhibition has afforded an opportunity of arranging specimens together, and bringing an illustration of how the elements of the existing floras come from the Tertiary. In the European Tertiary floras the ancestral types of the present indigenous flora are met with without doubt. This is seen in the genera Pinus, Alnus, Quercus, Fagus, Ulmus, Acer, &c, The transformation from some of the Tertiary species to the species of existing flora can be traced step by step through various formations. Coeval with the indigenous flora one finds, sometimes on the same slabs, representatives of the floras of America ; for instance, in the species of Sequoia, Taxodium, Myrica, Liquidambar, Carya, Tetrapteris, Robinia, &c. ; representatives of the Asiatic flora, such as Glyptostrobus, Planera, Cinnamomum, Engelhardtia, Ailanthus, &c.; of the African flora in the genera Widdringtonia, Callitris, Pterocelastrus, Rhus, &c. When these facts are considered with regard to the Australian flora, it seems probable that the ancestors of this flora are also contained in that of the Tertiary period. One cannot doubt this, since there are found the genera Casuarina, Leptomeria, Hakea, Dryandra, Banksia, Eucalyptus, &c. Now comes the question, how it happens that the floras of other quarters of the globe come to be found mixed together in Europe in Tertiary times. There are here but two courses open. Either their assemblage is the result of accident or they were indigenous. The first supposition is impossible, since no fallen leaves could have travelled so far, for even the most delicate and tender parts are well preserved. It is more probable that they were preserved on the spot where they grew. The European and exotic forms are all also in equally good preservation. which would not be the case had they come from different distances. From this we conclude that they grew where they are found, and formed a really indigenous European flora. It cannot be accepted that they have travelled from different parts, but rather that in the Tertiary time Europe and the other Continents were united,—for many of the plants could not have travelled by sea,—and that they have all sprung indigenously from the same spot. From these facts we deduce the following important conclusions as to the derivation of existing floras from the Tertiary flora. The elements or groups of types of ancient floras have been developed and spread over different parts of the globe in different ways ; generally in each direction one group has been developed more than the others, and the remainder appear scarcely to have survived. The elements which were best able to survive changing conditions have become characteristic floras, as in Australia, the Cape, &c. Since all the elements of the Tertiary flora had an equal development, we find that each recent flora has preserved the mixed character of its origin, as in the flora of Sumatra and most tropical countries.

The PRESIDENT : I am sure you will all have been glad to have heard this communication, of which, perhaps, it would be desirable to give as short an abstract as possible. He has alluded to that collection which he has shown in the neighbouring gallery, and mentions that you find in that collection the elements of the greater part of the flora of the present day. He shows that at that time there appears to have been no distinct separation between the European, the Asiatic, and African forms, but that all are to be found in the Tertiary remains which at present exist in Europe. He has cited certain trees of which not only the leaves but the fruits and flowers have been found—the *pinus*, for instance, of which he has specified two species. He then mentioned the discovery of the leaves of the oak, and also the chesnut, the older leaves of which approach most closely in character to those of the oak; so that it is almost doubtful whether in finding a leaf you have a leaf of the oak or the chesnut; but that it is a leaf of the chesnut rather than of the oak is proved by the presence of the flowers which he finds associated with the leaves. Those early leaves are not serrated to any extent, but as you come later down in time you find that the leaves in the higher deposits become more and more serrated, until at last in the pliocene time, you can hardly distinguish the fossil chesnut leaves from those existing in the present day. He then mentioned that in addition to these European forms, there are a certain number of forms which are now considered more characteristic of other regions; and with regard to America, he mentioned the twigs and fruits of the *sequoia*, some species of which are well known, such as the Wellingtonia, and the Taxodium. With respect to the Asiatic plants, there are the *Glyptostrobus* and the *Cinnamon*. With regard to the Australian, there are the Casuarnia and Dryandra. Then he showed that although all of these were present in the Tertiary plants in some special cases they appeared to be specialised, so that certain species become the dominating species in different regions. That is a general abstract of the paper with which he has favored us, and for a conclusion he has given me this, which I will now read:—From these facts we deduce the following important conclusions as to the derivation of existing flora: the elements of the creative type of all the highest floras have been developed and spread over different parts of the globe in different ways, generally in each direction, one group has been developed more than others, and the remainder appear scarcely to have survived. The elements which were best able to survive the change of conditions have become characteristic floras, as in Australia, the Cape, &c. Since all the elements of the Tertiary flora had an equal development, we find that each recent flora has preserved marks of its origin as in the flora of Sumatra and most tropical countries.

Having given you that brief abstract, I beg to return thanks to him for his very interesting communication. If no one has any remarks to make upon it, I will next call on Mr. J. S. Gardner.

ON THE EOCENE FLORAS OF THE HAMPSHIRE BASIN.

Mr. J. S. GARDNER: The deposits to which the notes I am about to read refer, are of fresh water origin and lower Bagshot age—that is, they occur between the marine eocene formation of the Bracklesham sands and the London clay.

The *district* embraced is situated on the southern borders of Hampshire and Dorset, extending from the Purbeck Hills to the Isle of Wight.

The *beds* consist of sands and clays. In the lower part they are of pure white pipe-clay, extensively worked for commercial purposes. In the upper part they consist of brick earths, sands, and grit beds. The leaves in the lower part are almost colourless, in the upper they preserve the brown or black tints of decaying vegetation.

These beds are extremely thick towards the west; nearly continuous beds of pipe-clay at Creech Barrow under the Purbeck Hills being upwards of 100 feet thick. They thin out rapidly to the east. At Alum Bay they are but fifteen feet thick, and at Whitecliffe Bay, in the east corner of the Isle of Wight, they are very *greatly* thinned out, sands alone being present with traces of leaves which, *there only*, bear every appearance of having been brought from a long distance.

The beds of clay and coarse quartzose sand are from a granitic source, and are the result of disintegration of high land to the west.

The horizon of both the Alum Bay and Bournemouth series is perfectly defined by the marine beds above and below it. The *age* of the beds is therefore not argued from the leaves as has been the case in many other instances. The collections made from these beds are thus of great importance as being of well ascertained geological age. They are further important as comprising the only *extensive* series of leaves of definite middle eocene age, except perhaps that from Monte-Bolca, the flora of which has not yet been described. The relative age of many deposits on the Continent, from which leaves have been collected and described, is not determined with certainty; but with the possible exception mentioned, and that of the deposit at Sezanne, few are older than miocene or upper eocene age.

The suites of specimens are sufficiently extensive to be taken as fairly representative of the flora of the period.

The only general attempt to describe the British eocene floras in print is a memoir by De la Harpe, published in the Bulletin de la Societé Vaudoise, 1856, and in the Mem. Geol. Survey, 1862. At the latter date only 300 specimens could be brought together from all sources, from these fifty species were determined. The Alum Bay and Bournemouth beds were, in this memoir, erroneously stated to be on *exactly the* same horizon.

It may in passing, be remarked that there seems reason for believe-ing that our eocene beds were deposited under different conditions from those which are thought to have prevailed during the deposition of beds on the Continent.

It is imagined that these conditions were, on the continent, archi-pelagic and similar to those now obtaining in the Pacific Isles; while the conditions we find prevailing in *these British eocene* beds, point to the conclusion that they were formed in a wide valley, occupied by the shifting bed of a torrent, somewhat analogous, for instance, to the Rhone Valley immediately above the Lake of Geneva.

The only evidence supposed to indicate the proximity of the sea, lies in pieces of drift wood having been found with teredo borings. Since the discovery of species of teredo living in fresh water in the Zambesi by Dr. Kirk; a circumstance mentioned with other instances by Gwyn Jeffreys, this evidence has become negative. No other organic remains, whatever, have been found, except a few unios, helices, elytra and insect wings, scale-insects and a solitary feather.

Each small patch of clay contains a flora almost peculiar to itself; that of Alum Bay, for instance, contains Aralia and Banksia, scarcely met with elsewhere; Castanea, Eucalyptus, fan and feather palm, each characterise localized deposits at Bournemouth; even the ferns, Asplenium, Pteris, Osmunda, Polypodium, &c., have particular cen-tres, outside which they are seldom ever met with. This extreme localisation of forms on the same horizon, indicates that the leaves have travelled no great distance, or they would have become mixed.

No attempt has yet been made to compare, systematically, the eocene and miocene leaf forms with regard to development; no work of reference exists for field workers. In fact the general opinion of *British* geologists, at all events, seems to be that the cretaceous, eocene and miocene floras present no distinct facies from each other. It is exactly the smaller differences, which *do* exist, that we have to appreciate, and this can only de done by minute study and comparison of the fossil plants from each of these ages. The less apparent the gradations of development at first seem to be, the more careful should our efforts be to become acquainted with them. In old volcanic districts, isolated patches of leaves occur, which would become of great value in helping determination of age.

As Baron Von Ettingshausen has kindly offered to undertake the determination of the British eocene floras, it would be out of place to make any statements as to their character, which would necessarily be founded on imperfect data, but it appears that although, at first sight the aspect of this flora resembles that of the miocene Continental beds, scarcely any of the species are identical. Another broad fact is that at Bournemouth, the immense majority of leaves are simple, linear, lanceolate, oval or ovate forms. It is probable that the average of simple forms of leaves is as thousands to one, as in all my work at Bournemouth, I have carefully preserved every fragment of lobed leaves, and they are all exhibited in the case; while on the other hand, for every *simple* leaf retained, thousands must have been rejected as they are often matted together in considerable thicknesses; specimens illustrating this may be seen in the cases in the adjoining room. A species of aralia, very common in individuals, found in the slightly older beds of Alum Bay—give that portion of the flora a somewhat different character. Whether this immense preponderance of simple form of leaves in these beds (every one of the numerous species of fig and oak being simple), has any significance, remains to be proved.

A lignitic bed, rich in seeds and fruits, such as Nipadites, occurs at Bournemouth, apparently on the same horizon; but exact measurements there are very complicated and difficult.

An important matter, to be taken into consideration in this study, is the differing condition which permitted distinct groups of plants to have lived almost side by side. A striking instance of this occurs at Bournemouth, where, within two miles, beds presumably on the *same horizon* present sub-tropical and temperate aspects. In the one we find abundantly large palms, Eucalyptus, Sequoia, a giant aroid, Osmunda, &c. In the other, all these plants are absolutely wanting, while small oak and willow leaves are the chief forms met with. The floras of these beds are so different, owing principally, we may suppose, to variety of soil and degree of moisture, that we should no doubt, were they from separate localities, have formerly described them as of a widely different age.

In conclusion, I would point out that merely naming these leaves, is not the only matter of consequence to us ; the greater number of forms *not being* associated with either flowers or seeds, may be, with equal reason, assigned to any of a number of genera ; what we require however, to know, is how many and what forms of leaves are present or absent relatively in cretaceous, eocene, miocene, and recent floras. To this end the first step is to carefully and accurately compare the leaf forms found in different parts of the world, with each other.

The PRESIDENT : I am sure you will all return thanks to Mr. Gardner for this most interesting communication, and we shall now be glad to hear anyone who has any remarks to offer. I hope there are some gentlemen present who have made botany their study.

Professor DUNCAN, F.R.S., President of the Geological Society : I could not remain silent after the very interesting papers we have heard. We first of all must congratulate ourselves on having had the presence of Baron von Ettingshausen, and seeing his magnificent collection, and now finally on listening to his interesting generalisations. These generalisations have been in the minds of English geologists for many years. They owe their origin to the researches of Dr. Hooker and Mr. Bentham years ago, when they illustrated the Australian, New Zealand, and Indian floras. Professor von Ettingshausen advances upon them from the other direction, for whilst these English philosophers worked from the present to the past, Baron von Ettingshausen has rather worked from the past to the present, and practically the two ends have met for the benefit of science. But it must be explained that after all these are but generalisations, and that when we use the term "Tertiary," and say that the present floras were existing generally during the tertiaries, we must understand that the word "Tertiary" is hardly applicable all over the world. The division of time into geo-. logical periods is a geological notion, an idea, an abstract notion altogether. For instance, the present flora really contains some relics by descent of far older floras than those of the Tertiary ; it contains relics of the carboniferous, of the oolitic, and of the cretaceous ; so that during all time there has been a gradual descent of species to the present. Again, if I have made this understood, you will see that in the southern hemisphere things were excessively quiet, and no great geological changes were progressing, when they were going on to a very considerable extent in the northern hemisphere. There were separate and many important movements in the crust of the globe in the northern hemisphere, while things remained quiet enough in the southern. Hence it is, when we talk of the eocene, pliocene, miocene, and glacial periods in the northern hemisphere, that we cannot do so satisfactorily of the south ; and we find ourselves constantly saying that we may be still in the mid-Tertiary period, so far as Australia and some other districts are concerned. All this must be taken into consideration. It is exceedingly difficult to understand, when we consider how the Australian floras are dispersed in Australia, how it can be unless we can admit that extraordinary changes in the physical geography of the southern hemisphere took place during the older periods of the world's history. We can infer those changes from the results of botany. There is a remarkable isolation, so to speak, of the flora of South Western Australia. It is isolated from that of south-eastern Australia. But it is most remarkable in its African affinities, necessitating a belief in the former existence of some land connection between Africa and Western Australia. Again, we find the flora of New Zealand connected with South America and with East Australia. Here, again, by inference we are led to believe that in long passed geological periods there were changes in physical geography ; but it is impossible to say when these occured, whether in the eocene, miocene, or pliocene times. I think we must again congratulate ourselves on having so distinguished a visitor as Baron von Ettingshausen.

Mr. WARINGTON SMYTH : I am sorry to occupy valuable time, even for a few minutes, but I cannot withhold my tribute of admiration to the work which has been brought before us, in some degree, because I happen, from accidental circumstances, to have seen how extremely valuable, in that particular part of Europe, have been the researches of Baron von Ettingshausen. It happened upwards of thirty years ago, and, therefore, about the time when the Baron was beginning his researches, that I was for some time in certain districts in Austria from which he has reaped so rich a harvest, and I was very much struck by the specimens which were obtained from various parts of Hungary, more particularly from Croatia, which were much discussed by Haidinger and others. One could not but be brought to the conclusion that there were here before us materials of the most interesting kind, showing the affinities, which the first glimpse could not but assure us of, between the Tertiary flora of those countries and that which now exists ; but I merely rose to point out that at that time all that we knew upon the subject was fragmentary, and, in great measure, confined to cases founded on a very small number of specimens, but by his long continued researches and studies, the Baron von Ettingshausen has thrown a light of a most important character, not only upon these various discoveries in South Eastern Europe, but also upon the recent flora of the whole world ; that, as you have seen, he has with wonderful perseverance, compared with these fragmentary findings from various places in Austria, the productions of all the four quarters of the earth ; that he has succeeded, at last, in bringing before us generalisations which must strike every geologist, and, I must say, every philosopher, and that in point of fact he has brought it to this point, that out of a state of confusion he has evolved order, that out a mere series of fragmentary guesses he has brought before us what really occupies the position of a reasonable theory.

The PRESIDENT : I will only again return the thanks of the Conference to the Baron von Ettingshausen for the great pains he has taken in making that magnificent collection and for attending here and giving us so valuable a communication. Unfortunately the auther of the next paper on the list is not present, but Mr. Judd will briefly call attention to Professor Szabó's

E E

METHOD OF DETERMINING THE SPECIES OF FELSPARS IN ROCKS.

Mr. JUDD : The distinguished Hungarian geologist, Dr. Szabó, is unfortunately unable to be present to read a communication which he desired to lay before this Conference, on the subject of the determination of the felspars in rocks. He has, however, prepared a series of tables and translated them into English, and accompanied them with such clear diagrams that I need do nothing more than simply direct attention to them. As they are on the wall at the end of the room and can be consulted by all present, I will confine myself to pointing out the great importance of the work which Professor Szabó has so well accomplished. Of late years no branch of geology has made greater progress than that of the application of the microscope to the study of rocks ; and we are fortunate in this collection to have a number of specimens brought together such as have never been seen in juxtaposition before. We have in the first instance the *classical* investigations of Mr. Sorby in this country well illustrated by his original specimens ; and side by side with these may be seen the collections sent by Dr. Boricky, Dr. Gümbel, of Bavaria, Messrs. Renard, Vogelsang, Dr. von Lasoulx, of Breslau, Dr. Szabó himself, and some others ; together with the results of chemical and physical researches like those of Professor Daubrée and Mr. Sorby. Of course with the microscope it is possible to discriminate the characters of minerals which make up rocks that appear to be of the most compact texture. Under the microscope these crystalline constituents are seen perfectly distinct, but there is one group of minerals—the most important group of all—in which the distinction of the various varieties is exceedingly difficult, by means of the microscope—I refer to the felspars. They consist of a series of silicates of alumina lime, soda and potash, and Dr. Szabó has well shown that when the proportion of the three bases, soda, potash, and lime can be determined the class of the felspar can also be stated ; and he has, therefore, after devoting attention during a great number of years to this question, devised a means by which the elements of a crystal as small as a grain of mustard seed, may be examined and the proportions of the bases in it determined, whereby he is able to tell the character of felspar in the rock.

In the chemical section you will see in a collection from Bonn the actual Bunsen-flame apparatus referred to in these tables, by which Professor Szabó carries on his researches; and in the Geological Gallery you will see a collection of the trachytes classified according to the felspars which have been determined by Professor Szabó's method I have seen this method of investigation carried on both in the University of Buda-Pesth and in that of Klausenberg by Professor Szabó, and his pupils, such as Professor Koch, and I must say that as a series of analytical researches carried on with care and skill there can be no doubt as to the value of the results. It is certain that the analyses of the ordinary methods carried on by Karl von Hauer at Vienna have in a number of instances strongly confirmed the results arrived at by the new method by Professor Szabó. The tables speak for themselves; and having said these few words and pointed out the extreme importance of the method, I need do no more than refer you to them to show how admirably that method is carried out.

The PRESIDENT : I am sure you will all return your thanks to Mr. Judd for this communication. If no one has any remarks to make upon it, I will next call upon Mr. Rowley for his paper

ON THE CONSTRUCTION AND USE OF A TRANSIT THEODOLITE FOR MINE SURVEYING AND OTHER PURPOSES.

Mr. WALTER ROWLEY, C.E., F.G.S. Leeds.: The great object to be attained in mine surveying is perfect accuracy, and the two things most essential to success are—a careful surveyor, and an instrument properly constructed for the purpose. This is now so generally admitted, that it appears unnecessary to dwell upon its importance, but rather to limit this paper to a simple description of the construction and use of the instrument now exhibited, omitting any reference in detail to the ordinary Circumferenter or Magnetic Compass—the instrument in general use and best understood by mining surveyors.

The want of the latter directed the special attention of the writer to the subject now proposed to be submitted for consideration; and while not claiming any novelty of invention regarding the theodolite, would venture to hope that its application to mine surveying in the manner now intended to be described will be a step in advance of the crude and imperfect instrument generally in use. Having repeatedly experienced the unsatisfactory results of trusting wholly to the magnetic needle in underground surveying, I have had the instrument now exhibited constructed from my own designs by the eminent opticians, Messrs. Troughton and Simms. It is a six-inch transit theodolite, fitted with supplementary telescope and other appliances, specially adapting it for underground surveying; and, as will be readily observed, differs widely in construction from the ordinary six-inch transit theodolite.

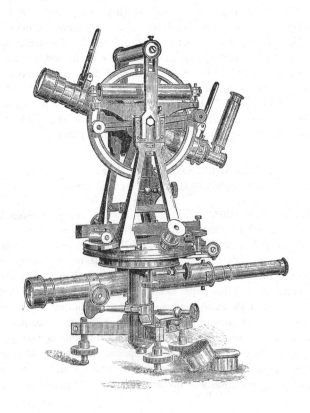

The axes of the transit telescope of this instrument are hollow, and a lamp, which requires to be cased in copper gauze, is fitted to illuminate the centre of the telescope.

This telescope is fitted with a diagonal eye-piece for the purpose of observing directly into the zenith.

There is also attached to it a magnetic needle of unusual length and delicacy.

The lower limb is divided as in ordinary theodolites: it is also numbered from zero right and left to ninety degrees and from 180 degrees right and left to ninety degrees and by the aid of the verniers it reads to ten seconds of a degree.

The vernier plate is fitted with reading lenses in tubes six inches long, in order to prevent the heat from the lamp used in reading off the angles underground scorching the face of the surveyor.

The instrument is finally fitted with four tripods, or sets of legs—one, an ordinary strong set for surface surveying, and general engineering purposes, and three other tripods adapted in size and character to the conditions under which they are used in underground surveying.

As I have already observed, underground surveys dependant upon the magnetic needle are extremely unsatisfactory; so much so, indeed, that in the whole course of my experience I have never seen two surveys (the magnetic bearings of which had been taken with different instruments) agree exactly. In order therefore to attain accuracy, I abandon altogether the use of the magnetic needle in underground surveying, and adopt the following method :—

I place the instrument at the bottom of the shaft, in the best position for at once observing directly in the zenith, and also for getting the longest sight possible in the main road leading from the shaft to the workings, or the other varied passages of a mine.

Having adjusted the instrument, I direct the telescope into the zenith, where, by the aid of a couple of assistants, a line is ranged across the mouth of the shaft, dividing it as nearly as possible in semicircles: this line is made to coincide with the vertical hair in the telescope of the instrument; the line is then ranged across the entire estate under which the workings are to extend, and permanently pegged out so as to be found at any future time. It is then measured and accurately laid on the plan to be used as a datum for all future underground surveys. Having observed that the line is truly coincident with the hair in the telescope, I unclamp the upper limb, and direct the telescope on a light placed on a tripod at as great a distance from the shaft in the main road as it can be seen; if the roof be rock, holes ought to be drilled into it, to indicate the line of survey; the angle contained between this line and the surface line is then carefully taken, read off, and entered in a book arranged in a convenient form for that purpose; by this process the underground workings are connected with the surface plan, to do which accurately is of the greatest importance, for however carefully the underground survey may be made, unless an equally accurate system of connecting the two is adopted the object will not have been fully attained.

The upper limb being securely clamped, the instrument is then taken off the tripod (care being observed not to move the tripod in the slightest degree) and placed on the tripod on which the light sighted in the last observation stood.

The transit telescope is then turned over, and directed towards a light now placed on the tripod on which the instrument last stood at the bottom of the shaft.

In order to adjust the vertical hair of the telescope on this light, it will be necessary to release the whole instrument, taking care not to touch the clamp fastening the upper limb.

Having adjusted the transit telescope on the back light, as above directed, release the supplementary telescope by slackening its clamp, and bring it also to bear on the back light. Having adjusted it to cut the light, clamp it firmly, then turn over the transit telescope towards the light placed in advance; release the clamp fastening the upper limb, and bring the hair to cut the light; clamp the upper limb, and read off, and enter the angle. If the supplementary telescope still remains undisturbed and fixed on the back light, it will be evident that the observation has been correctly performed.

This done, the instrument is taken off the tripod, and placed on that in advance on which the light last observed stood. A light is now placed on the tripod on which the instrument last stood, and the telescopes are again adjusted on this light as in the previous observation.

Having sent another tripod in advance, and placed a light upon it, the transit telescope is directed on the light in advance, as before; the angle taken, read off, and entered in the same manner as in previous observations; and this process is continued throughout all the devious roads of the mine.

The reason for numbering the instrument in quadrants will now be obvious, the angles being read and entered as ordinary compass bearings, calling, for the sake of convenience, the surface line a meridian, which in reality is not a meridian at all.

All the angles taken can now be placed or laid down from an ordinary traverse table, as by this method of surveying every angle is measured from the surface line, and hence the sines and cosines of the angles will be the difference of latitude and departure, or the Northing and Southing, and Easting and Westing, of the ordinary traverse tables used for nautical purposes. Should there at any time be a difficulty in obtaining sufficiently accurate traverse tables, the sines and cosines of the angles can be computed for the purpose of plotting.

The elegance and accuracy of this method of surveying and plotting an underground survey, will, it is thought, at once be admitted by every competent surveyor, insuring as it does, what is most required, an absolutely correct plan of the underground workings, and an accurate method of connecting the same with the plan of the surface ; and for the atter purpose the instrument is used for ascertaining the necessary angles involved in a systematic trigonometrical survey, the only truly reliable, althought no universal method, adapted for securing a correct plan to form the basis of all future underground surveys, for without such a system of triangulations greater cost of time and expense must be incurred, by not being able to fix the position of the stations independently from one base line, in the laying out of which the greatest care and discretion are required, for if accurately done it affords an absolute check against errors in chaining, the locality of which would otherwise be difficult to ascertain.

As regards the perfect accuracy of this method of underground surveying, there can, I think, be no question, and its infinite superiority to that generally practised by unskilful surveyors, with the old circumferenter or miner's dial, is beyond a doubt, for with the latter instrument, owing to its construction, a magnetic bearing cannot be reliably taken nearer than one-eighth of a degree, independent of the variation which the needle is daily subject to, and its liability to attraction, &c.

This instrument is admirably adapted for setting out drifts or roads to be driven for the purpose of connecting shafts or localities in mines a great distance apart, and which without such an instrument could not be constructed, except at a sacrifice of needless time and expense, and of which the writers own experience affords convincing evidence, as well as for surface surveying and general engineering work, and will also answer for solar observations.

In conclusion, the writer trusts he has not erred in venturing to introduce into a section of this museum (with which there are so many distinguished names in mining and geological science associated) a subject so thoroughly practical in character, and one that he fears will have lacked the general interest attached to some of the previous papers; but those who are interested in mining operations know well its importance, and the disastrous results to life and property that have so often arisen for want of possessing a more reliable instrument than the ordinary dial to guide them in laying out the underground workings in the most advantageous manner, and keeping the same within their proper boundaries, thereby avoiding frequent and costly litigation, insuring the greatest economy of time and capital, and the most perfect security against floods and other evils resulting from not possessing a thoroughly reliable plan; and if the description of an instrument constructed for the writer's own use directs, more than hitherto, the attention of those interested in mines and mineral property to the importance of the subject, and assists in bringing into more general use an instrument calculated to secure by its application such valuable results, his object will have been fully attained.

The PRESIDENT: We are all glad to hear a paper which directs attention to a matter which is so thoroughly practical. There is no doubt whatever of the great value of mining instruments and of the extreme importance of accuracy being obtained, and Mr. Rowley appears, I think, to have thoroughly appreciated this. I do not pretend to be a judge of these matters, but it is most important to take some given line for surveys as a base rather than a magnetic line. I think we have some gentlemen amongst us well versed in mining matters who will perhaps be able to offer some opinion on this instrument.

Mr. WARINGTON SMYTH : There is no question of the very great importance of the principle which has been laid down here, but we must not express our satisfaction with what has been told us without remembering that the same thing has been said for the last fifty years by a great number of experienced surveyors and of ingenious constructors of instruments, and the best proof of that will be seen in the admirable collection of instruments down stairs, not by any means so numerous as it might have been, but at all events showing that for the last half century it has been the object of surveyors desirous of obtaining full accuracy to dispense as far as possible with the use of the needle and to substitute for it a definite fixed line, and from that to work into the interior of the mine. For this purpose the instruments have been more or less different. I should be very sorry indeed to see these beautiful instruments taken down into a great number of mines with which I am connected in the south-west of England, because as the author is no doubt perfectly aware there is a great deal of rough work to be done there, and what with water falling in all directions in naked shafts, and the levels being sometimes so low that you have to creep in the most uncomfortable positions, with rocks above and below you and on each side you have to get hold of an instrument of the very simplest and roughest description to do the work. As the author has pointed out to us, it is a matter of very high importance that we should as often as possible dispense with the use of the needle, which is so apt to lead us into difficulties, and that every surveyor should seek, whatever the instruments are, to carry out his work with the greatest amount of accuracy. I have no doubt, however, having had some experience in mines, that a great number of the accidents and loss of property to which he has referred have been not so much in consequence of the mistakes to which we have been brought by the use of the magnetic needle, as from the disuse or misuse or the forgetting altogether to use the instruments which ought to be employed in surveying, and by substituting a sort of guess-work for accurate instrumental observation.

Mr. ROWLEY : I am exceedingly pleased that the reading of this paper has induced our distinguished friend Mr. Smyth to make these remarks, which of itself is a valuable result of my effort ; but it necessitates my making one explanation, namely, that this instrument was not constructed for a mine of less thickness than four feet. Mr. Smyth is perfectly correct in saying that in what we call thin mines of coal this would not be a practically available instrument.

Mr. SMYTH: I refer more especially to metalliferous mines.

Mr. ROWLEY: Quite so; this was intended for what we call the thick sections of our coal measures, in which I happen to be located. Mr. Smyth made one remark with regard to the principle of the instrument being to a great extent old. No doubt the theodolite is an old instrument; but it is no less true, from my own experience and observation, that (although there have been scientific men who have constructed these instruments for the use of mining surveyors), such an instrument is still very far from being in general use, and in some of our colliery districts is practically unknown, so far as its application to coal mines is concerned, and think it would be an advantage to all interested in mining operations (whether as an owner of mineral property, or a lessee endeavouring to work the same to the greatest advantage), that a somewhat more accurate and reliable instrument than the circumferenter, or ordinary miner's dial, which, as used at present, shows very little improvement in character since [its first construction, in the early period of coal mining, and when the workings were principally confined to basset operation, should be brought into more general use.

The PRESIDENT: The last paper on our list is one of the Rev. Nicholas Brady, on the desirability of a uniform national notation for crystallography. It is a somewhat technical matter, but I am glad we have some crystallographers present who will take an interest in what is to be said.

ON A SERIES OF CRYSTALLOGRAPHIC MODELS AND DIAGRAMS.

Together with a suggestion that the Conference should recommend one out of the many systems of Crystallographic Notation for universal use.

———

Rev. NICHOLAS BRADY, M.A., Rector of Wennington: The chief characteristics by which we classify the numerous natural substances met with in the inorganic world, and which we call minerals, are chemical composition, or the internal, invisible molecular combination of elements, and crystalline form, or the external, visible molecular shape. The former requires various and often complicated processes of analysis to be employed before we can determine it ; but the latter may frequently be distinguished at once, or by accurate measurement of the mutual inclination of some of the planes or faces of the crystal. It is therefore most necessary and helpful to the mineralogist to know the laws which govern the formation of these flowers of the mine and of the rock, to investigate their wonderful, yet simple symmetry, and to learn to decipher the principal simple forms and combinations at sight. A thorough knowledge of these laws requires, no doubt, high mathematical attainments ; but a vast amount of practical information may be gained by patient observation and diligent use of the goniometer (a very fine series of these instruments may be seen in the adjoining gallery), since the relations of one face to another, and one form to another have been already mathematically worked out by many observers, and their angles have been recorded in treatises on crystallography. Diagrams, and models, however, are of the very greatest use in teaching the general shape of a crystal, and in exhibiting its relations to other forms ; for, not only do they assist the student in enabling him to realize the form, so greatly lightening the labour of the professor, but they also express qualities and facts which cannot be so easily conveyed in any other way, and, therefore, I hope that you will not think your time wasted if I explain, as shortly as possible, the diagrams and models which I have selected from my collection for exhibition in this museum, and which I have constructed with my own hands, especially as without a little explanation they would possibly convey but few ideas even to some few of the members of this Conference : how much more would this be the case with those who have little or no scientific education to help them. A crystal may be drawn and its characters exhibited in several ways : first, we may depict it as it is really seen by the eye that is in perspective, and the kind of perspective generally employed is that known as isometrical perspective, in which the lines preserve their parallelism, there being no vanishing points as in ordinary perspective (this system is shewn in the first column of the series of diagrams, and in fig. 1), and I should like to point out a beautiful series of eighteen by Professor Von Rath, of Bonn, on the gallery wall. A much more instructive diagram, however, is produced if we employ a method known as projection, and two kinds of projection are used by crystallographers, viz.—one a projection

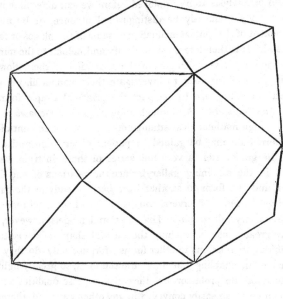

Fig. 1

of the faces in a plane parallel to the paper, and perpendicular to one of the principal axes of the crystal, the other a projection of the poles of each face on great circles of a circumscribing sphere, the poles being normals to the faces drawn from the point where the axes intersect, the first of these, or the projection in plan or section as we may call it, (examples of which are given in the second column of the diagrams, and in fig. 2), shews the appearance of the crystal as we see it if we look along the axis which is perpendicular to the plane of the paper, the faces inclined to that axis are grouped around the centre, which is the position of the axis, and faces parallel to the axis, should there be any, are seen merely as boundary lines ; but more than this, the laws of symmetry are such that a number of faces, or their edges, generally lie in the same zone or belt, and in this projection these zones are shewn by the plane of the paper, and lines drawn through the centre of the figure which divide it symmetrically, thus in fig. 2, which is a projection of one of the varieties of the four-faced cube,

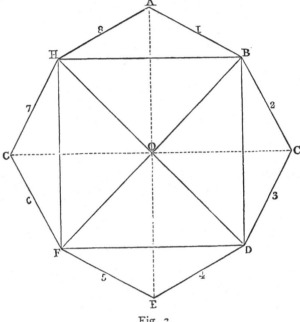

Fig. 2.

the same simple form as fig. 1; the boundary lines 1 to 8 signify faces perpendicular, to the plane of the paper, and parallel to the axis chosen, the dotted lines being the other two axes, and lines passing through the centre joining any two opposite letters indicate the zones— by constructions of this kind the angles of a crystal may very often be most easily calculated. A still better form of projection is that of the poles of the faces on great circles of a sphere, fig. 3 (as portrayed in the third column of the series of diagrams, in the last diagram, in which the positions of all the chief zone circles of the cubic system are exhibited with the poles of all the faces of the system drawn on them, except those of the greater number of the varieties of the six-faced octahedron which lie one in each of the triangles formed by the zones).

I may remark in passing that, the second series of diagrams on the wall comprise perspective drawings of all the observed varieties I am acquainted with of the simple forms of the cubic system, and examples of all three kinds of drawings of one variety of each species. Turning

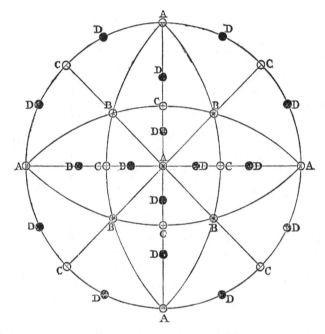

Fig. 3.

now from the diagrams to the models, I have on the table a large wire model which is, so to speak, a summary of the cubic system, coloured exactly to match the diagrams, and a large series of pasteboard models, in all of which the positive and negative hemihedral forms are denoted by dark and light tints respectively of the same colour. In this wire model, the lines meeting in the centre, are the crystallographic axes cubic red, octahedral blue, and dodecahedral green. The edges, or the line of junction of any two adjacent faces, is shewn by a wire coloured according to the simple form to which it belongs. The forms are of course arranged according to their relative positions, or are properly orientated, as the phrase is. Thus by concentrating ones attention on two or more sets of colours, we can see what shape would be taken by them when combined together, and what edges and what solid angles would be replaced or truncated. In this model are comprised a circumscribing and three inscribed cubes 100. The positive and negative tetrahedron + and—K 111. The octahedron 111. The dodecahedron 110. And one each of the varieties of the other simple forms, viz.— The positive and negative twelve-faced scalenohedron + and—K 122. The three-faced octahedron 122. The positive and negative three-faced tetrahedron + and—K 211. The twenty-four faced trapezohedron 211. The positive and negative irregular pentagonal dodecahedron + and—K 210. The four-faced cube 210. The positive and negative six-faced tetrahedron + and—K 531. The positive and negative irregular twenty-four faced trapezohedron + and—K 531, and the six faced octahedron 531, in all, nineteen separate simple forms. Outside we have the whole of the chief zone circles in flat steel spring, with the position of all the poles which lie in them, painted upon them. On the table will be seen the solid models of the forms included in the diagrams and wire model, and in the room beyond is a large series of models of all the chief varieties of the system as far as I have, as yet, had time to complete it, made to a scale of three inches for the circumscribing cube, and fitting into a hollow glass cube to demonstrate their orientation ; together with a series of combinations of the principal varieties of each simple form with the cube tetrahedron, octahedron, and dodecahedron, fully illustrating the passage of one form to another, so that we have, what may be called, the complete morphology of the system. I should not, however, omit to mention the very beautiful series of wire models of Professor Story Maskelyne. The one representing the cubic system I have ventured to have placed on the table, and you will see that, that whereas in my model, the crystal forms themselves are exhibited, in his the planes of the zone circles, or the planes of symmetry of the system are shewn, on which by coloured worsted or threads of any kind, the crystal which he desires to demonstrate can be built up. Then again there is a fine series of models made from glass plates by W. Apel, of Gottingen, many of which shew by a model inside it, the formation of its hemihedral form by the developement of half its faces, and then lastly, Professor O'Reilly has contributed a most ingenious model which can be arranged to suit the parameters of any of the six systems of crystallography; the crystallographic axes are indicated by wooden rods, which can be inclined at any angle to each other, and the edges of any face, are indicated by elastic cords stretched at given distances from parameter to parameter, by hooks and eyes, usually employed to fasten ladies dresses, and by strings crossing these elastics and stretched by means of small leaden weights.   I also wish to bring under your notice a very ingenious model, the Raumgitter, invented by Professor Sohncke, of Carlsruhe, which, by means of joints, can express the inclination of the faces of the cube and all the varieties of the rhombohedron, the right prism and the oblique prism.   This communication has already occupied too much of your valuable time, but before I close I should like to throw out a suggestion of a practical nature, and to bring before you a subject which is very suitable for discussion in an International Conference, viz., the desirability of a uniform international notation for crystallography, and the steps by which such a system of notation might be attained— could this be achieved a great difficulty would surely be removed from the student's path, and many more would be likely to be attracted than now to study the wonderful symmetry and properties of crystals—the need of such a course of action is patent to everyone who merely glances over the pages of a mineralogical library even if it only contain works or memoirs in his native tongue, and if a sudent desires to become acquainted with the labours of the chief authors in this difficult branch of science, he must learn not one, but many of the dialects of crystallography.   Indeed, the catalogue of this magnificent collection is itself a commentary on the desirability of uniformity in this matter in as much as more than one system of notation is spoken of in its pages. Of course the relation of a face to its parameters is the same under every system of notation, and the symbols expressing that relationship are mathematically convertible into the symbols of each system, yet they require translation into the language which the student first learned, and which is, so to speak, his mother tongue. This is much to be regretted, for the secrets of rock, and vein, and mine, are hard enough to unravel without putting any unnecessary difficulty in the way.   It is easy however to see how this variation of language has arisen if we look at the terminology of any of the cognate sciences. The first efforts are vague and crude, and invention finds its place in devising means of expressing the required qualities in symbol or by word. Several pioneers it may be are traversing the same field of research. They arrive at the same result; but by different routes and by different modes of thought or expression, and finally meeting they interchange ideas, and select for future use the shortest and most pithy expressions, the neatest and most simple symbols as their foundation for future work. To such a point in the history of science, the crystallographer may now be supposed to have arrived. The conditions of a mathematical analysis of crystalline form have long been made known, and we may well agree to cast aside those old languages which have done good work in their day, but which are now obsolete, or have been succeeded by simpler and therefore better formulæ and signs. It was an immense step when Weiss, about 1814, laid the foundations of the six systems of crystallography, calling them the regular system—the four membered, the two and two membered, the three and three membered, the two and one membered, and the one and one membered; but with the exception of the first, none of them are in general use at the present day, at any rate in England. Such barbarous terms could not possibly survive, the systems however are but six in number, and though each of them is known by several different names, any confusion of terms is not likely to be very onerous to the earnest student, but it is far otherwise, when we descend from what may be styled natural orders or families to the symbols expressing the simple forms, or the genera and their varieties or species with the numberless combinations of them which occur. I need not enlarge upon the characteristics of the notations. Amongst others there is Levy's modification of the system of Réné Just Haüy, the founder of the modern school of crystallography; as the late Master of Trinity College, Cambridge, calls him in his admirable history of the inductive sciences, whose contact goniometer is one of the most prized relics of this marvellous exhibition, and this system is the one adopted by Mons. Des Cloizeaux in his splendid " Manuel de Minéralogie," then we have the systems of Weiss and Gustav Rosé, Naumann, Professor Miller, Mitchell, and James Dwight Dana's modification of Naumann. Crystallographers know these systems well, and their relative advantages and disadvantages, but the table in the diagram (figure 4) giving the symbols of each of these systems for one of each of the species of cubic crystals, will be quite sufficient to illustrate how diverse they are, and to shew what a boon it would be for the student of the future to be able to read crystallographic treatises in all languages expressed in one universal system of notation; in the same way as Greek is well nigh the universal language upon which the terminology of science generally is founded.

Fig. 4.

Examples of different notations of each species of simple form in the cubic system of crystallography :—

	Cube.	Octahedron.	Dodeca-hedron.	3 faced Octahedron.	24 faced Trapezohedron.	4 faced Cube.	6 faced Octahedron.
Levy Brooke and Des Cloizeaux	p.	a^1	b^1	$a^{\frac{1}{2}}$	a^3	b^2	$b^1\,b^1/_3$ $b^1/_5$
Weiss and Rose	a:∞a:∞a	a:a:a	a:a:∞a	a:a:2a	a:a:½a	a:2a:∞	a:$^1/_3$a:$^1/_5$a
Naumann	∞ O ∞	O.	∞ O	2.O	2O2	∞O2	$5.0.^5/_3$
Whewell and Miller.	100	111	11∞	122	211	210	531
Mitchell	1∞∞	112	11∞	112	122	12∞	$1^5/_3 5$
Dana	O.	1.	2'.	·2	2-2	2'-2	$5—^5/_3$

This day is, however, far distant, for the books now in use will continue for a longer or shorter time, according to their relative value, but it will come, if we can induce *all* crystallographic authors to use for the future one International notation, and if we can persuade *all* professors in our great schools and universities to gradually adopt the system chosen, and to teach it, at any rate, with that system employed in the text books they recommend. It must take time for new text books and revised editions to be prepared, but we may be encouraged by the way a new system of notation has been and still is being adopted by chemists without any great derangement of their methods of teaching. The fact being admitted, as I should suppose is generally the case, that a universally adopted notation is most desirable. The great question is what system shall be recommended. Shall it be one of those now in use, a mixture of two, a modification of one, or a new one altogether? We all of us probably have our own ideas on this subject, and I for one, consider that the notation of my old master, Professor Miller—to whom I owe so much, and whose kindness to me when I was an undergraduate attending his lectures, I shall never forget—is by far the simplest and best that I know of; and I have lately heard, that in his new edition, Naumann has adopted Professor Miller's notation. We shall not, I fancy, be willing, off hand, to decide upon the relative merits of the several systems of notation, neither would it be wise for us in our corporate capacity to recommend the adoption of any of the systems under consideration; we could, however, affirm the desirability of a uniform system of crystallographic notation for general use, and perhaps some agency might be devised for bringing this question before the learned societies and eminent crystallographers in Europe, America, and elsewhere. An International commission might be appointed to fully discuss this matter, and make what recommendation they think fit to the bodies concerned. If I might make a further suggestion, it would be to this effect, that this Conference, recognising the desirability of a uniform International crystallographic notation, should request our chairman or Professor Miller, or Mons. Des Cloizeaux, who we are most proud to have welcomed amongst us, to introduce this subject to the persons and learned societies I have mentioned, and ask them to nominate a commission to discuss this matter and report to them the recommendation they think fit.

The PRESIDENT: I am sure we are all much obliged to Mr. Brady for bringing this matter before an International Conference such as this, though I am sorry to say the foreign element is not so fully represented as I could wish. There is one point, however, with regard to it, on which I must make a correction, and that is, where he suggested that the matter should be referred to me or some other eminent crystallographer; because of all things in the world I certainly am not a crystallographer, and cannot pretend to any knowledge whatever of that branch of science. At the same time it does appear to me that the question which has been raised by Mr. Brady is one well worthy of consideration, because there can be no doubt that a considerable loss of time and energy arises from there being such diversity of notation, whether in chemistry, geography, geology, crystallography, or any other science, and the method which he has propounded of referring it to some International Committee, is one which has been in some instances adopted, and, I believe, is like to prove successful. When there was a Congress on pre-historic archæology two years ago at Stockholm, the question of an international system of symbols to be used on pre-historic maps, was in that manner referred to a committee, consisting of one or two members of each different nationality, the general committee consisting of some fourteen or fifteen, and the result has been that a report has been agreed upon, and a series of symbols has been devised, which I think is likely to be adopted generally for such a purpose; and it would appear that it is possible for the purpose which has been suggested by Mr. Brady, that some similar course of action must be adopted. I am glad, however, to see that, although we have not many distinguished foreign crystallographers amongst us, we have some here from France and Belgium, and I am not sure whether Germany is not also represented; and possibly some one will express an opinion on the question which has been raised. At any rate I see Mr. Maskelyne has just come in, and I feel sure that he will be able to offer us some valuable remarks upon this subject.

Mr. MASKELYNE : I feel much embarrassed, sir, on coming into the room to be immediately called upon to address so distinguished an assembly on a discussion of which I really have heard nothing. I can only divine what Mr. Brady has said ; but I am quite sure if what Mr. Brady has said is what I divine he has, he has said a good thing, and told you that we have already a system of crystallographic notation, which is an admirable system, and, I believe myself, the most simple system that can be devised.

The PRESIDENT : Will you mention which you consider the best ? I will not say which Mr. Brady has recommended.

Mr. MASKELYNE : I do not wish to occupy your time, but though I am in the presence of the distinguished foreigners who, I am quite sure, are more or less wedded to good systems of their own, I am bound to say that I think the most simple system is that which has been called by one friend of mine the English system, namely, that of our distinguished professor of mineralogy at Cambridge, W. H. Miller. There is no system, in fact, that represents the *face* of a crystal so completely, and is so simply, and so readily converted into a mathematical expression, as that which he has given us. I, of course, having an official position in England, have amongst my friends mineralogists of considerable distinction as crystallographers abroad, and I have had the great pleasure of talking over this subject, amongst others, with M. Vom Rath, and with a distinguished French gentleman now present, M. Daubrée, who stands at the head of the Ecole des Mines, and M. des Cloizeaux, a gentleman who, in his valuable memoirs and books, has used the French system of Haüy, modified by Levy. And I have talked with each of these gentlemen separately, on the merits of their different systems. I have generally found the consensus of different opinions was that, if you had to begin over again, and to start a new system of crystallography for the world, they might be willing to accept the Millerian system. But we cannot begin *de novo*. There are our French friends, and it is very hard to ask a gentleman like M. des Cloizeaux, when he sits down to write a great book on mineralogy, to throw aside a method in which he has become a master, and to begin over again from a new point of view. He gives us the book, and he says : " Here is my work ; I give it to you in the manner in which I have learned my subject. It is for you to interpret it." It is for us to interpret it, and, to those familiar with the subject, there is only the trouble involved in doing so. But the question is, in the long run, how is a desirable uniformity to be established, and what is to be the solution of this difficulty? I think the answer again is, that we shall find the system of Miller will gradually be accepted by everybody. I am confident it will be accepted in Germany. It has already been translated—and translation means selection—as being the best system, by persons no less eminent than that very distinguished Frenchman M. de Sénarmont. It has been translated also by another gentleman who, in early life, was devoted to this science, but who, unfortunately for science, has been lately beguiled into the troubled paths of political life, Signor Sella in Italy, and finally it was translated by Grailich into German. It is now the crystallographic language of Southern Germany, and it is gradually coming into use in Northern Germany. In a book which Professor Groth has published, he says, although he uses Naumann's system, that he recognises in Miller's the true system for geometers and mathemati-

cians. Therefore, we have got a long way in Berlin, and I think when
Berlin has accepted Miller's system, the world will accept it, all but
our friends in Paris.   They occupy a very peculiar position on this
subject.   Haüy, if not the father of crystallography, was the great
originator of it as a science, and he left behind him a mode of
treating the subject of the relation of the faces of crystals, which was
in itself a remarkably simple and beautiful one, and which has
been developed by Lévy since.   That is the French system, and
although Monsieur de Sénarmont has translated Miller's system, at
present there is no very advanced step made in France to accept it.
This much, however, has been done; that you recognise it as necessary to the work of English and Germans who employ Naumann's
method, that in dealing with the question of zones (that is, the relation
of the faces of crystals which lie with their faces all parallel to the
same axis)—in dealing with those zones every crystallographer has
to come to what are really symbols expressed in the language of
Miller's system.   He has to reduce his symbols into a form in which
he can handle them, as Miller handles them directly.

That is the prospect, I think, of our subject, and I believe by
degrees it is making its way.   It is now being partly introduced into
our own School of Mines, where the students are beginning to learn
that system, and I have not the slightest doubt in the course of a
few years it will be the system acknowledged not only in South
Germany, but in North Germany, and I venture to hope, that if
another edition of Des Cloizeaux is written, he may see good to
adopt it also in France.   One point which gives Miller's system its
enormous superiority over every other is this, that by three single
letters or figures you are able to treat the whole subject of the faces
of a crystal, and represent every conceivable face that a crystal can
present by the interchanges of those letters, and of the signs attached
to them; you have plus or minus attached to each, and you have the
relative positions which those three figures can represent.   It is a very
singular thing that with the one *form* which contains the greatest
number of faces that any crystal can by possibility present that are
symmetrically connected, a form, namely, of the cubic system which
has forty-eight faces, those forty-eight exactly represent the number of
interchanges that you can make between three figures, putting them

in different order, if you change the plus or minus to every one. You thus get forty-eight different interchanges, and those exactly represent the possible number of faces of the crystallographic *form* which has the maximum number of faces ; so that Miller's system answers all the demands you can make upon it, and completely fulfils the requirements of presenting you with a separate symbol for every *face* upon the most complicated *form* that a crystal can present. I think everybody can see at once that a system which can do that has an enormous power, and at the same time has a remarkable simplicity. I will not detain you by discussing this in detail, but if you take those figures which Miller has given, the three indices as he calls them, you get the most beautiful representations of the relations of faces and groups of faces of crystals with respect to one another. By performing on those figures an operation of cross multiplication (that is the simplest way in which I can put it before you), you can deal with them as with determinants. In short, you are able by this treatment to get the most simple grasp upon all the relations first, of all the faces belonging to the same zone, and then of the zones that have a face in common, and so on. Thus you are able to treat the whole of the subject of the geometry of the crystal simply from the point of view of the algebraic treatment of those three simple letters which give the co-efficients in the equations of the planes that represent the faces of the crystal.

Those are the main points of Miller's system. It is a very difficult thing to express what is really a complex mathematical idea in simple language. I do not pretend to have done it, but I hope I have pointed out how these particular merits of Miller's system must place it pre-eminently in the front as a system capable of meeting all the requirements of mathematicians as well as of practical crystallographers. It is most simple to learn, much easier to learn than Naumann's, and therefore I do not think I have put the claim for it too high when we contrast it with what other systems can do. You will find that none of them are capable of performing algebraic operations of this sort, except by transforming them first into forms equivalent to Miller's symbols. It is, therefore, a mark of his pre-eminent superiority, that he begins where the others leave off, that their symbols have to be converted by a complex method to get them into the forms in which you are enabled to treat them as Miller's symbols.   Therefore, for simplicity, for grasp, for the general handy purpose of daily life, there is no system comparable to Miller's.   I believe that that fact is becoming so rapidly recognised all over the world that we shall certainly have Miller's system in another twenty years thoroughly established.   I feel confident of it myself, because I know it only has to have its merits brought forward to meet with general consent.   I hope I have convinced you, at any rate, that we in England may with confidence sustain the system of our very eminent Professor at Cambridge, and I trust .before long he will be recognised everywhere, as he is in a great many important centres of education in Europe as the true leader, in our time at least, of this really great and beautiful science.

Mons. DAUBRÉE said that after the elaborate paper read by Mr. Brady, and the further expositions which had been given by so eminent an authority as Professor Maskelyne, it did not become him to say anything, but he might just add that the fact of Mons. Sénarmont having translated Miller's work was really a proof of the great esteem in which he held him.

The Rev. N. BRADY : Professor Maskelyne spoke of the disadvantage of not having heard what had gone before, but I must say that he could not have expressed more strongly my own feelings with regard to Professor Miller than he has done with regard to the great simplicity and beauty of his system of notation.   My object in bringing forward this question was not so much to treat of the relation of the different systems one with another as to direct the attention of the crystallographers of different countries to the fact that it is desirable as far as possible that we should get to one common language, that we should express our ideas and our thoughts in one common dialect.

The PRESIDENT : I think we must be glad not only that we have had this interesting paper from Mr. Brady, but that it has led to so interesting a discussion.   After what we have heard as to the translation of Professor Miller's works into various foreign languages, and as to the high esteem which it has in this country, we may also trust to some process of natural selection for its general adoption throughout the civilised world.

I have now only to tell you that we have concluded the series of Conferences which will be held in connection with this Loan Exhibition of Scientific Apparatus. I have not had an opportunity of attending those in the other sections, but I must express my satisfaction at the number of subjects that have been brought before this section and the manner in which they have been treated, and I think and I hope it will be generally felt that these Conferences have to a great extent carried out the wishes of the committee in bringing before those who come to see this collection the principal objects of interest which are here exhibited, and in drawing attention to the historical points which are of the greatest interest in connection with those objects. I can only now thank you for the kind manner in which you have borne with me during the time I have had the honour of presiding over you.

Mr. WARINGTON W. SMYTH : I think we should be scarcely fulfilling our duty or inclination if we were not, before we part—and I am afraid it is for a long time—to bear testimony to the admirable manner in which the duty of chairman has been fulfilled by Mr. John Evans. Without intruding on your time, I beg, therefore, to propose a cordial vote of thanks to him for his admirable conduct in the chair.

The proposition having been carried by acclamation,

The PRESIDENT said : I am much obliged to you for this kind expression of feeling, and I would only say that it was a post in which I was placed by a series of accidents rather than by any seeking of my own, and that I have done my best to fill it as far as my ability would allow me.

THE END.

Printed in the United States
By Bookmasters

.